Springer Series in Biophysics 5

Editor: P.M. Bayley, London

Springer Series in Biophysics

Volume 1 **Structure, Dynamics and Function of Biomolecules**
Edited by A. Ehrenberg, R. Rigler, A. Gräslund and L. Nilsson (1987)

Volume 2 **Biophysics and Synchrotron Radiation**
Edited by A. Bianconi and A. Congiu Castellano (1987)

Volume 3 **Cytoskeletal and Extracellular Proteins**
Edited by U. Aebi and J. Engel (1989)

Volume 4 **Electron Probe Microanalysis**
Edited by K. Zierold and H.K. Hagler (1989)

Volume 5 **Biophysics of the Cell Surface**
Edited by R. Glaser and D. Gingell (1990)

R. Glaser D. Gingell (Eds.)

Biophysics of the Cell Surface

With 77 Figures

Springer-Verlag Berlin Heidelberg New York
London Paris Tokyo Hong Kong

Professor Dr. Roland Glaser
Humboldt Universität zu Berlin
Sektion Biologie
Wissenschaftsbereich Biophysik
Invalidenstraße 42
1040 Berlin, DDR

Professor Dr. David Gingell
Department of Biology as Applied to Medicine
Middlesex Hospital
Medical School
London W1P 6DB, Great Britain

ISBN-13: 978-3-642-74473-0 e-ISBN-13: 978-3-642-74471-6
DOI: 10.1007/978-3-642-74471-6

Softcover reprint of the hardcover 1st edition 1990

2131/3145(3011)-543210 – Printed on acid-free paper

Preface

It is common practice to publish conference papers in books or monograph series. This gives some advantage to those who did not have the opportunity to attend the meetings, but it irritates and disappoints others who may have hoped for a set of closely related reviews.

With this book we have tried to find a compromise. It presents a selection from the topics which have been discussed in a series of international symposia entitled "Biophysics of Cell Surface", held in 1976, 1978, 1981, 1985 and 1988 in the GDR, and subsequently published in the journal STUDIA BIOPHYSICA (volumes 56, 74, 90, 110, 127). Nearly all the contributors to this book participated in one or more of the meetings. We hope that our choice of topics selected for this book manages to reflect the variety and interest of the broad range of subjects which fall within the scope of membrane biophysics, without taking on the randomness of a scientific car-boot sale.

We would like to express our thanks to all colleagues and organisations who helped to realize the conferences and particularly this book. Financial support for the symposia of 1985 and 1988 was provided by the IUPAB. A number of topics, reflected in this book, resulted in international cooperations, supported by various organisations. We are especially grateful for the support of UNESCO research project on biophysics in this respect. The European Bureau (ROSTE) of UNESCO supported the editorial work of this book.

London, July 1989 — Roland Glaser, David Gingell

Preface

Contents

Introduction

Membrane biophysics has become one of the most rapidly developing fields in biological science. Membranes are everywhere, surrounding the cell, and separating numerous compartments inside it. They act as barriers separating different phases as well as being surfaces which can carry arrays of functional proteins, like elements on an electronic current board. Artificial membranes can be made and studied in the laboratory. This plethora of membranes and functions together with the diverse and often complex methods which have been developed to study them, is responsible for the vast scope of the subject.

The title of this series of symposia and, consequently of this book, was chosen in order to concentrate attention on a particular area, namely the role of membranes in regulating cellular processes and the interactions of cells with their environment. To make this task managable we have narrowed the field still further by introducing the word "biophysics" into the title. This knife cuts rather sharply, since "biophysics", even handled with care (taking into account the difficulty of finding an acceptible definition), generally excludes the important field of immunological reactions at the cell surface.

What are the factors that influence the cell from outside? There are a number of molecular influences, including peptide messages and toxins, as well as more "physical" influences, like contacts with other cells, viruses, or non-cellular surfaces. Viscous shear forces and electromagnetic fields form another category.

These external influences are partly non-specific, partly specific. In the latter case, receptor systems transform them into cytoplasmic messages, which sequentially control cellular functions. External stimuli may also modify membrane transport, altering the molecular composition, or mosaic organization of membrane molecules, and many other properties. Thus the study of the action of external factors on cell surfaces merges naturally into the study of control processes.

Knowledge of such mechanisms can provide a deeper understanding of cellular functions. It can also provide the means of influencing them by drugs of by physical factors. For this the membrane must be considered as part of a complex dynamical system, tightly connected with the cell and with cellular metabolism.

The papers of this book try to indicate some of the recent progress in this field and point to questions which remain as yet unsolved.

Translocation and distribution of phospholipids across eukaryotic plasma membranes

A.Herrmann*, P. Devaux and A. Zachowski
Institut de Biologie Physico-Chimique
13, rue Pierre et Marie Curie
F-75005 Paris
France

Introduction

A major task of current cell biology is to elucidate those mechanisms which generate and maintain the molecular composition and topology of different cell compartments. According to the 'fluid-mosaic' model (Singer and Nicolson 1972), the lipid bilayer phase of biological membranes was considered as a homogenous 'floor' serving only as a building block of biological membranes. Intrinsic and extrinsic proteins were considered as responsible for the functional features of biological membranes. Much attention was given to their inhomogenous lateral and asymmetric transverse arrangement in biological membranes. However, increasing evidence for the involvement of lipids in complex cellular processes, where they may serve as substrates and chemical mediators, has been presented in recent years (Hanahan and Nelson 1984). For example, the involvement of phosphoinositides in signal transduction (Berridge 1987) and in lipid specific regulation of the activity of membrane bound enzymes (Rando 1988) has given important impetus to lipid biochemistry and biophysics. With the awareness that the different membrane properties caused or mediated by lipids cannot always be explained by bulk physical properties of the lipid phase, much information on the lateral and transverse inhomogenous organization and dynamics of the lipid phase has also been accumulated in the past decade. The heterogenous lipid composition typical for biological mem-

*permanent adress: Humboldt-Universität, Sekt. Biologie, Bereich Biophysik, Invalidenstr. 42, 1040 Berlin, G.D.R.

branes (Curatola and Bertoli 1987) is accompanied by a non random distribution of lipids in the plane of the membrane triggered by ions, temperature, and/or specific interactions between lipids and proteins.

The present chapter will focus on the translocation and transverse distribution of phospholipids in plasma membranes. In particular, attention is given to those mechanisms responsible for the generation and maintenance of phospholipid asymmetry. The transverse diffusion and distribution of cholesterol (see for a review Schroeder 1985) as well as lipid topology in membranes other than plasma membranes will not be dealt with herein (see for review Zachowski and Devaux 1989).

Phospholipid asymmetry in mammalian plasma membranes

Since the pioneer work of Bretscher (1972), a general picture of the transbilayer distribution in mammalian plasma membranes has emerged. Several techniques have been used to determine the transmembrane topology of endogenous lipids. Briefly, the lipid localization was inferred from the selective modification of these molecules either by exogenous phospholipases or by chemical reagents. Monitoring lipid exchange kinetics between two populations of membranes catalyzed by a soluble exchange protein offers another way to establish the topology of phospholipids. These techniques have been described and critically discussed in previous reviews (Op den Kamp 1979, Etemadi 1980). In general, these approaches do not allow to follow rapid translocation of phospholipids across the membranes with half-times of the order of several minutes because the time required to assess lipid localization often exeeds the time of phospholipid translocation.

The study of phospholipid asymmetry in plasma membranes requires a difficult isolation procedure since eukaryotic cells possess internal membranes which corresponds up to 85 % of total cell membranes. Fortunately, this is not the case for mammalian erythrocytes comprising only a single membrane.

Consequently, the erythrocyte plasma membrane is the most highly characterized system with respect to lipid asymmetry. It is now well established by different techniques and laboratories that the zwitterionic phospholipids phospatidylcholine (PC) and sphingomyelin (SM) are enriched in the outer monolayer, while the aminophospholipids phosphatidylserine (PS) and phosphatidylethanolamine (PE) are preferentially located in the inner monolayer (Op den Kamp 1979, Van Deenen 1981). Recently, using spin-labeled lipid probes, which give insights on the time dependent reorientation of lipids (see below), it has been shown that the equilibrium distribution of these exogenous phospholipids is in agreement with the topology of endogenous phospholipids from which they derive (Devaux et al. 1988).

With few exceptions (Session and Horwitz 1983), an asymmetric transverse distribution of the four main phospholipids (PC, SM, PS and PE) similar to erythrocyte membranes was found to be a general feature of plasma membranes. This conclusion is based on data of platelets, enterocytes, murine LM fibroblasts, chick embryo fibroblasts, B-16 melanoma variants and LM metastatic cell lines (Op den Kamp 1979, Schroeder 1985, Zachowski and Devaux 1989). A more complex situation arises in the polarized cells of simple epithelia where the apical and basolateral plasma membrane domains have different lipid compositions (Simons and van Meer 1988). A high content of glycolipids was established in the outer leaflet of the apical membrane, whereas the lipid composition of the basolateral membrane is similar to that of plasma membranes of unpolarized cells (Brasitus and Schachter 1980). The so-called tight junctions which were identified as the permeability barrier between the cells (Diamond 1977) are considered as the boundary between the apical and basolateral domain of the plasma membrane. Interestingly, these differences in lipid composition of epithelial cell plasma membrane domains are restricted to exoplasmic lipids since it was shown that the free diffusion of lipids between the apical and basolateral part was only prevented on the outer leaflet by the tight junctions but not on the cytoplasmic layer (Spiegel et al.

1985). However, as with non-polarized cells choline-containing lipids are preferentially on the outer leaflet, whereas aminophospholipids are enriched on the cytoplasmic leaflet (Simons and van Meer 1988). It was suggested that tight junctions might enable epithelia to segregate at least some lipids to the apical or basolateral membrane, thus, maintaing an asymmetric lipid distribution (Dragsten et al. 1981).

Phospholipid asymmetry in subcellular membranes of eukaryotic cells is not so firmly established and is still uncertain. The literature contains conflicting conclusions. For example, contradictory results were obtained for the transversal topology of aminophospholipids in a microsomal fraction from hepatocytes (see Zachowski and Devaux 1989). A more homogenous picture has emerged in the case of muscle sacroplasmic reticulum where PC is symmetrical, PE preferentially outside and PS preferentially inside (Herbette et al. 1984).

Spontaneous transmembrane diffusion of phospholipids

In unilamellar phospholipid vesicle membranes, the characteristic half time of phospholipid transverse diffusion is of the order of days. For a review see Op den Kamp, 1979. Modification of the phospholipid head group has little influence on the rate of spontaneous translocation, although SM has a slower rate than most other phospholipids (Zachowski et al. 1985a). By contrast a diacylglycerol molecule is capable to diffuse between two membrane leaflets at a very high rate i.e. with a half-time of a few seconds (Ganong and Bell 1984). It was proposed that the rate limiting step of transverse diffusion is the occurrence of corresponding defects in both leaflets due to statistical fluctuations (Homan and Pownall 1987). It is likely, that the probability of those defects can be enhanced by reconstituting integral membrane proteins and, consequently, would accelerate the flip-flop. De Kruijff et al. (1978) established an increase of the transbilayer movement of PC of at least two orders of magnitude after incorporation of glycophorin into sonicated PC-vesicles.

The presence of integral membrane proteins might be responsible for the elevated translocation rate of PC across human erythrocyte membranes in comparison to large unilamellar vesicles made with total lipid extract (Zachowski et al. 1985a). However, PC diffuses relatively slowly through the red blood cell membrane. The half-time is of the order of ten hours or more, the transfer being faster if the acyl chains are shorter (Fujii et al. 1985) and if unsaturations are present (Middelkoop et al. 1986).

In summary, results obtained with erythrocytes are very similar to the results obtained with artifical membranes. But this similarity is only true for PC. The other major phospholipids behave differently. Indeed the aminophospholipids (PS and PE) are transported from the outer to the inner layer by a specific transmembrane carrier protein which explains the origin of phospholipid asymmetry in eukaryotic plasma membranes (Devaux 1988).

The aminophospholipid translocase in the human erythrocyte membrane

Although the phenomenon of phospholipid asymmetry in mammalian plasma membranes is now well established, the origin and regulation is only in its beginning of understanding. Besides chemical attack after vesicle formation (Hope and Cullis 1987), artifical lipid vesicles can be made asymmetric by transverse lipid segregation due to high curvature (small unilamellar vesicles) or by a transmembrane pH gradient. However, there is no report that these mechanisms are sufficient to establish such a pronounced and stable asymmetric distribution of mixed lipids observed in plasma membranes.

The phospholipid asymmetry in the plasma membrane is **de novo** generated by the cell. Lipid asymmetry in plasma membranes could be the result of vectorial membrane biogenesis, combined with a low transverse diffusion rate. However, for cells, in particular erythrocytes, which have a low metabolic activity and slow lipid metabolism, this stable asymmetric topology must last much longer than the half-time of flip-flop. The normal life span of erythrocytes is about 120 days. Even in cells where plasma membranes undergo a continously recycling by endo- and exocytosis one would expect that at least a fraction of lipids is redistributed between both leaflets. It has been suggested that a specific interaction of the membrane skeleton proteins with aminophospholipids, in particular PS, could be responsible for the stable lipid asymmetry of erythrocytes and, probably, other eukaryotic plasma membranes (Haest and Deuticke 1978). PC and SM would be enriched in the outer bilayer simply by counterbalancing the inner monolayer. But there is now compelling evidence that the principal origin of the lipid asymmetry is a lipid specific ATP-dependent aminophospholipid translocase, so-called flippase. Originally, Bretscher has proposed a 'specific flippase' in 1973.

The aminophospholipid translocase continously pumps the aminophospholipids from the outer to the inner leaflet (Seigneuret and Devaux 1984) much faster in comparison to the mere diffusion of the choline-containing lipids. The latter would then accumulate on the outer monolayer as a result of the aminophospholipids occupying the inner leaflet.

Protein-mediated lipid translocation in human erythrocytes was discovered in 1984 using spin-labeled phospholipids with a short β-chain bearing a nitroxide radical (Fig.1A) (Seigneuret and Devaux 1984). The partial water solubility of these molecules offers an easy way to incorporate them in cell membranes without the participation of a phospholipid exchange protein. Spin-labeled phospholipids localized in the outer monolayer can be readily extracted back by bovine serum albumin (BSA) (Fig. 1B) as can be lysophospholipids and fatty acids (Haest et al. 1981). Thus, the flip-flop of spin-labeled lipids and their transmembrane steady-state distribution can be monitored.

When spin-labeled analogues of PS and PE are incorporated in the outer membrane leaflet at 37^0C, they rapidly translocate to the inner monolayer with half-times of 3-5 min for PS and 30-45 min for PE, respectively (Fig. 2). It is evident that the steady-state distribution of the spin-labeled analogues is identical to that of endogenous phospholipids (values refer to the percentage located on the inner leaflet): 95% for the spin-labeled PS, 87% for spin-labeled PE, 30 % for spin-labeled PC and about 10 % for spin-labeled SM (values for PC and SM being interpolations).

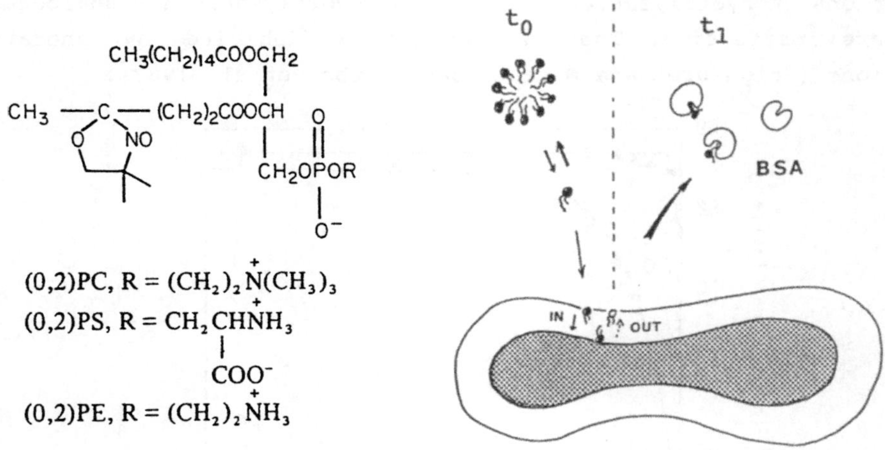

A B

Fig. 1 Technique for monitoring the inward and outward translocation of spin-labeled phospholipids (A) or the corresponding SM in plasma membranes. In an aqueous solution, these lipids form micelles in equilibrium with monomers. In the presence of large excess of membranes, they incorporate spontaneously and very rapidly on the outer monolayer of biological membranes (time t_0). Spin-labeled analogues which reside on the outer leaflet can be readily extracted by exposing the membranes to defated BSA. Hence, the fraction of spin-label outside (at t_0 100%) can be measured by Electron Spin Resonance. Addition of ferricyanide allows to re-oxidate reduced probes. It has been shown that reduction of probes takes place on the cytoplasmic layer. Thus, the presence of reduced probes of PS and PE on the outer layer provide evidence that the steady-state distribution results from a continous inward and outward motion (Bitbol and Devaux 1988).

Essentially the same translocation kinetics were obtained by Roelofsen and collaborators (1987) using long chain radioactive phospholipids which were introduced via a non specific exchange protein (Fig. 2). Their transmembrane distribution was assessed by the phospholipase A_2 technique. This technique does not allow the determination of the inital rate of phospholipid inward motion because of the time required for incorporation and transmembrane localization. However, Fig. 2 shows that the two techniques give very consistent results in the time domain where they overlap.

The selective inward motion of aminophospholipids in human erythrocytes requires the presence of millimolar concentrations of hydrolyzable Mg-ATP as non-hydrolyzable ATP analogues are ineffective. The translocation is inhibited by vanadate ions (Seigneuret and Devaux 1984, Bitbol et al. 1987).

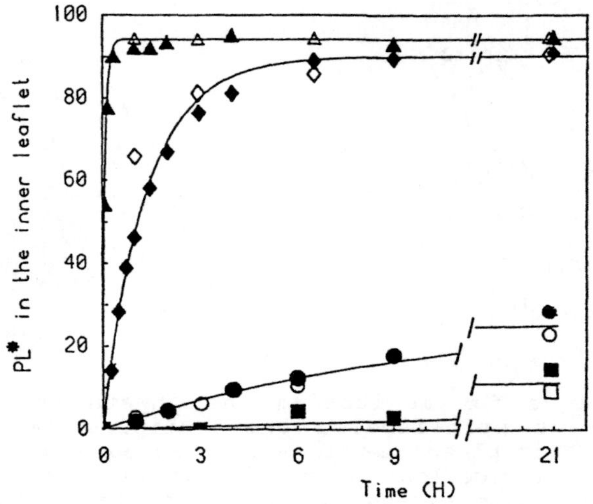

Fig. 2 Spontaneous passage at 37^{0}C in human erythrocytes from the outer to the inner layer of spin-labeled phospholipids (closed symbols (Morrot et al. 1989)) and long chain radioactive phospholipids (open symbols (Roelofsen et al. 1987)). PS: △ and ▲ ; PE: ◇ and ◆ ; PC: O and ● ; SM: □ and ■ .

The existence of a transport protein specific for aminophospholipids is confirmed by the sensitivity of fast inward motion of PS and PE to protein reagents such as NEM either at low (Zachowski et al. 1986) or physiological temperature (Fig. 3). Competition experiments demonstrated that the aminophospholipid translocase recognizes both PS and PE, but the apparent affinity of the latter molecule is approximately tenfold lower than that of the former one (Zachowski et al. 1986). Systematic investigation of the substrate specifity of the flippase established a high sensitivity to all chemical modification of the aminophospholipid headgroup. Also, the glycerol backbone and at least a short β-chain are required for fast inward motion via the aminophospholipid translocase. The affinity of lysoderivatives is very low and, as a result, they are not rapidly transported (Morrot et al. 1989). The slow diffusion of PC and SM is ATP-independent and NEM insensitive suggesting that PC and SM cross the membrane by a different mechanism. This view is sustained by a higher activation energy of PC diffusion when compared to the activation energy of PS and PE transport (Zachowski et al. 1986).

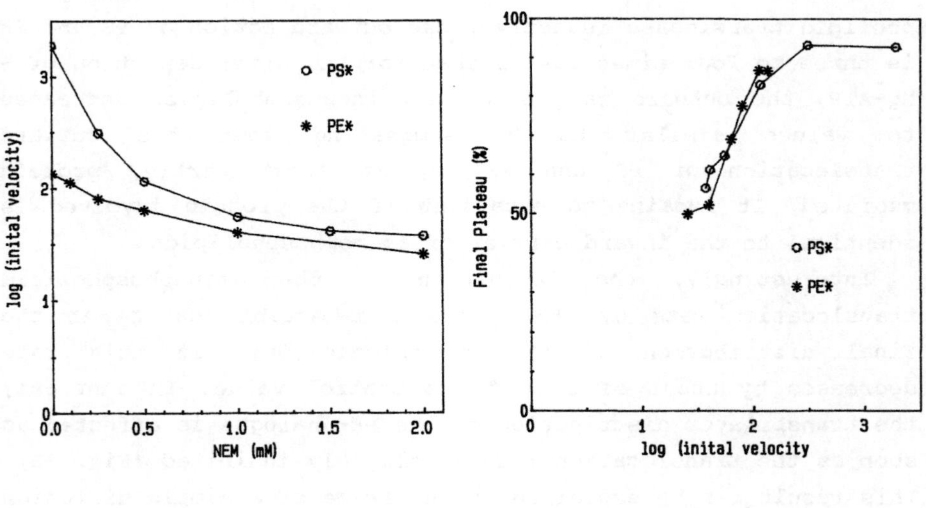

Fig. 3 Inhibition of the aminophospholipid translocase after preincubation of human red blood cells in the presence of NEM for 15 min at 37^{0}C. The inital velocity of inward motion [% of lipid analogues/min] for PS is much more affected by NEM in comparsion to that of PE (A) suggesting that both the activity as well as the specifity of the carrier are inhibited by blocking of SH-groups. In Fig. 3B the final plateau of lipid distribution [% inside] is plotted as a function of the inital velocity (3A) for a given NEM-concentration.

The transport is inhibited by micromolar amounts of intracellular Ca^{2+} (~1 µM), while extracellular Ca^{2+} has no effect (Zachowski et a. 1986, Bitbol et al. 1987). Both ligands, ATP and lipid, bind independently to the flippase, however, they need to be loaded at the same time (Devaux et al. 1988).

As suspected from theoretical consideration (Herrmann and Muller 1986) the equilibrium distributions for all lipids are in fact steady-states corresponding to the balance between continuous inward and outward translocation of lipids. The kinetic parameters of the outward transport were determined recently with spin-labeled phospholipids (Bitbol and Devaux 1988, Sune et al. 1988). Under conditions were the aminophospholipid translocase is active, the outward motion of PS and PE is three to four times faster than for PC. After depletion of -Mg-ATP, the outward rate of both aminophospholipids decreases to values similar to PC suggesting that the outward translocation of PS and PE is, at least partly, protein mediated. It remains to estabilsh if the protein involved is identical to the inward carrier of aminophospholipids.

Interestingly, the inhibition of the aminophospholipid translocation rate by NEM yields a detectable change in the final distribution of the PS-analogue only if this rate decreases by about of 80% of its control value. In contrast, the transbilayer distribution of the PE-analogue is affected as soon as the translocation rate is slightly inhibited (Fig. 3B). This result can be explained in the frame of a simple diffusion model, which as a first approximation describes the steady-state distribution of phospholipids as a result of an equality between the number of molecules transported inwards and the number of molecules transported outwards (Herrmann and Muller 1986, Bitbol and Devaux 1988). Although the model does not take into account the dependence of the inward motion on ATP and lipid concentration (Zachowski et al. 1986), it gives reasonable estimates for the inward and outward translocation rates k_i and k_0, respectively, as well as for the transversal equilibrium distribution q. According to the model, when $k_i/k_0 \simeq 1$, q is very sensitive to any variation of either k_i or

k_0. q is almost invariable when k_i/k_0>>1. These predictions are consistent with recent measurement of the inward and outward translocation of PS and PE, where k_i/k_0 is about 19 for the PS-analogue whereas it is about 6.7 for spin-labeled PE.

The ATP-dependent, protein-mediated translocation of aminophospholipids in red blood cells was confirmed by several independent techniques (see also Fig. 2). Connor and Schroit (1987), using resonance energy transfer, showed that exogenously supplied NBD labeled fluorescent PS was selectively transported to the inner leaflet, whereas PC remained in the exoplasmic leaflet. Taking into account the difference of viscosity between the two leaflets of erythrocyte membrane, Morrot et al. (1986) established by photobleaching measurements that the fluorescent long chain PS molecules are enriched on the inner leaflet. Daleke and Huestis (1985) took advantage of cell morphology as an assay for exogenous phospholipid distribution and concluded that the selective inward motion of aminophospholipids was mediated by a Mg^{2+}- and ATP-dependent protein.

Several years ago Haest et al. (1978) suggested that spectrin, the main cytoskeleton component, could be involved in maintaining phospholipid asymmetry. This hypothesis was supported by experiments on model systems showing an interaction between spectrin and PS and PE (Mombers et al. 1980). However, this interaction is strong only at acidic pH (below pH 6.5). Furthermore, it was demonstrated that the PS-spectrin interaction does not seem to be strong enough to account for the observed asymmetry (Maksymiv et al. 1987; Bitbol et al. 1989). Recently, we have shown that in spectrin-depleted vesicles obtained from erythrocytes, aminophospholipids are accumulated in the inner leaflet, as in native cells, provided that the vesicles contain ATP (Fig. 4) (Calvez et al. 1988). The translocation of the spin-labeled analogues PS and PE in heat-treated erythrocytes was still very effective even after 0.5 h incubation at 50^0C. The heating process (15 min at 50^0C) is accompanied by a partial (50%) denaturation of spectrin (Brandts et al. 1977).

Another argument in favor of a minor role of the

cytoskeleton in maintaining lipid asymmetry is the fact that the lipid transmembrane equilibrium is not a fixed state with a fraction of lipids anchored to spectrin but rather a steady-state with continous inward and outward motion of aminophospholipids across the cell membrane (see above). Further evidence for the essential role of the aminophospholipid translocase in generating and maintaining lipid asymmetry is given by the influence of NEM on the carrier activity. Taking the inital velocity of the inward transport of PS and PE as a measure of the carrier activity, the relation between carrier activity and transmembrane equilibrium shows a reasonable coincidence between PS and PE (Fig. 3B). In the presence of specific interaction between spectrin and PS curves should deviate from each other. This suggests that under these conditions the final distribution of aminophospholipids can be explained solely by the activity of the aminophospholipid translocase.

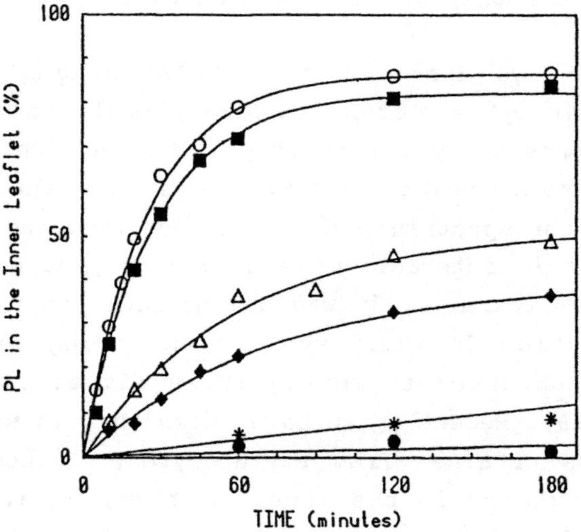

Fig. 4 Transmembrane reorientation at 37^{0}C of the different phospholipid analogues in heat-induced vesicles containing two different ATP-concentrations. Vesicles were formed by incubating red blood cells at 50^{0}C for a few minutes. PS with 0.6 (O) or 0.3 (■) mM ATP. PE with 0.6 (Δ) or 0.3 (◆) mM ATP. PC (*) and SM (●) with either ATP concentration (Calvez et al. 1988).

Recently, Middelkoop et al (1989) followed transmembrane equilibration of glycerophospholipids using Friend erythroleukemic cells (FELC) corresponding to an early stage of erythropoiesis which does not yet contain a complete membrane skeleton. Besides a rapid transverse equilibration they found a symmetric distribution for PE and a less pronounced asymmetric distribution for PS in comparison to mature cells. On the basis of these results they concluded that both an ATP-dependent translocase and spectrin were required to maintain the observed asymmetry in mature erythrocytes. This, however, appears to be a rather indirect demonstration which implies that in FELC spectrin is the only membrane protein which is not fully developed. Even, when compared with the red cell membrane, the enhanced reorientation of PC in FELC indicates other membrane alterations.

In conclusion, at the present time, it appears that the available data is more supportive of the exclusive role of the aminophospholipid translocase in the generation and maintenance of lipid asymmetry in red blood cells.

Aminophospholipid translocase in the plasma membrane of other cells

Using the spin label assay the same active transport of aminophospholipids as well as the asymmetric steady-state distribution was found in red blood cell membranes of other mammals: cows, sheeps, rats (Devaux et al. 1988) and guinea pigs (Sune et al., 1988). Interestingly, the temperature dependence of phospholipid translocation can be very different from one system to another. For example at 4^0C, in sheep, there is no measurable translocation of the PS-analogue while in pigs and in humans PS is transported rapidly to the inner layer (Devaux et al. 1988). The selective translocation of spin-labeled analogues of PS and PE across other plasma membranes has been demonstrated for platelets, reticulocytes and lymphocytes, as well as a bidirectional translocation of these lipids (Zachowski et al. 1985b, 1987; Sune et al. 1987, 1988). Similar to erythrocytes, a competition for a putative aminophospholipid translocase of the two analogues has been observed (Zachowski et al. 1987). Shape change analysis of platelets and reticulocytes confirmed the conclusion of a selective translocation of aminophospholipids (Daleke and Huestis 1985, Sune and Bienvenue 1988; Sune et al. 1987, 1988). By following the redistribution of exogenous incorporated fluorescent phospholipid analogues within the internal membranes of cells, evidence was obtained for a protein-mediated and ATP-dependent selective translocation of aminophospholipids in the plasma membrane of cultured fibroblasts. In contrast the transmembrane movement of the corresponding fluorescent PC across the plasma membrane of the same cells was not detectable (Sleight and Pagano 1984, 1985: Martin and Pagano 1987). Thus, there is now strong experimental evidence for the concept of a rather universal ATP-dependent membrane bound enzyme which can selectively translocate phospholipids and, consequently, control their transmembrane distribution in the plasma membrane of eukaryotic cells. It remains to establish whether such an enzyme also exists in prokaryotic cells. It should be mentioned that with cells containing internal membranes, the transmembrane steady-state distribution is more difficult to establish because of the flow of lipids toward the organelles.

Possible physiological function of the lipid asymmetry

The selective accumulation of aminophospholipids on the inner leaflet of plasma membranes is generated and maintained by the aminophospholipid translocase at the expense of ATP consumption. Thus, one should expect that such non-equilibrium distribution is important for cell homeostasis. However, the biological significance of this phenomenon is poorly understood. Several hypotheses have been put forward.

It has been shown that the membrane shape of erythrocytes and platelets is changed during the transverse diffusion of spin-labeled phospholipids incorporated on the outer leaflet (Seigneuret and Devaux 1984, Daleke and Huestis 1985, Sune and Bienvenue 1988). These results are consistent with the bilayer couple hypothesis of Sheetz and Singer (1974), which assumes that bilayer curvature is imposed by the relative areas of the inner and outer leaflets. Thus, regulation of the phospholipid asymmetry by the aminophospholipid translocase might also maintain cell shape together with cytoskeletal proteins imposing some topological constraints.

Recently, it has been proposed that asymmetric viscosity of the lipid phase contributes also to the determination of cell shape (Zachowski and Devaux 1989). There are several indications that the asymmetric lipid distribution is accompanied by an asymmetric fluidity in plasma membranes due to differences in unsaturation of fatty acid chains of both layers. The outer monolayer of erythrocytes was found to be less fluid (Seigneuret et al. 1984, Morrot et al. 1986) whereas the opposite asymmetry was found in the case of LM fibroblasts (Wisnieski and Iwata 1977).

Both asymmetric lipid distribution and fluidity are considered to be important in modulation of the activity of membrane bound enzymes and receptors. For example, the physical contact and the functional linking of the two separate proteins of the receptor-adenylate cyclase complex are influenced by the fluidity of each leaflet (Sweet and Schroeder 1988). It was suggested that the vertical displacement of membrane proteins, in particular the exposure of receptor proteins, is affected by alterations of the trans-bilayer lipid distribution which involves changes in lipid fluidity (Schroeder 1985). The activity of integral membrane enzymes such as the Ca^{2+}-ATPase (Niggli et al. 1981) or the protein kinase C (Rando 1988) are shown to be differentially modulated by negatively charged phospholipids.

Lipid asymmetry may be crucial in triggering membrane-membrane interaction involving recognition, adhesion and fusion of cells and/or subcelluar organelles. It has been shown that the presence of PS on the outer leaflet of erythrocytes, and the loosening of lipid packing, will increase their binding by macrophages and, subsequently, their uptake due to phagocytosis (Schroit et al. 1985). Thus the control of lipid asymmetry is one of the requirements of cell survival in the blood stream. This is underlined by the activation process of platelets which involves the redistribution of PS in the outer leaflet of the plasma membrane (Zwaal 1988).

The important role of aminophospholipids in membrane fusion provides potential clues concerning the rationale of lipid asymmetry. It was suggested that the preferential distribution of PE and PS in the outer plasma membrane leaflet of myoblasts may be related to plasma membrane fusion to form myotubes (Session and Horwitz 1983). Thus, cell-cell fusion which is normally a rare event may be prevented by exposing PC and SM on the outer leaflet as it is known that PC reduces fusion competence. Schlegel and Williamson (1988) postulated that the fusion competence of erythrocyte membranes depends on the packing of phospholipids in the outer leaflet which is controlled by the transverse lipid distribution; membranes with tightly-packed phospholipids are fusion-incompetent, whereas those with loosely-packed phospholipids are fusion-competent. Thus, phospholipid asymmetry and the aminophospholipid translocase may be important in exocytosis and endocytosis processes. The enrichment of aminophospholipids on the inner leaflet of most plasma membranes and on the outer layer of cytoplasmically located membrane vesicles, such as chromaffin granules (Buckland et al. 1978), may facilitate exocytosis. On the other hand, endocytosis requires local invaginations of cell surface which must be accompanied with an accumulation of lipids on the inner leaflet (Devaux 1989). Consequently, the activity of the aminophospholipid translocase could be involved in creating an excess of aminophospholipids in the inner leaflet. In this respect, it is interesting to note that (1) the translocation of PS to the inner leaflet is much higher in plasma membranes of lymphocytes than in erythrocytes (Sune et al. 1988); (2) high levels of ATP can induce red cells to endocytose (Ben-Bassat et al. 1972).

Acknowledgement

The authors thank Dr. M. Clague for correcting our English.

References

Berridge MJ (1987) Inositol triphosphate as a second messenger in signal transduction. Annu. N.Y. Acad. Sci. 494: 38-51.

Ben-Bassat I, Bensch RG, Schrier SL, (1972) Drug induced eryhtrocyte membrane internalization. J. Clin. Invest. 51: 1833-1840.

Bitbol M, Fellmann P, Zachowski A, Devaux PF (1987) Ion regulation of phosphatidylserine and phosphatidylethanolamine outside-inside translocation in human erythrocytes. Biochim. Biophys. Acta 904: 268-282.

Bitbol M, Devaux PF (1988) Measurement of outward translocation of phospholipids across human erythrocyte membrane Proc. Natl. Acad. Sci USA 85: 6783-6787.

Bitbol M, Dempsey, C, Watts, A, Devaux PF (1989) Weak interaction of spectrin with phosphatidylcholine-phosphatidylserine multilayers: a ^{2}H and ^{31}P NMR study. FEBS lett. 244: 217-222.

Brandts J, Erickson C, Lysko K, Schwartz AT, Taverna RD (1977) Calorimetric studies of the structural transitions of the human erythrocyte membrane. The involvement of spectrin in the A transition. Biochemistry 16: 3450-3454.

Brasitus TA, Schachter D (1980) Lipid dynamics and lipid-protein interactions in rat enterocyte basolateral and microvillus membranes. Biochemistry 19: 2763-2769.

Bretscher M (1972) Phosphatidyl-ethanolamine: Differential labelling in intact cells and cell ghosts of human erythrocytes by an impermeable membrane reagent. J. Mol. Biol. 71: 523-528.

Bretscher M (1973) Some general principles of membrane structure. In: Cell surface in development (Moscona AA, ed.) pp. 17-27. Wiley, N.Y.

Buckland RM, Radda GK, Shennan CD (1978) Accessibility of phospholipids in the chromaffin granule membrane. Biochim. Biophs Acta 513: 321-337.

Calvez JY, Zachowski A, Herrmann A, Morrot G, Devaux PF (1988) Asymmetric distribution of phospholipids in spectrin-poor erythrocyte vesicles. Biochemistry 27: 5666-5670.

Connor J, Schroit AJ (1987) Determination of lipid asymmetry in human red blood cells by resonance energy transfer. Biochemistry 26: 5099-5105.

Curatola G, Bertoli E (1987) Microheterogenity of biological membranes. In: Structural and functional implications in biomembranes and receptor mechanisms. (Bertoli E, Chapman D, Cambia A, Scapagnini V, eds.) pp. 143-161. Fidia Res. Series ed. 7, Liviana Press, Padova.

Daleke DL, Huestis WH (1985) Incorporation and translocation of aminophospholipids in human erythrocytes. Biochemistry 24: 5406-5416.

De Kruijff B, Van Zoelen EJ, Van Deenen LLM (1978) Glycophorin facilitates the transbilayer movement of phosphatidylcholine in vesicles. Biochim. Biophys. Acta 509: 537-542.

Devaux PF, Morrot G, Herrmann A, Zachowski A (1988) Protein involvement in plasma membrane lipid asymmetry. studia biophysica 1-3: 183-191.

Devaux PF (1988) Phospholipid flippases. FEBS lett. 234: 8-12.
Devaux PF (1989) The aminophospholipid translocase. A transmembrane lipid pump - Physiological significance. News Physiol. Sci., in press.
Diamond JM (1977) The epithelial junction: Bridge, gate, and fence. Physiologist 20 (1):10-18.
Dragsten PR, Blumenthal R, Handler JS (1981) Membrane asymmetry in epithelia: is the tight junction a barrier to diffusion in the plasma membrane? Nature 294: 718-722.
Etemadi AH (1980) Membrane asymmetry. A survey and critical appraisal of the methodology. II. Methods for assessing the unequal distribution of lipids. Biochim. Biophys. Acta 604: 423-475.
Fujii T, Tamura A , Yamane T (1985) Trans-bilayer movement of added phosphatidylcholine and lysophosphatidylcholine species with various acyl chain length in plasma membrane of intact human erythrocytes. J.Biochem 98: 1221-1127.
Ganong BR, Bell RM (1984) Transmembrane movement of phosphatidylglycerol and diacylglycerol sulfhydryl analogues. Biochemistry 23: 4977-4983.
Haest CWM, Deuticke B (1978) Spectrin as a stabilizer of the phospholipid asymmetry in the human erythrocyte membrane. Biochim. Biophys. Acta 509: 21-32.
Haest CWM, Plasa G, Deuticke B (1981) Selective removal of lipids from the outer membrane layer of the human erythrocyte without hemolysis. Consequences for bilayer stability and cell shape. Biochim. Biophys. Acta 649: 701-708.
Hanahan DJ, Nelson DR (1984) Phospholipids as dynamic participants in biological processes. 25: 1528-1535.
Herbette L., Blasie JK, Deffon S, Fleisher S, Bick RJ, Van Winkle WB, Tate CA, Entmann ML (1984) Phospholipid asymmetry in the isolated sarcoplasmic reticulum membrane. Arch. Biochem. Biophys. 234: 235-242.
Herrmann A, Müller P (1986) A model for the asymmetric lipid distribution in the human eryhtrocyte membrane. Biosc. Rep. 6:185-191.
Homan R, Pownall HJ (1987) Effect of pressure on phospholipid translocation in lipid bilayers. J. Am. Chem. Soc. 109: 4759-4760.
Hope MJ, Cullis PR (1987) Lipid asymmetry induced by transmembrane pH gradients in large unilamellar vesicles. J. Biol. Chem. 262: 4360-4366.
Maksymiv R, Sui SF, Gaub H, Sackmann E (1987) Electrostatic coupling of spectrin dimers to phosphatidylserine containing lipid lamellae. Biochemistry 26: 2983-2930.
Martin OC, Pagano RE (1987) Transbilayer movement of fluorescent analogs of phosphatidylserine and phosphatidylethanolamine at the plasma membrane of cultured cells. J. Biol. Chem. 262: 5890-5898.
Middelkoop E, Lubin BH, Op den Kamp JAF, Roelofsen B (1986) Flip-flop rates of individual molecular species of phosphatidylcholine in the human red cell membrane. Biochim. Biophys. Acta 855: 421-424.

Middelkoop E, Coppens A, Llanillo M, Van der Hoek EE, Slotboom AJ, Lubin BH, Op den Kamp JAF, Van Deenen LLM, Roelofsen B (1989) Aminophospholipid translocase in the plasma membrane of Friend erythroleukemic cells can induce an asymmetric topology for phosphatidylserine but not for phosphatidylethanolamine. Biochim. Biophys. Acta 978: 241-248.
Mombers C, De Gier J, Demel RA, Van Deenen LLM (1980) Spectrin-phospholipid interaction. A monolayer study. Biochim. Biophys. Acta 603: 52-62.
Morrot G, Cribier S, Devaux PF, Geldwerth D, Davoust J, Bureau JF, Fellmann, Herve P, Frilley B (1986) Asymmetric lateral mobility in the human erythrocyte membrane. Proc. Natl. Acad Sci. USA 83: 6863-6867.
Morrot G, Herve P, Zachowski A, Fellmann P, Devaux PF (1989) Biochemistry, in press.
Niggli V, Adunyah ES, Crafoli E (1981) Acidic phospholipids, unsaturated fatty acids and limited proteolysis mimic the effect of calmodulin on the purified erythrocyte Ca^{2+}-ATPase. J. Biol. Chem. 256: 8588-8592.
Op den Kamp JAF (1979) Lipid asymmetry in membranes. Annu. Rev. Biochem. 48: 47-71.
Rando RR (1988) Regulation of protein kinase C activity by lipids. FASEB J. 2: 2348-2355.
Roelofsen B, Op den Kamp JAF, Van Deenen LLM (1987) Structural and dynamical aspects of red blood cell phospholipids, featuring phosphatidylcholine. Biomed. Biochim. Acta 46: S10-S13.
Schlegel RA, Williamson P (1988) Role of phospholipid asymmetry in cellular membrane fusion. In: Molecular mechanisms of membrane fusion. (Ohki S, Doyle D, Flanagan TD, Hui SW, Mayhew E, eds). pp. 289-302. Plenum Press. N.Y., London.
Schroeder F (1985) Fluorescence probes unravel asymmetric structure of membranes. In: Subcellular Biochemistry, Vol 11 (Roodyn DB, ed.) pp. 51-101. Plenum Publishing Corporation.
Schroit AJ, Madsen JW, Tanaka Y (1985) In vivo recognition and clearance of red blood cells containing phosphatidylserine in their plasma membranes. J. Biol. Chem. 260: 5131-5138.
Seigneuret M, Devaux PF (1984) ATP-dependent asymmetric distribution of spin-labeled phospholipids in the erythrocyte membrane: Relation to shape changes. Proc. Natl. Acad. Sci. USA 81: 3751-3755.
Seigneuret M, Zachowski A, Herrmann A, Devaux PF (1984) Asymmetric lipid fluidity in human erythrocyte membrane: New spin-label evidence. Biochemistry 23: 4271-4275.
Session A, Horwitz AF (1983) Differentiation related differences in the plasma membrane phospholipid asymmetry of myogenic and fibrogenic cells. Biochim. Biophs. Acta 728: 103-111.
Sheetz MP, Singer SJ (1974) Biological membranes as bilayer couples. A molecular mechanism of drug-erythrocyte interactions. Proc. Natl. Acad. Sci. USA 71: 4457-4461
Simons K, Van Meer G (1988) Lipid sorting in epithelial cells. Biochemistry 27: 6197-6202.

Singer SJ, Nicolson GL (1972) The fluid mosaic model of the structure of cell membranes. Science 175: 720-731.
Sleight RG, Pagano RE (1984) Transport of a fluorescent phosphatidylcholine analog from the plasma membrane to the Golgi apparatus. J. Cell Biol. 99: 742-751.
Sleight RG, Pagano RE (1985) Transbilayer movement of a fluorescent phosphatidylethanolamine analogue across the plasma membrane of cultured mammalian cells. J. Biol. Chem. 260: 1146-1154.
Spiegel S, Blumenthal R, Fishman PH, Handler JS (1985) Gangliosides do not move from apical to basolateral plasma membrane in cultured epithelial cells. Biochim. Biophys. Acta 821: 310-318.
Sune A, Bette-Bobillo P, Bienvenue A, Fellmann P, Devaux PF (1987) Selective outside-inside translocation of aminophospholipids in human platelets. Biochemistry 26: 2972-2978.
Sune A, Vidal M, Morin P, Sainte-Marie J, Bienvenue A (1988) Evidence for bidirectional transverse diffusion of spin-labeled phospholipids in the plasma membrane of guinea pig blood cells. Biochim. Biophys. Acta 946: 315-327.
Sune A, Bienvenue A (1988) Relationship between the transverse distribution of phospholipids in plasma membrane and shape change of human platelets. Biochemistry 27: 6794-6800.
Sweet WD, Schroeder F (1988) Lipid domains and enzyme activity In: Advances in membrane fluidity. Vol. II (Aloia RC, Curain CC, Gordon LM, eds.) pp. 17-42. Alan R. Liss Inc. N.Y.
Van Deenen LLM (1981) Topology and dynamics of phospholipids in membranes. FEBS lett. 123: 3-15.
Wisnieski BJ, Iwata KK (1977) Electron spin resonance evidence for vertical asymmetry in animal cell membranes. Biochemstry 16: 1321-1326.
Zachowski A, Fellmann P, Devaux PF (1985a) Abscence of transbilayer diffusion of spin-labeled spingomyelin on human erythrocytes. Comparison with the diffusion of several spin-labeled glycerophospholipids. Biochim. Biophys. Acta 815: 510-514.
Zachowski A, Craescu CT, Galacteros F, Devaux PF (1985b) Abnormality of phospholipid transverse diffusion in sickle erythrocyte. J. Clin. Invest. 75: 1713-1717.
Zachowski A, Favre E, Cribier S, Herve P, Devaux PF (1986) Outside-inside translocation of aminophospholipids in the human eryhtrocyte membrane is mediated by a specific enzyme. Biochemistry 25: 2585-2590.
Zachowski A, Herrmann A, Paraf A, Devaux PF (1987) Phospholipid outside-inside translocation in lymphocyte plasma membranes is a protein-mediated phenomenon. Biochim. Biophys Acta 897: 197-200.
Zachowski A, Devaux PF (1989) Bilayer asymmetry and lipid transport across biomembranes. Comm. Molec. Cell. Biophys., in press.
Zwaal RFA (1988) Scrambling membrane phospholipids and local control of blood clotting. News Pysiol. Sci. 3: 57-61.

Protein-lipid interactions at membrane surfaces

Anthony Watts
Department of Biochemistry,
University of Oxford,
South Parks Road,
OXFORD, OX1 3QU,
United Kingdom

INTRODUCTION

Specific molecular interactions that determine many of the functions of a biomembrane have a high probability of occurring at the surface of that membrane. However, unlike for those interactions which occur in the hydrophobic core, the interactions between lipids and proteins which may occur at the polar-apolar interface of biomembranes have been somewhat neglected experimentally. The chemical heterogeneity and complexity (Ohshima and Kondo, 1986; 1987) encountered in all components at the membrane surface, makes a simple but comprehensive molecular description of their mutual molecular interactions (such as electrostatic, hydrophilic, allosteric, etc) very difficult. In addition, probing the membrane surface often involves a perturbation of those determinants, such as lateral pressure (Cevc, 1987, 1988), which are being studied and are so important in determining the molecular architecture of the membrane surface.

Plasma membrane surfaces are the initial site of contact for a cell with the external environment. They are involved in cell-cell adhesion, cell fusion and other cytotic mechanisms, recognition for requisite as well as "foreign" molecules, absorption and transport of ions and metabolites, the binding site for numerous structural and enzymatic proteins, for example myelin proteins and those involved in second messenger cascade sequences and other trigger events leading to such diverse effects as cell death on the one hand to cell replication on the other. Indeed all chemically distinguishable biomolecules, including nucleic acids, interact at some stage with either plasma or intracellular membranes and it could be argued that all aspects of cellular activity involve, fairly closely, a membrane of one sort or another. Furthermore, macroscopic alterations in membrane structure involving fusion and cytotic related processes (De Kruijff and Cullis, 1980) affect the membrane surface and may be initiated or controlled through surface interactions. Traditional methods for obtaining detailed structural information about biomolecules, including diffraction methods, electron microscopy and more recently high-resolution 2D nuclear magnetic resonance techniques, are still inappropriate for all but the simplest of membrane problems, although progress towards this goal is accelerating

(Campbell, 1988). Sequencing methods are powerful in detailing linear arrays of polypeptide chains in membrane proteins, but to translate this information into three dimensional structural organization for the sequence is less than straightforward. In none of these methods is it possible to extend the information obtained to explain the way in which proteins and lipids dynamically interact at the membrane surface to affect either any conformational or any functional consequences of those interactions.

Membranes are not static structures and therefore structural information must be considered together with dynamic information. The dynamic range for molecular interactions within biomembranes is very large. To define a dynamic benchmark related to function is not easy; the initial events in photoreception occur in picoseconds and bulk membrane turnover occurs in hours or days depending upon the mechanism. Spectroscopic methods for studying molecular motions in membranes also have their own intrinsic time-scale sensitivity. To understand fully membranes in detail from a *functional* viewpoint, therefore requires a comprehensive *chemical* description, the two being linked by structural and dynamic details; this linking is the major task of the membrane biophysicist.

Both integral and peripheral proteins perturb, at least structurally and dynamically, the phospholipid head groups of membranes (Figure 1) (Israelachivili *et al.*, 1980). Similarly, proteins are affected, perhaps allosterically in some cases (Sandermann, 1986), by the lipids with which they interact. The important question is "do these surface interactions have any significance functionally for the cell". In this chapter, a comprehensive review of the literature will not be presented. Some basic principles will be described and supporting experimental indications for these principles presented, when available. Also, the discussion will be confined to naturally derived proteins and peptides, rather than synthetic ones.

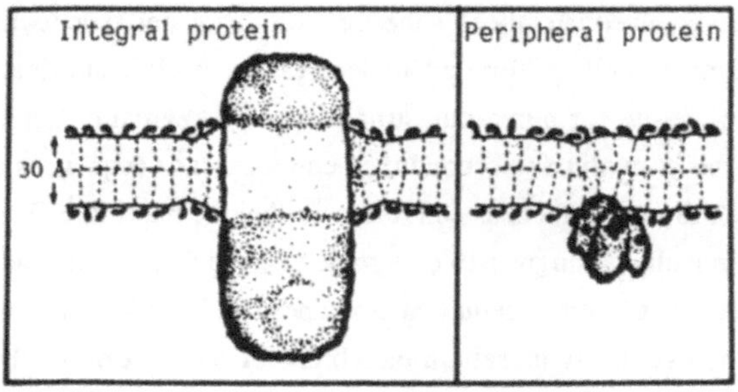

Figure 1: Sketch showing how both integral and peripheral proteins can perturb the membrane lipid bilayer architecture and dynamics, possibly by necking or menisus formation at the hydrophobic lipid-protein interface (Hippe and Lüth, 1986). Conversely, protein structure can be modulated by interaction with lipid polar head groups. The types of interaction possible include electrostatic, hydration, hydrophilic and allosteric. (adapted from Israelachvili *et al.*, 1980).

SURFACE-ACTIVE PEPTIDES

The major energetic contribution to the binding of peripheral (or extrinsic) proteins to biological membranes is that provided predominantly through electrostatic and hydrophilic interactions (Cevc, 1987). Such interactions could occur at the membrane surface between basic amino acids and anionic phosphates of phospholipids (usually 50% or more of the total membrane lipid) or between acidic amino acid side chains and primary amino groups of, for example, phosphatidylserine and phosphatidylethanolamine. Partial penetration of peripheral peptides and proteins does take place, to varying degrees depending upon the protein, but the hydrophobic contribution to binding is not likely to be great when compared with the hydrophilic and electrostatic forces. Since the insertion of integral proteins into membranes in the initial stages involves the association and then penetration of the bilayer a small leading signal peptide of between 15 and 30 amino acids, this initial sequence being highly basic, any information regarding small peptide associations with membranes is likely to be of some assistance in understanding the energetics and mechanism of insertion of any protein into a membrane. In addition, because of the lack of structural information on integral proteins, studies have been carried out with small peptides in the anticipation that results of peptide-membrane interactions may be applicable to integral proteins; this may well not be the case however.

Polymyxin-B

Of physiological interest is the antibiotic peptide polymyxin-B which has a hydrophobic side chain and polar cyclic amino acid ring which can electrostatically associate with lipid head groups. This peptide is cationic and binds to the outer and cytoplasmic membranes of Gram-negative bacteria causing highly increased membrane permeability, release of cytoplasmic material and finally cell death. It has been suggested that this peptide has a high affinity for anionic lipids (HsuChen and Feingold, 1973; Sixl and Galla, 1981; El Mashak and Tocanne, 1980; Hartman *et al.*, 1978).

Phosphorus-31 NMR and ^{2}H-NMR studies of specifically deuterated phospholipids in bilayers have been carried out to try and give an insight into the molecular associations between polymyxin-B and anionic phospholipids. Deuterium-NMR in particular, is capable of monitoring, in a quantitative way, the electrostatic changes which occur at a membrane surface by using deuterated lipid head groups as sensors at the bilayer surface (Watts and Poile, 1986; Seelig *et al.*, 1987). This method was therefore used to study the effects of polymyxin-B with charged and zwitterionic lipid bilayers of phosphatidylglycerol with dimyristoyl chains, perdeuterated ($-O-C(\alpha)^2H_2-C(\beta)^2HOH-C(\gamma)^2H_2OH$) in its head group in single and mixed bilayers with dimyristoylphosphatidylcholine, which was also deuterated in its polar head groups. On the addition of the antibiotic, the spectrum from the labelled phosphatidylglycerol molecules is very similar both in the mixed and the single lipid bilayers. A converse experiment in which the dimyristoylphosphatidylcholine is labelled shows that the perturbing effect of phosphatidylglycerol on the conformation or amplitude of motion of the phosphatidylycholine head groups (measured from the deuterium quadrupole splittings, $\Delta\nu_q$ in hertz; Sixl and Watts, 1982) is reversed after addition of polymyxin-B. Thus, on binding of polymyxin to the mixed lipid bilayers, the phosphatidylcholine head groups give ^{2}H-NMR spectra with quadrupole splittings which are the same as for phosphatidylcholine bilayers in the absence of phosphatidylglycerol molecules (Sixl and Watts, 1982), implying that lateral phase separation of peptide-charged lipid complexes had occurred.

Both these independent observations show that the peptide induces lateral phase separation of the phosphatidylglycerol molecules into phosphatidylglycerol-polymyxin-B complexes which become separated from the dimyristoylphosphatidylglycerol-dimyristoylphosphatidylcholine bilayers of ideally mixed lipids in the absence of the peptide (Sixl and Watts, 1982) (Figure 2). This demonstrates that the ^{2}H-NMR method is capable of identifying lateral lipid phase separation, where the lifetime of the separated regions is shorter than 0.1ms, which is the time needed to average the quadrupolar anisotropy. It may well be the case, therefore, that the negatively charged phospholipids are laterally separated from the rest of the lipids in the bacterial membrane bilayer leading to destabilization, possibly at the peptide-complex interface, finally leading to membrane rupture and the release of cell contents. The acyl chain motional characteristics of phosphatidylglycerol lipids in bilayers can also be switched isothermally from the gel to the liquid crystalline phase by pH titration of the lipid phosphate (Watts *et al.*, 1978) and such changes almost certainly also take place on peptide or ion binding.

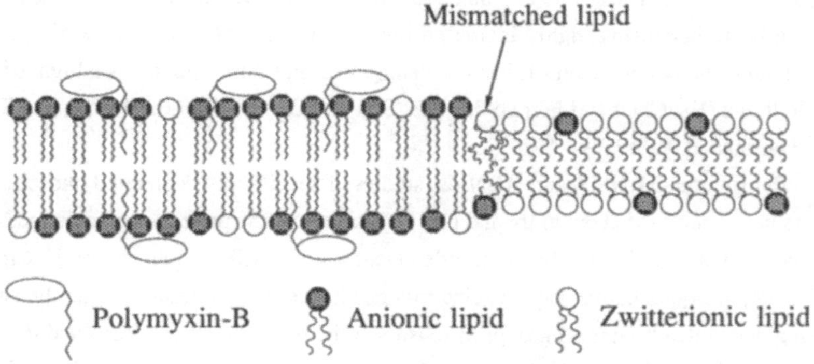

Figure 2: One possible way in which polymyxin-B could induce lateral phase separation of anionic lipids from uncharged lipids in a mixed bilayer. The interaction between peptide and lipid would cause a change in bilayer thickness through modulation of the acyl chain motions. The mis-match which then occurs between separated regions may be a point of instability in the cell membrane and a site at which membrane rupture takes place. The peptide antibiotic could, by this mechanism, cause release of cell contents and cell death.

To obtain information about the dynamics of the head-group segments, spin-lattice (T_1) relaxation time measurements were also performed (Sixl and Watts, 1985) to give information about the fast carbon-carbon bond motions in the head group, typically with τ_c of less than 10ns (Brown *et al.*, 1979). Polymyxin-B reduces slightly the fast segmental motions of the head group, especially the remote γ-C^2H_2 segment of the glycerol head group (Sixl and Watts, 1985). This would correspond to an increase in surface viscosity for the bilayer upon peptide binding.

The ^{2}H-NMR approach for studying directly the surface of membranes can therefore give useful information about rather subtle dynamic changes induced on ion (Seelig *et al.*, 1987) and small peptide associations with the membrane surface.

Melittin

Melittin (3.4kDa) has been widely studied in crystals, free solution and in its interactions with membranes. Being a surface-active peptide, its mode of association with bilayers has been of special interest but unfortunately the peptide, when prepared, is often contaminated with phospholipase A_2 (PLA_2). Recently, however, PLA_2-free melittin (Dasseux *et al.*, 1984) has been studied in its interaction with the surface of phospholipid bilayers by ^{2}H-NMR methods (Dufourc *et al.*, 1986; Dempsey and Watts, 1987). These studies have shown that the previously reported elevated phase-transition temperatures of saturated acyl chain phospholipid bilayers (Lavialle *et al.*, 1980) are not due to the peptide alone, but may be induced by a combination of the effects of melittin and PLA_2-generated hydrolytic products, namely fatty acids and mono-acyl phospholipids (Dassuex *et al.*, 1984; Dempsey and Watts, 1987).

Melittin is a surface-active peptide but it induces gross molecular organization of membrane bilayers and ^{2}H-NMR, in common with ^{31}P-NMR (de Kruijff *et al*, 1985a), can be used to identify these macroscopic phase changes of a phospholipid dispersion as well as under ideal conditions, the time scale for their formation (Dempsey and Watts, 1987). This arises because the broad, spherically averaged NMR spectra (typically kHz wide) recorded for extended bilayers, due to partial but incomplete averaging of the axial anisotropic magnetic interactions of ^{2}H and ^{31}P nuclei (Seelig, 1977; Seelig and Seelig, 1980; Davis, 1983), can be isotropically averaged by rapid motion of the nuclei with respect to the applied magnetic field in small complexes which are less than about 100nm in diameter (Burnell *et al.*, 1980) to give narrow (less than about 100Hz) spectral lines. Narrowed ^{2}H- and ^{31}P-NMR spectra are also recorded for some other lipid phases, designated by X-ray studies as cubic or rhombic phases, which also have tight radii of curvature around which the molecules can laterally diffuse quickly and average out the spectral anisotropy. Other lipid complexes, for example micelles and small vesicles, additionally tumble quickly themselves in solution to average out the spectral anisotropy.

This spectral sensitivity can be used to indicate the macroscopic reorganization of bilayers. It has been shown from measurements of the ^{2}H-NMR spectra of acyl chain-labelled dipalmitoyl phosphatidylcholine (Dufourc *et al.*, 1986) and head-group-labelled dimyristoylphosphatidylcholine (Dempsey and Watts, 1987) that melittin induces the formation of small micellar phospholipid particles, probably discs (Dufourc *et al.*, 1986) (Figure 3), below the dimyristoylphosphatidylcholine phase-transition temperature of about 21-23°C. Above the phase-transition temperature, purified melittin induces reformation of large (with diameters greater than ca. 100nm) extended complexes of liquid crystalline bilayers.

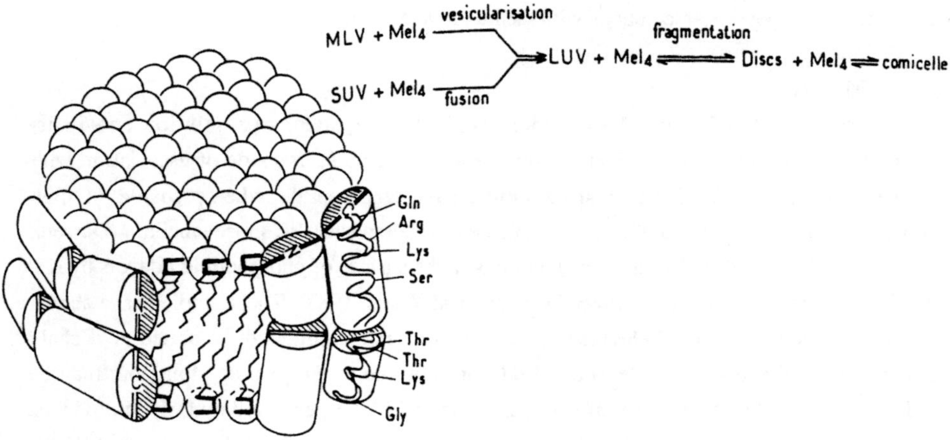

Figure 3: The bilayer to disc transition induced by melittin in bilayers on going from above to below the acyl chain liquid crystalline to gel phase transition. The small particles formed below the phase transition temperature of the lipid give rise to narrow, single line ^{2}H-NMR and ^{31}P-NMR spectra through averaging of the spectral anisotropy. (taken from Dufourcq *et al.*, 1986)

An interesting aside in this work is the effect that melittin, which is thought to reside in a near helical conformation on the bilayer surface, may reinforce the orientational influence of the applied magnetic field on bilayers in a strong magnetic field, such as that used in high field NMR experiments. This orientation, due to the diamagnetic anisotropy induced by the parallel packing of the acyl chains, is observed as a distortion in the expected spherically averaged ^{2}H-NMR powder pattern by an increased intensity of the spectral contribution from the CD-groups in the 90°-orientation, that is, the outer wings of the spectrum (Seelig and Seelig, 1985). In the ^{2}H-NMR spectra for head group deuterated dimyristoylphosphatidylcholine recorded after re-warming the bilayer dispersion above the phase transition, such a spectral distortion was quite clearly observed. Thus, the extended bilayers were oriented by the magnetic field during reformation as they were warmed in the presence of melittin so that the field direction was parallel to the bilayer plane and the peptide molecular axis aligned with the magnetic field. The potential is now available for describing the orientational behaviour of, say, a labelled peptide bound to or embedded in a membrane (see below).

With the suggestion that phospholipid head groups can act as sensors of membrane surface charge (Seelig *et al.*, 1987), it follows that their orientational and conformational properties may sense the binding of peptides if that binding perturbs the surface electrostatics. Melittin has been shown to bind to phosphatidylcholine bilayers in the melittin concentration range of 0-100μM, with a partition coefficient of $(2.1 \pm 0.2) \times 10^3\ M^{-1}$. As in a previous study (Dempsey and Watts, 1987), the lipid head groups were sensitive to the peptide binding in the same way as for other positive membrane surface charges in general. Indeed, quantitatively, melittin was found to be one of the most efficient head group modulators, the efficiency per unit charge being comparable to that of charged local anaesthetics (Watts and Poile, 1986; Seelig *et al.*, 1988) or hydrophobic ions (Altenbach and Seelig, 1985).

From titration ^{2}H-NMR experiments of melittin into bilayers of head group deuterated phosphatidylcholine and phosphatidylserine (with dimyristoyl acyl chains), it has been possible to reverse the perturbation of phosphatidylcholine head groups by melittin through specific binding of the peptide to the anionic lipid head groups (Dempsey *et al.*, 1989). No long-lived (with a life-time of greater than ms) lateral phase separated regions of either lipid were observed in the presence of the peptide. Also, it was found that the anionic lipid greatly stabilized the phosphatidylcholine bilayers with respect to the previously reported melittin-micellization (see above) indicating that this property of melittin may not be important for the hæmolytic activity of the peptide. In this study with mixed lipid bilayers, however, it was shown that the perturbation of the lipid head groups on peptide binding could not be simply accounted for exclusively by electrostatic effects. It was suggested that some conformational changes are induced which are not charge related and may possibly be due to surface dehydration as shown for phospholipase A_2 binding (Jain and Vaz, 1989) and suggested from considerations of the various contributions to membrane surface pressure (Cevc, 1987; 1988) in determining polar head group conformations.

PERIPHERAL PROTEINS

Mechanisms whereby peripheral proteins can associate with membrane surfaces are often studied by changing the bulk pH or ionic strength; it is these changes which are usually used to release the protein from the membrane surface during purification processes. The effects on either the molecular motions at the bilayer surface or within the hydrophobic core, are then monitored. Indeed, the effects of pH and ionic strength cannot be readily separated since lipid-lipid interactions are controlled at the membrane surface and have corresponding repercussions in the core due to the changes in bilayer packing and chain motion cooperativity.

Cytochrome *c* and apocytochrome *c*

Cytochrome *c* binds avidly to liposomes of particular composition at physiological pH and ionic strength in the absence of calcium. Nine lysine residues exist along one edge of the protein (M_r 12,300) and are available, in principle, for binding to a charged bilayer surface. For example, with bilayers of cardiolipin (which could provide a possible binding site in the mitochondrion) the stoichiometry has been shown to be 43 lipids per protein from binding studies (De Kruijff and Cullis, 1980). Phosphorus-31 NMR experiments have been used to demonstrate that such protein-lipid complexes, with single lipid types only, always consist of bilayers but these may coexist with a population of non-lamellar lipid with mixed lipid dispersions. Such dispersions, on the addition of calcium, form non-bilayer, hexagonal H_{II} phase structures which can also occur in the absence of the protein (Rand and Sengupta, 1972; Cullis *et al.*, 1978).

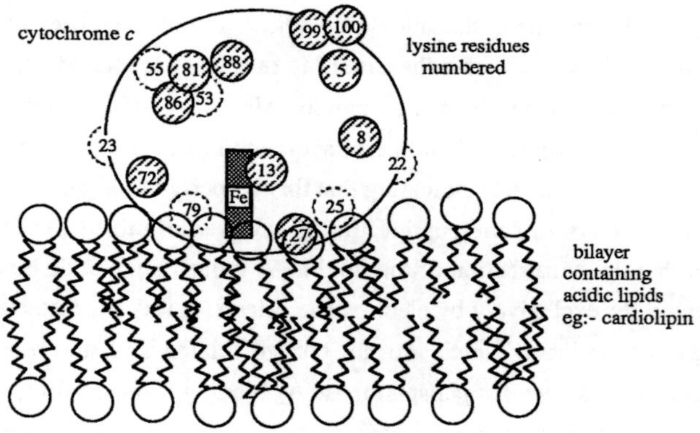

Figure 4: Representation of cytochrome *c* showing the approximate location of the basic lysine residues which could potentially bind to the head groups of anionic lipids, such as cardiolipin, at bilayer surfaces.

It would be interesting to study the interaction of cytochrome *c* with cardiolipin head groups, in view of a possible essential rôle for this molecular specificity in the mitochondria, using the non-perturbing ^{2}H-NMR method and labelled lipid, preferably in the glycerol head group. Such a lipid has been produced biosynthetically by incorporating deuterated glycerol into the growth medium of an E. coli mutant which was enhanced in its capability to produce cardiolipin (Allegrini *et al.*, 1984). However, this expensive experimental approach has not been pursued further and an efficient complete synthesis of cardiolipin, as well as a number of other phospholipids, must be resolved before more minor lipids can be studied realistically using biophysical methods.

In the absence of deuterated cardiolipin, cytochrome *c* has been studied in its interaction with bilayer surfaces by NMR methods in particular, using the natural cardiolipin. Phosphorus-31 NMR studies have suggested that the addition of cytochrome *c* to negatively charged phosphatidylserine or phosphatidylglycerol liposomes was found to cause no conformational or spectrum of these lipids (De Kruijff and Cullis, 1980). Similarly, the protein did not induce changes in the ^{31}P-NMR bilayer spectrum of mixed phosphatidylcholine/cardiolipin liposomes, unlike phosphatidylethanolamine/cardiolipin liposomes which did show indications of macroscopic phase changes to non-bilayer dispersions (Cullis *et al.*, 1980).

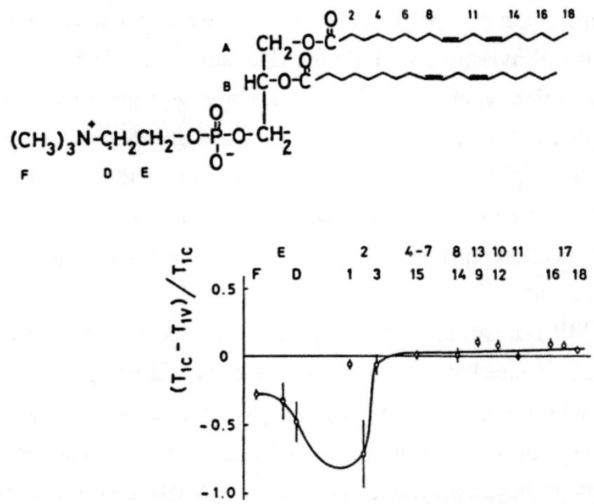

Figure 5: Values of $(T_{1c} - T_{1v})/T_{1c}$, corresponding to a fractional change in *effective mobility* of a particular lipid segment determined from ^{13}C-NMR T_1 relaxation times following binding of cytochrome *c* to bilayers of bovine heart cardiolipin/hen egg yolk phosphatidylcholine, 1:4 (w/w). Negative values indicate reduced mobility and T_{1c} and T_{1v} are the spin-lattice relaxation times for the segments with and without protein respectively. Numbers and letters in the lower part of the diagram refer to the acyl chain segments and head groups respectively, as shown in the upper formula of phosphatidylcholine (adapted from Brown and Wüthrich, 1977).

Proton and ^{13}C-NMR studies have been used to identify specific ionic associations between ferricytochrome *c* and head groups of phospholipids in small rapidly tumbling vesicles of 30nm diameter composed of bovine heart cardiolipin and egg phosphatidylcholine (1:4 by weight) (Brown and Wüthrich, 1977). The lipid in these bilayer vesicles gave high-resolution ^{1}H- and ^{13}C-NMR spectra and it was suggested that considerable slowing of the overall protein motion occurred on binding to the vesicles. Relaxation time measurements of the proton resonances were made to construct an "effective mobility" profile across the bilayer on binding of cytochrome *c* (Figure 5). Only near the bilayer polar-polar interface, that is, up to the third methylene segment, was the chain mobility decreased on protein binding, with the maximum perturbation being at the glycerol backbone of the lipids. The changes in the long-range order in the upper part of the bilayer may be due to an increased lateral pressure at the bilayer interface as reported from monolayer studies (Quinn and Dawson, 1969). Very little perturbation of the mobility in the interior of the bilayer by the protein was observed, as shown also by ESR spin-label studies (Van and Griffith, 1975). However, the losses of intensity of both ^1H and ^{13}C-NMR resonance lines from the lipids were interpreted as being indicative of a small population of phospholipids, probably cardiolipin, preferentially associating with cytochrome *c* at the bilayer surface thereby causing a non-random distribution of lipids in the membrane.

More recently, deuterium NMR studies have been reported on the binding of cytochrome *c* to extended bilayers of phosphatidylcholine and phosphatidylserine in equimolar proportions, both with dimyristoyl acyl chains (Devaux *et al.*, 1986). Both phospholipids were perdeuterated in their acyl chains and the serine lipid was specifically deuterated in the α-position to the phosphate of the polar head group. No changes were detected in the order of the acyl chains for either lipid on the addition of cytochrome *c* under conditions of maximum protein binding to the mixed gel phase or liquid crystalline lipid bilayers, as deduced from measurements of quadrupole splittings of spectral moments which give an average of the orientational order of the acyl chains.

Deuterium NMR relaxation time determinations were used to show that no long-term specific association takes place between the protein and individual lipid molecules, rather that a Coulombic interaction between the lipid head groups and protein slows down the overall lipid reorientation (membrane plus water) in which the lipids move. This suggestion is made from the assumption that the rather small changes detected by NMR are not due to large-scale steric or conformational alterations in the lipids but much more subtle mobility changes. It is not clear, however, how one lipid type, the serine lipid, can detect these subtleties but not the choline lipid in equimolar and ideally mixed bilayers in which no lateral phase separation on any time scale takes place. In direct contrast to the results from parallel experiments but using chain-deuterated lipids, the slow motions ($\tau_c \sim 10^{-4}$ s) of the serine head group contributing the measured spin-spin relaxation times remained unchanged on cytochrome *c* binding, but spin-lattice (T_1) relaxation times, which are sensitive to fast ($\tau_c \sim 10^{-9}$ s) segmental motions, increased by 50% for the serine lipid head group; this was also indicated from the ^1H-NMR studies (Brown and Wütrich, 1977).

To investigate any *structural* perturbations of the membrane surface on cytochrome *c* binding, equimolar mixed bilayers of phosphatidylcholine and phosphatidylserine were made in which the serine lipid contained two deuterons at the serine α-carbon chiral center of the head group (Devaux *et al.*, 1986). At the one protein-lipid ratio examined, an increase in the quadrupole splitting, $\Delta\nu_q$, (and thus the amplitude and/or conformation of the CD-group) for one deuterated resonance from the head group of phosphatidylserine was observed. For bilayers containing various proportions of two lipids (1:1, 5:1 and 10:1; choline:serine lipids) very little difference (no more than 10%) in the values of $\Delta\nu_q$ for the head group segments was seen on protein binding at similar protein concentrations at the bilayer surface. In no samples was "immobilized" lipid detected which would give rise to broad spectral components.

It therefore appears that motional rates, but not amplitude or conformation of phospholipid head-group segments, can be induced by cytochrome *c* on the surfaces of bilayers composed of the limited and perhaps not so appropriate, lipids studied. Also, cytochrome *c* binds and partially penetrates into cardiolipin/phosphatidylcholine bilayers as well as induces lipid lateral phase separation, as shown by high resolution NMR studies (Brown and Wüthrich, 1977), but from the more recent deuterium work the protein does not appear to penetrate phosphatidylserine-phosphatidylcholine bilayers or laterally separate the lipids (Devaux *et al.*, 1986), the differences most probably being due to the dissimilar nature of the bilayer surfaces in their affinity for cytochrome *c* for the different lipids studied.

The degree of acyl chain perturbation was found to correspond to the binding ratio of apocytochrome *c* to phosphatidylserine bilayers, using the nitroxide spin-labelled phosphatidylglycerol as a reporter. As shown in Figure 6 (left hand side), the extent of protein binding decreases in a similar way to the binding ratio of apocytochrome *c* with increasing ionic strength (Görrisen *et al.*, 1986). In parallel with the screening effects for these bilayers of salt on bilayer surface potential (Cevc *et al.*, 1980), little binding is observed above 0.4M NaCl. Similar surface binding behaviour is observed for lysozyme at pH 6 (Mateu *et al.*, 1978) or for cytochrome *c* (Görrisen and Marsh, 1988) to negatively charged bilayers.

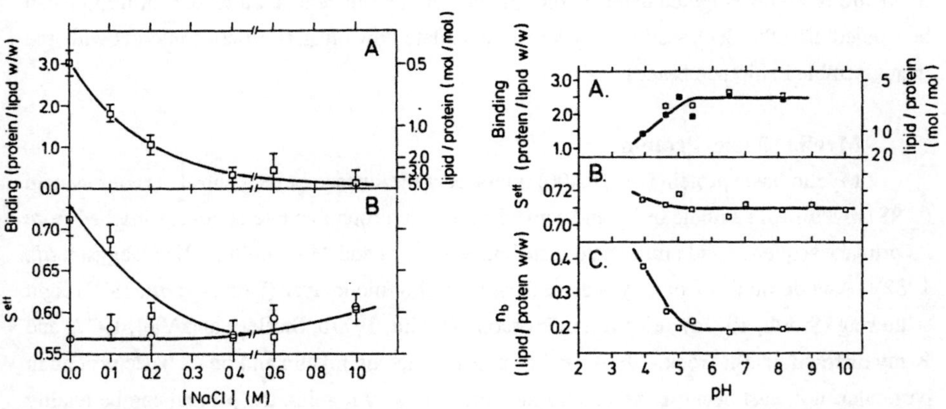

Figure 6 (left hand side): Salt dependence of **A**: apocytochrome *c* binding and **B**: the effective order parameter (overall amplitude of lipid chain motion), S^{eff}, at 30°C, of the phosphatidylglycerol nitroxide spin-labeled on the C5 position of the *sn*-2 chain in dispersions of bovine phosphatidylserine at pH 7.0; □., in the presence, and, O in the absence of saturating amounts of apocytochrome *c*. (from Görrisen *et al.*, 1986).

Figure 6 (right hand side): pH dependence of **A**: apocytochrome *c* binding; **B**: effective order parameter, S^{eff}, at 30°C of the same label as above, and **C**: the number of motionally restricted lipids, n_b, for the same spin-label but with the nitroxide on the C12 position of the *sn*-2 chain in dispersions of dimyristoylphosphatidylglycerol. (Görrisen *et al.*, 1986)

To demonstrate the titratable protein side chains participate in peripheral protein binding to charged lipid bilayers, the pH dependence of apocytochrome c has been studied, again using spin-labelled phospholipids (Görrisen *et al.*, 1986). Protein binding decreases significantly between pH 4 -5 (Figure 6; right hand side), a region where the phosphatidylglycerol bilayers have no titratable groups (Watts *et al.*, 1978). Possible contributing amino acid side groups are carboxylic acid side groups in this pH range since the surface lysine amino groups would titrate at pH 9 - 10. In contrast, the acyl chain motion, as reported by the spin-label, does not change significantly in this range suggesting that the apparent change in binding may correspond to a change in the number of proteins bound rather than a wholesale displacement of the protein from the bilayer surface.

Electron microscopy has been used to show that cytochrome *c* can induce the formation of lipidic particles in bilayers which are also revealed in ^{31}P NMR studies as narrower isotropic spectral lines superimposed upon the more usual spherically averaged powder pattern from extended bilayers (De Kruijff and Cullis, 1980). Such lipidic particles may be due to inverted micelles which have the potential to mediate the passage of fairly large (up to 7nm in diameter) hydrophilic particles, such as a protein, across otherwise impenetrable bilayers. This interesting suggestion could explain how such a hydrophilic protein as cytochrome *c* can permeate mitochondrial membranes which contain a significant proportion of negatively charged lipids such as cardiolipin. Such approaches to cytochrome *c* and apocytochrome *c* interactions with bilayers will clearly take an interesting turn when the same kinds of interactions are investigated using cardiolipin containing bilayers in which the cardiolipin can be studied directly. Such studies may help to elucidate how integral proteins interact with and then assemble in biomembranes.

Myelin Basic Protein

Myelin basic protein (M_r 18,400) interacts at the surface of the central nervous system (CNS) myelin with anionic lipids and some degree of hydrophobic interaction, as implied from its primary sequence and amino acid distribution (Boggs and Moscarello, 1978; Boggs *et al.*, 1982). A wide range of biophysical studies, including monolayer (Demel *et al.*, 1973) light scattering (Smith, 1977a) , electron microscopy (Smith, 1977b; Stollery and Vail, 1977) and X-ray diffraction (Blaurock, 1986) have been performed on myelin and on its basic protein in particular, not least because the myelin membrane is very regular, the protein can be readily purified and it is relatively stable.

The high affinity of myelin basic protein for acidic lipids was first demonstrated in a solvent complex (Palmer and Dawson, 1969) in which the relative association with bound lipids decreased upon acylation of the protein lysine residues. The protein has 31 basic amino acids (Lys, Arg) and 11 acidic residues (Asp, Glu) at physiological pH of a total of 170 amino acids thereby giving the protein its basic character (Boggs and Moscarello, 1978). The 52%

apolar residues, of which 24% are hydrophobic, suggest that this protein not only binds to membrane surfaces but also penetrates the hydrophobic core. The protein may take up some secondary structures, such as α-helix or β-sheet, on binding to bilayers of certain lipid types (Boggs *et al.*, 1982). Also phase diagrams have been constructed from DSC studies of the bilayer phase transition temperatures of mixed lipids and basic protein to show that 27-34 molecules of acidic lipids bind to the protein. ESR studies of Mn^{++}-binding to protein residues not accessible to lipid binding, also confirmed that around 31 protein residues can bind to acidic bilayer lipid head groups (Boggs *et al.*, 1981).

Since myelin basic protein partially penetrates lipid bilayers (Figure 7), it has been possible to monitor this hydrophobic interaction using nitroxide spin-labels, with the probe on the lipid acyl chain (Sankaram *et al.*, 1989a). Although the major interaction in stabilizing the protein association is electrostatic and hydrophilic in nature, any surface interactions mediated at the membrane interface should be observed in the membrane interior by such labels.

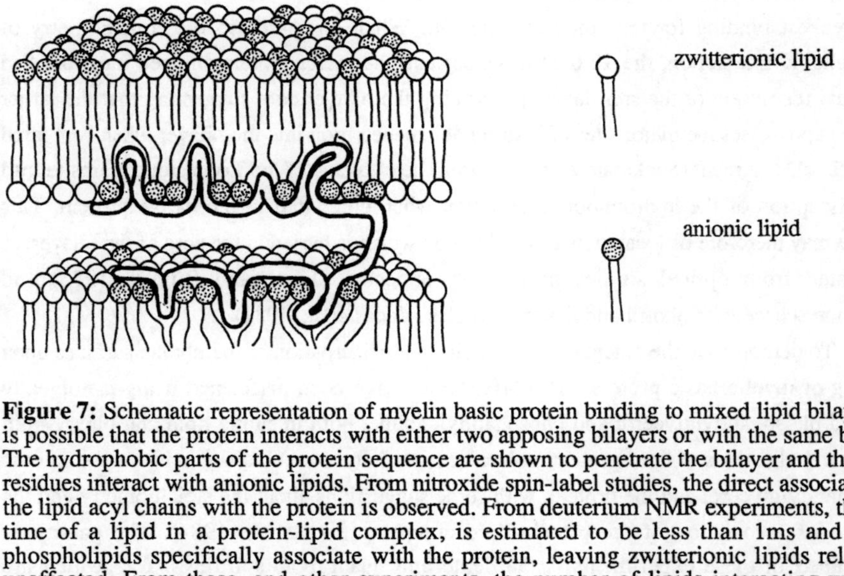

Figure 7: Schematic representation of myelin basic protein binding to mixed lipid bilayers. It is possible that the protein interacts with either two apposing bilayers or with the same bilayer. The hydrophobic parts of the protein sequence are shown to penetrate the bilayer and the basic residues interact with anionic lipids. From nitroxide spin-label studies, the direct association of the lipid acyl chains with the protein is observed. From deuterium NMR experiments, the life-time of a lipid in a protein-lipid complex, is estimated to be less than 1ms and acidic phospholipids specifically associate with the protein, leaving zwitterionic lipids relatively unaffected. From these, and other experiments, the number of lipids interacting with the protein monomer at any one time is between 25 and 36; the protein has 31 basic amino acids (Lys, Arg) which are capable of undergoing specific binding at the membrane surface. (adapted from Boggs *et al.*, 1982).

A series of phosphatidylglycerol spin labels, with the nitroxide on the C5, 7, 10, 12, 14, and 16 positions of the *sn*-1 acyl chain, have been incorporated separately into bilayers of dimyristoylphosphatidylglycerol. All the labels were shown to be interacting with the protein penetrating segments and quantitation of the interaction showed a maximum degree of association of 36 lipids per protein monomer. The hydrophobic interaction was monitored as far as the C12 and C14 positional isomers of the spin-labels, giving some indication about the depth of protein penetration.

From these spin-label studies, the salt titrations and pH changes made with protein-lipid complexes implied that electrostatic interaction of the protein with the lipid head groups was necessary for protein binding and incorporation into the bilayers. Above 1M NaCl, no protein was bound to the membrane, as expected for an electrostatically bound, peripheral protein. At pH 10 which is above the pI of 9.6 for this basic protein, the protein was again seen to be significantly reduced in association with the bilayer surface.

Further spin-label studies on the two tryptic cleavage fragments of myelin basic protein, into 12.6 and 5.4kD fragments, showed that the saturation binding against salt concentration of the association with phosphatidylglycerol bilayers was additive with 11, 24 and 36 lipids per monomer for the 5.4kD, 12.6kD segments and intact protein respectively. The strength of binding for the fragments was in the same order, with strongest binding for the intact protein and weakest binding for the smallest fragment. When compared to the stoichiometry of hydrophobic interaction, the 12.6kD fragment and the intact protein showed most marked motional restriction of the spin-labels penetrating the bilayer core suggesting that the larger fragment possesses the major sites of hydrophobic penetration into the bilayer when compared to the 5.4kD fragment (Sankaram *et al.*, 1989b). The cleavage of the Try residue seems to lead to a disruption of the hydrophobic interaction when compared to the native protein. This residue may therefore be located in a folded region which is buried in the core of the bilayer, as suggested from optical studies of the Try emission (Jones and Rumsy, 1975) and phosphorescence with phosphatidylserine vesicles (Vadas *et al.*, 1981)

To demonstrate the extent of direct molecular perturbation of membrane surfaces upon binding of myelin basic protein, ^{2}H-NMR studies have been performed using isotopically labelled phosphatidylglycerol and phosphatidylcholine, both in single lipid and binary lipid bilayers. Indications of lipid-protein molecular specificity were obtained between the anionic phosphatidylglycerol and the protein, with some suggestions about the stoichiometry and the lifetime of association of the protein-lipid complex (Sixl *et al.*, 1984). The parameter determined in these experiments is the deuterium quadrupole splittings, $\Delta\nu_q$, which are sensitive to the amplitude of motion or conformation of the CD_2-group. Values of $\Delta\nu_q$ in hertz, were analyzed in lipid-protein titration experiments assuming fast exchange of lipids within the bilayer into and out-of the protein-lipid complex. The observed value of $\Delta\nu_{q(observed)}$ was then the weighted mean of the proportion of labeled lipids in a complex (n_c) and the proportion in the bulk phase (n_f) away from the complex. Thus:

$$\Delta\nu_{q(observed)} = n_c . \Delta\nu_{q(complex)} + n_f . \Delta\nu_{q(free)}$$

where $\Delta\nu_{q(free)}$ is the value for protein free bilayer unperturbed by the protein and can be determined directly. This can be rearranged to give:

$$\Delta\nu_{q(observed)} = n_c(\Delta\nu_{q(complex)} - \Delta\nu_{q(free)}).\frac{1}{n_t} + \Delta\nu_{q(free)}$$

The *stoichiometry* (number of lipids in a lipid-protein complex, n_c) can therefore be obtained in principle from the slope of the dependence of $\Delta\nu_{q(observed)}$ with $1/n_t$, the protein:lipid mole ratio. If it is assumed that the lipid head groups are totally disordered (or at least give a ^{2}H-NMR spectrum with no measurable quadrupole splitting, $\Delta\nu_q \sim 0$) then a value for n_c can be deduced; a value of $n_c \sim 21 - 35$ lipids per protein was estimated from four independent sets of experimental results, those from three deuterated segments of the glycerol head group as well as the phosphorus spectral changes.

Similarly, the *exchange rate* (ν_{ex}) for phospholipids into and out-of a protein-lipid complex can be estimated from the difference in the frequency of the NMR spectra measured for deuterated phospholipid head groups when protein-free and when in a protein complex ($\nu_{ex} = \Delta\nu_{q(free)} - \Delta\nu_{q(complex)}$), giving at least a fast exchange limit for this motion of >1kHz; that is, the life-time for a lipid in a protein-lipid complex is < 1ms. The observation of:

(i). single component ^{2}H-NMR and ^{31}P-NMR powder spectra and the absence of broad spectral components which could be due to "immobilized" (on the spectroscopic time-scale of deuterium NMR quadrupole anisotropy averaging; Watts, 1981) lipid as seen in spin-label ESR experiments on similar systems, and;

(ii). a linear dependence of $\Delta\nu_{q(observed)}$ from the resonant lines with the protein:lipid ratio from all labelled positions in the head group of the glycerol phospholipid, confirmed the fast exchange of lipids onto and off the protein interface.

The changes in the observed NMR quadrupole splittings from head-group deuterated phospholipids induced on the addition of myelin basic protein were rather large, in percentage terms, for the charged phosphatidylglycerol lipid, whether alone, or mixed in a equimolar proportion with phosphatidylcholine. Similar effects were observed in the ^{31}P-NMR chemical shift anisotropy in the bilayer powder patterns. No changes were observed for choline lipid bilayers alone and all these studies were performed with lipid bilayers containing dimyristoyl acyl chains in the liquid crystalline phase. Although some spectral broadening was observed (no spin-spin relaxation time determinations were made) quadrupole splitting changes could be measured over a very wide range (from 1000:1 to 22:1, lipid:protein mole ratio) of protein concentrations. It appears, from the earlier determinations of available basic amino acid residues for phospholipid head group association from Mn^{++} binding studies (see above), that there is a one-for-one molecular association between a lipid and an amino acid to facilitate binding of this protein to the membrane surface (Figure 7).

Although myelin basic protein binding was not extended to saturation limits and the analysis is perhaps not complete, the study was instrumental in showing the power of non-perturbing ^{2}H-NMR methods for attempting a quantitation of *exchange rates*, *selectivity* and *stoichiometry* of lipid-protein interactions at membrane surfaces. The method shows that the order and/or conformation of the membrane surface is significantly perturbed on interaction with the surface binding protein. Having now formulated a direct approach to the study of protein-lipid interactions by ^{2}H-NMR, it will be interesting to see whether other proteins are as amenable to similar study. In particular, hydrogen bonds could be formed between lipid head groups and proteins and the ^{2}H-NMR method may be a way of quantifying such associations. It will be therefore be interesting to see if any other independent non-perturbing experimental approaches can give the kind of detail the ^{2}H-NMR seems potentially able to do and whether other proteins conform to a *general* mechanism for binding to a membrane surface.

Spectrin

Spectrin is the major protein of the erythrocyte cytoskeleton. It forms heterodimers composed of two non-identical subunits (α chain, 240000 daltons, and β chain 220000 daltons). Each subunit is composed of several homologous triple helical segments (Speicher and Marchesi, 1984). Spectrin heterodimers participate in head-to-head association to form tetramers (Marchesi, 1983). Spectrin heterodimers or tetramers also associate with actin filaments and band 4.1 to form a meshwork. This meshwork is linked to the plasma membrane by association of spectrin with ankyrin, which, in turn, is associated to band 3 (Bennett, 1989). Besides its interaction with a number of other cytoskeletal proteins, spectrin is thought to have a direct association with the cytoplasmic surface of the lipid bilayer. The polar head specificity of this association is however, still a matter of controversy. Electron microscopic studies have demonstrated that spectrin heterodimers bind to phosphatidylserine liposomes, at multiple sites of the protein chains (Cohen *et al.*,1986). It was also shown, using a monolayer technique at constant lateral pressure, that spectrin penetrates specifically between phosphatidylserine molecules (Mombers *et al.*,1983). This phenomenon was attributed to direct electrostatic interaction between negatively charged phosphatidylserine polar head and positive charged segments of spectrin.

Some authors have questioned the suggestion that specific spectrin-phosphatidylserine interaction occur at bilayer surfaces. Analysis of intrinsic fluorescence of spectrin in the presence of various mixed lipids liposomes did not show any difference between the effect of pure phosphatidylserine vesicles and phosphatidylcholine ones. Phosphatidylethanolamine seemed to be more efficient in quenching spectrin fluorescence than both phosphatidylcholine and phosphatidylserine (Sikorski *et al.*, 1987). More recently, binding of spectrin dimer was found to be similar for pure dimyristoylphosphatidylcholine (DMPC) liposomes and DMPC:DMPS (1:1) liposomes. No effect of bound spectrin of the ^{2}H-NMR quadrupolar splitting of DMPC deuterated in the polar head group was observed. Also, both the phase transition temperature of the DMPC:DMPS mixed bilayers and the spin-lattice (T_1) relaxation time (reflecting very fast head group motions with τ_c < ns) of the deuterated DMPS head-group, was unaffected by spectrin (Bitbol *et al.*, 1989). Clearly, spectrin might provide a good protein with which to study protein-lipid interactions at membrane surfaces but much more work is still required to resolve the molecular specificity and the extent of its interactions.

INTEGRAL PROTEINS

Integral proteins, surprisingly perhaps, significantly perturb the *structure* and *dynamics* of the molecular components of the membrane surface. It may well be that such perturbations are involved in the lateral communication which can take place along membranes in, for example, nerve cells. Similarly, surface electrostatics, hydrogen bonding (Hauser *et al.*, 1981), hydration levels (Cevc, 1987; 1988) and so on, may be changed by the integral proteins interacting with a membrane surface. Deglycosylation of integral viral proteins with enzymes (neuraminidase) induces significant changes in the motional properties of the acyl chains in the hydrocarbon chains (Ohyashiki *et al.*, 1987), probably through alterations in the surface hydration which causes changes in lipid lateral packing and thus bilayer volume. It cannot, however, be expected that a general description for lipid-protein interactions which occur at the membrane interface, both at the hydrophobic core and at the polar-apolar interface where the interactions will be strongest, can be formulated; every protein will be different.

The functional significance of lipid-protein interactions are now well documented (Sandermann, 1986). Ganglioside GM1 which is the receptor for cholera toxin, triggers the regulatory cascade involving adenylate cyclase and protein kinase (Moss and Vaughan, 1979) and gangliosides have more recently been shown to regulate cell growth by binding to an allosteric site on the epidermal growth factor receptor (Hakomori, 1979). Ion transport is also lipid modulated, as shown for the Na^+/K^+-ATPase reduced activity in hydrolysing ATP in the absence of either a specific number (~ 60) of lipids or negative phospholipids (de Pont *et al.*, 1978). It has been suggested that kinetic cooperativity between lipids and protein activity can occur with two possible mechanisms, (i) allosteric control resulting from interaction at lipid-protein binding sites (which could be at the membrane surface) and (ii) non-allosteric mechanism requiring a lipid solvation shell around the protein (see Table 3 in: Sandermann, 1986). Selectivity of such interactions have a high probability of occurring at the membrane surface, between lipid head groups and amino acid residues at the polar-apolar interface. Both low-specificity and a multiple lipid binding is accommodated in these models. The traditional way of observing such interactions has been by probing the hydrophobic motional changes which occur upon association between lipid and protein, even though such interactions are probably determined at the polar-apolar interface.

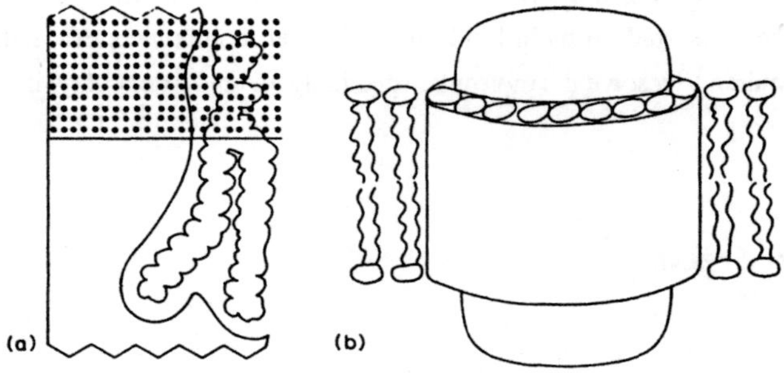

Fiure 8. Schematic representation showing a possible arrangement for protein-lipid interactions in membranes which can occur at the membrane surface. (a). Specific lipid association with a protein may occur at a single binding site. Activation may occur through alignment of the hydrated lipid polar group with a hydrated amino acid side-chain (or groups of chains) and the mobile fatty acid chains with non-polar amino acids. The dotted areas symbolize hydration. (b). Illustration of the multiple lipid binding sites of the *boundary lipid* or *solvation layer* around an integral protein. Non-boundary or sites may exist within a cleft in a protein, or at the interface between proteins which are self-associated; the exchange for these lipids would then be governed by the life-time for protein-protein interaction (Adapted from Sandermann, 1986).

The methods available for describing the energetics of lipid-protein associations and specificity give rather low resolution and the contributions from hydrophobic and electrostatic interactions, hydrogen bonding, entropic and steric factors, are not clear. In addition, the types of activation of membrane bound enzymes by lipids, is still an area where more information is required.

On extracting integral proteins from biomembranes, numerous different lipid types (defined in their head group chemistry) are found to be associated with the isolated protein. This alone does not indicate specific lipid-protein interactions, simply that the extracted complex reflects some characteristics about the proteins and lipids in response to the detergent used. One good and perhaps the only way at the moment, to demonstrate molecular specificity between lipids and proteins is to isolate a protein entirely free of endogenous lipids and then test its affinity by a biophysical approach against added lipids of different types. This is best done in a reconstituted membrane in which the functional activity of the protein can be tested. Structural *and* functional selectivity need to be demonstrated for any protein-lipid molecular specificity to be convincing.

The first indication of a structural phospholipid-integral protein molecular specificity was obtained with cytochrome *c* oxidase from yeast using ESR spin-label probes (Knowles *et al.*, 1981). Here, cardiolipin and phosphatidic acid spin-labels were shown to be more strongly associated with the protein than other charged or zwitterionic phospholipid labels, with residency times at the protein interface of 5- and 2-times longer than other labels respectively, assuming that the number of lipids around the protein is the same for each lipid; this may not actually be the case if a lipid is required to stabilize oligomers of a protein. This pioneering work has now been substantiated with this protein (Powell *et al.*, 1985) and other further experiments and examples (see Marsh and Watts, 1982; Devaux and Seignereut, 1985; Marsh, 1986; Marsh and Watts, 1988 for reviews).

The ESR spin-label method gives experimentally distinguishable ESR spectra for labels at the protein interface and those in the bulk phase away from the interface, since the labels exchange slowly on the ESR hyperfine anisotropy averaging time-scale (Watts, 1981). The number of labels restricted in motion by the protein, n^*_b, can be quantitated by spectral subtraction methods, to give the equilibrium association equation (Marsh, 1986):

$$n^*_f/n^*_b = n_t/(N_1.K_r^{av}) - 1/K_r^{av}$$

where n^*_f/n^*_b is the ratio of labelled lipids free in the bilayer to those associated in the first shell sites on the protein and is obtained by spectral subtraction, assuming that all the labels are observed in the spectrum and K_r^{av} is the average relative association constant defined by:

$$K_r^{av} = \sum_{i=1}^{m} n_i K_i/N_1$$

and N_1 is the total number of first shell association sites on the protein. By titrating different protein concentrations into a bilayer and quantifying the "boundary" lipid fraction, values of n^*_f/n^*_b against the lipid:protein mole ratio gives both the number of lipids at the protein-lipid interface (n^*_b) but also K_r^{av}. Extending this approach, it is possible to define any selectivity between a particular labelled phospholipid and a protein by incorporating different label types into a complex and quantifying the changes in n^*_b at the same lipid/protein ratio in the complex. Now, assuming that the number of any lipid types surrounding the integral protein, N_1, does not change:

$$(n^*_f/n^*_b)^{L_2}/(n^*_f/n^*_b)^{L_1} = K_r^{av}(L_1)/K_r^{av}(L_2)$$

where $K_r^{av}(L_1)/K_r^{av}(L_2)$ is the ratio of the average relative association constants for a particular spin-labelled lipid, L_1, to that for spin-labelled lipid, L_2; often phosphatidylcholine spin-label is chosen for L_2 since it is a common and major component of eukaryotic membranes and shows little selectivity for integral proteins.

This approach to analyzing spin-label experiments, designed to give information about lipid-protein interactions in bilayer membranes, allows both the *stoichiometry* and *selectivity* of the associations to be determined.

A measure of the *stoichiometry* (the number) of lipids at any one time in the protein-lipid interface, enables estimates to be made for the hydrophobic interface for an integral protein and hence its average diameter. For example, bovine rhodopsin has a hydrophobic diameter of 30 - 35Å which is capable of accommodating ~25 phospholipids, a number which is measured in ESR spin-label experiments (Watts *et al.*, 1979). Stoichiometries have now been determined for a wide range of readily available proteins in bilayers (see Table 9-1 in Marsh and Watts, 1988). It is important in this kind of treatment to know the aggregation state of the protein and the analysis only applies in its simplest form for monomeric proteins. Aggregation, which may be promoted or stabilized by lipids through interactions at the membrane surface, clearly reduces the lipid-protein interaction by using the interface available for lipid associations to form protein-protein contacts (Figure 9). Deviations from a linear dependence of n^*_f/n^*_b versus n_t from above is an indication that protein aggregation occurs as its concentration in the bilayer increases, as observed with cytochrome c oxidase (Knowles *et al.*, 1979). This is an important consideration and needs to be understood in more detail, especially if ways of producing regular molecular arrays are to be devised for diffraction studies of membrane proteins.

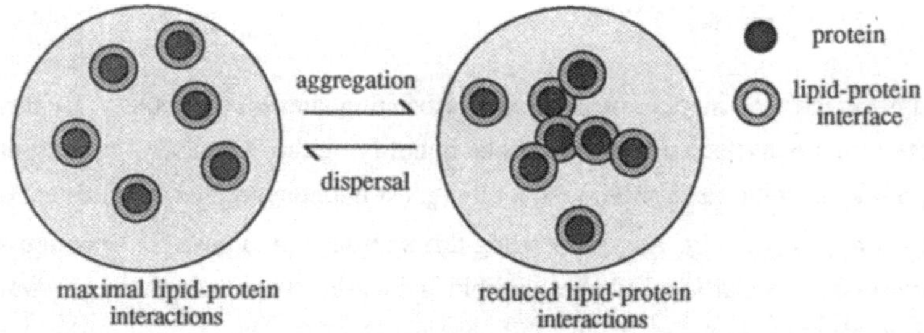

Figure 9: Aggregation of integral membrane proteins can be either mediated through protein-protein associations or mediated by lipids, with such interactions having a high probability of taking place at the polar-apolar membrane interface. As examples of these, bacteriorhodopsin appears to require charged lipids to promote formation into trimers (Sternberg *et al.*, 1989) and band 3 from erythrocytes aggregates independently of the lipid surroundings (Dempsey *et al.*, 1986). In both cases, lipid-protein interactions are conserved by protein-protein contacts, thereby reducing the experimental observation of these interactions.

Selectivity of lipid-protein associations may either be as a result of changes in N_1 or as a result of different K_r^{av} for some lipids. Table 1 shows some of the information available on some integral proteins. The selectivity, which is probably determined at the membrane apolar-polar interface, is therefore reflected in a longer residency time of the selected lipid at the protein interface. In terms of K_r^{av}, the ratio of the exchange rate constants is then given by:

$$n^*_f/n^*_b = \tau_b^{-1}/\tau_f^{-1} = n_t/(N_1.K_r^{av}) - 1/K_r^{av} \qquad \text{and;}$$

$$(n^*_f/n^*_b)^{L_2}/(n^*_f/n^*_b)^{L_1} = (\tau_b^{-1}/\tau_f^{-1})^{L_2}/(\tau_b^{-1}/\tau_f^{-1})^{L_2} = K_r^{av}(L_1)/K_r^{av}(L_2)$$

where τ_b^{-1} and τ_f^{-1} are the exchange rates for the fluid and bound lipid label components respectively. For the myelin-proteolipid protein recombinants it was found that the on-rate (τ_f^{-1}) for lipids to the protein interface is diffusion controlled and the same for all lipid types. It is the off-rate (τ_b^{-1}) from a protein interface which is reflected in the selectivity of a particular lipid type. The selectivity for different lipids is given in Table 2 for the myelin-proteolipid protein. Such selectivity is clearly related, for this example, to the electrostatic interactions possible between the lipids and protein, since the charged labels have a stronger association then the zwitterionic lipids. Such selectivity is not always so straightforward since some proteins select for charged as well as neutral lipids in preference to zwitterionic lipids which usually display a lower specificity for proteins than other lipid types, as shown in Table 1.

Protein	Order of lipid selectivity
Myelin proteolipid	SA > PA > CL $\gtrsim$ PS > PG ≈ SM ≈ PC > PE
(Na^+, K^+)-ATPase	CL > PS ≈ SA $\gtrsim$ PA > PG ≈ SM ≈ PC ≈ PE
Cytochrome oxidase	CL > PA ≈ SA > PS ≈ PG ≈ SM ≈ PC ≈ PE
Acetylcholine receptor	SA > PA > PS ≈ PC ≈ PE
Ca^{2+}-ATPase	CL > PS ≈ SA $\gtrsim$ PA $\gtrsim$ PG ≈ SM ≈ PC ≈ PE
Rhodopsin	CL ≈ PA ≈ SA ≈ PS ≈ PG ≈ SM ≈ PC ≈ PE

Table 1: Order of selectivity of spin-labelled lipids for association with integral memrbane proteins. CL, cardiolipin; PA, phosphatidic acid; PS, phosphatidylserine; PG, phosphatididylglycerol; PC, phosphatidylcholine; PE, phosphatidylethanolamnie; SM, sphingomyelin. (from Marsh and Watts, 1988).

Lipid (L)	$\tau_b^{-1}(PC)/\tau_b^{-1}(L)$	$K_r^{av}(L)/K_r^{av}(PC)$
Phosphatidic acid	3.5	3.0
Phosphatidylserine	1.4	1.2
Phosphatidylethanolamine	0.7	0.6
Stearic acid	5.8	5.3
Phosphatidylcholine	1.0	1.0

Table 2: Relative association constants (related to PC) and exchange rate constants (related to PC) for the selectivity of lipid spin-labels in myelin proteolipid protein-DMPC lipid recombinants at 30°C. (from Horváth *et al.*, 1988).

Deuterium NMR studies of isotopically labelled phospholipids have been performed to gain an insight into lipid-protein interactions (see Seelig *et al.*, 1982; Bloom and Smith, 1985; Devaux and Seigneuret, 1985; Watts, 1986;1987c; Marsh and Watts, 1988; Watts, 1989 for reviews). In many cases, however, extensive titrations of protein concentration have not been produced, or alternatively the functional state of the protein has not been monitored. In many studies, it was initially assumed that, since only single component NMR spectra were obtained, all lipids were in fast exchange throughout the bilayer membrane and that ESR studies were incorrect and this error was often attributed to the perturbing nature of the nitroxide. However, more recently it has been realized that the time-scale sensitivity for the two spectroscopic methods show fast lipid exchange by deuterium NMR but slow exchange by ESR of labelled lipids into and out of the protein-lipid interface (Watts, 1981). The situation is not quite so simple, since exceptions can be and have been found to every rule. However, this field has led to some close examination of the use and meaning of different spectroscopic methods in biological systems.

The head-group *conformation*, *amplitude* and *rate* of motion have all been shown to be altered by integral proteins using ^{2}H-NMR and ^{31}P-NMR methods utilizing deuterated phospholipids.The approach has however been used to demonstrate the temperature-induced aggregation state in dimyristoylphosphatidylcholine bilayers of band 3, the anion transport glycoprotein (85 - 100 kDa) from human erythrocytes (Dempsey *et al.*, 1986) and that rhodopsin from bovine retinal disc membranes appears to be monomeric at all temperature and protein concentrations studied (Ryba *et al.*, 1986). The results were analyzed in a similar way as described above for myelin basic protein and it was suggested that the labelled lipids were in fast exchange throughout the bilayer between the bulk and the lipid-protein interface. The perturbation of the surface structure and dynamics of bilayers by cytochrome *c* oxidase has also been studied by ^{2}H-NMR where the head-group segments of phosphatidylcholine were deuterated. NMR observation of the deuterons, ^{14}N of the choline group and phosphate, all showed relatively small differences in their splittings whether protein was present or not. Some line broadening did occur in the spectra although no complete protein-lipid titrations were performed (Tamm and Seelig, 1983).

One interesting situation has been observed for the small integral protein, the coat protein (M_r ~ 5,000) from M13 bacteriophage. About 3,000 of these proteins are arranged spirally around the mature phage which buds after a non-pathological infestation of *E.coli.* When reconstituted into phosphatidylcholine bilayers deuterated in the polar choline head group, the mature coat protein appears to induce a second ^{2}H-NMR spectral component, indicative of trapped (or "immobilized" lipids) on the time scale for deuterium quadrupolar anisotropy averaging (Van Gorkom *et al.*, 1989). Such lipids are though to be arranged in between parallel aggregates of the M13 coat protein which aligns itself in this way in preparation for extrusion of the naked phage. Spin-label ESR experiments also show boundary (or trapped) lipids but there are about 25% more boundary lipids indicated by ESR experiments (Wolfs *et al.*, 1989) than ^{2}H-NMR experiments. Here the time-scale differences of the two methods are important since ESR monitors both trapped lipids and lipids at the interface of aggregates whereas ^{2}H-NMR monitors only the trapped lipids. This implies that the lipids at the protein aggregate interface are exchanging faster (ν_{ex} ~ MHz) than the trapped lipids (ν_{ex} ~ kHz) with the bulk of the lipids. This also gives an estimate of >1ms for the shortest life-time possible of a protein aggregate, although it could be much longer.

However, any molecular selectivity between phospholipids and integral proteins which might be expected to occur at the bilayer surface, has yet to be convincingly demonstrated using this the NMR method.

ISOTOPICALLY LABELLED PROTEINS

Despite the ubiquity of protons and ^{13}C nuclei, membrane proteins are not readily amenable to study by ^{2}H-NMR or ^{13}C-NMR, at least not when interacting with large, extended membranes. Some progress has been made recently in the understanding of ^{1}H- and ^{13}C spectral line-shapes which are complicated because of the restricted, anisotropic slow motion inherent in membranes and the relatively large dipolar couplings possible among various spin 1/2 nuclei in a membrane (Deese and Dratz, 1986). In particular, the dipolar spin-lattice relaxation mechanisms lead to broadening of the spectra, rendering them relatively featureless and not readily interpretable.

Elegant solutions to these problems have been devised, as demonstrated for phospholipase A_2 in which the aromatic residues in phospholipase A_2 have been deuterated and the dynamics of the protein side chains determined in solution and micelles (Allegrini *et al.*, 1985). The next stage is clearly the study of protein-lipid interactions from observation of the protein and any conformational changes which may occur upon membrane surface association.

Another approach to the NMR study of membrane proteins is to incorporate, into a protein, nuclei whose anisotropic magnetic interactions can be exploited to give useful information. As a start of the study of surface-active proteins, we have successfully deuterated melittin at four amino sites by chemically methylating the protein adjacent to the three lysines and the *N*-terminal glycine (Dempsey *et al.*, 1987). The ^{2}H-NMR spectra from the deuterons confirm the occurrence of the macroscopic changes induced by melittin and seen in the ^{2}H-NMR spectra from the bilayers of deuterated phospholipids (described above). The assignment from such spectra is not simple since it appears that two of the deuteromethylaminolysines have the same quadrupole splittings. However, the spectra do indicate that the peptide binds to varying degrees on the surface of bilayers made of different types of phospholipid with tighter binding and reduced peptide intramolecular motion (giving broadened spectra lines) when deuterated melittin is bound to bilayers with a surface negative charge from phosphatidylserine-containing bilayers or erythrocyte ghosts.

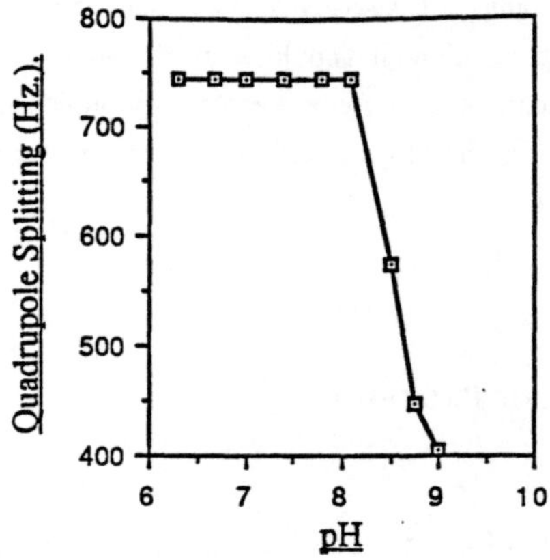

Figure 10: Deuterium NMR quadrupole splittings shown as a function of bulk pH for melittin specifically deuterated at Lys 21 and 23 whilst bound to large, multibilayers of dimyristoylphosphatidylcholine. The apporoximate pK_a for these two peptide groups suggests that they are exposed to the aqueous phase at the bilayer surface. (Dempsey, Lemon and Watts, unpublished).

In addition, pH titrations of titrateable groups, a frequently measured parameter for soluble proteins by ^{1}H-NMR, can be determined in principle from membrane surface-bound melittin. The variations in $\Delta\nu_q$ values with pH for melittin (Figure 10) (Dempsey, Lemon and Watts, unpublished) were measured directly, demonstrating the potential of the method for determining individual side group pK_a's for a protein actually on the bilayer surface, or inside a membrane for integral proteins. The accessibility of protein residues to the extramembraneous environment and bilayer surface as well as surface electrostatics, can therefore now be studied not only for phospholipids isothermally in bilayers (Watts and Poile, 1986) but also for proteins. Therefore the binding, function and stability of proteins in response to pH can now be studied directly. This approach clearly has much potential, especially if the proteins can be labelled biosynthetically, although this will be impractical for mammalian systems without some genetic manipulation.

CONCLUSIONS

Biological function is controlled by highly specific molecular interactions, many of which may occur in biomembranes although their strength and significance may be limited (Lee, 1987). Membrane surfaces are clearly well placed to be the sites for such molecular specificity and in the biological situation and so many cellular functions and consequences are triggered at this important region of the membrane. Lipid-protein interactions have been implied in:

(i). changes in the macroscopic organization of the membrane;

(ii). the control, either through providing solvation for a protein or specific allosteric associations, of the kinetics and activity of membrane bound enzymes;

(iii). the perturbation of the structure, conformation and dynamics of membrane surfaces;

(iv). modulation the surface electrostatics and hydration with concomitant changes in the motional properties of the membrane hydrophobic core.

In view of the complexity of the membrane surface, chemically, structurally and dynamically, much more information is needed before the influence of lipids on the function of proteins and eventually the cell as a whole, can be described in detail. Only when atomic resolution for membrane proteins and complete catalytic mechanisms have been described, with any influence of surrounding lipids, can we be sure that the involvement of the interactions which occur at the membrane surface are fully resolved and understood; this is a goal which is clearly many years away.

ACKNOWLEDGMENTS

This work was supported by grants from SERC, MRC, The Multiple Sclerosis Society of GB and Northern Ireland and the European Economic Community.

REFERENCES

Allegrini PR, Pluchke G and Seelig, J (1984) *Biochemistry* **23**, 6452-6458
Allegrini PR, van Scharrenburg GJM, Slotboom AJ, de Haas GH and Seelig J (1985) *Biochemistry* **24**, 3268-3273
Altenbach Ch and Seelig J (1985) *Biochim Biophys Acta* **818**, 410-415
Bennett V (1989) *Biochim Biophys Acta* **988**, 107-121
Bitbol M, Dempsey CE, Watts A, and Devaux PF (1989) *FEBS lett* **244**, 217-222
Blaurock A (1986) In: *Progress in Protein Lipid Interactions* (Watts A and De Pont JJHH, eds), Elsevier, Amsterdam, Vol 2, Chapter 1

Bloom M and Smith ICP (1985) In: *Progress in Protein Lipid Interactions* (Watts A and De Pont JJHH, eds), Elsevier, Amsterdam, Vol 1, Chapter 2
Boggs JM and Moscarello MA (1978) *J Membr Biol* **39**, 75 96
Boggs JM Clement IR and Moscarello MA (1981) *Biochemistry*, **20**, 1065-1073
Boggs JM, Moscarello MA and Papahadjopoulos D (1982) In: *Lipid-Protein Interactions* Jost PC and Griffith OH, eds), Wiley-Interscience, New York, Vol 2, Chapter 1
Brown LR and Wüthruth K (1977) *Biochim Biophys Acta* **468**, 389-410
Brown MF, Seelig J and Haberlen U (1979) *J ChemPhys* **70**, 5045-5053
Burnell EE, Cullis PR and De Kruijff B (1980) *Biochim Biophys Acta* **603**, 63-69
Campbell ID (1988) in: *Transport through membranes: Carriers, Channels and Pumps* 91-101
Cevc G (1987) *Biochemistry*, **26**, 6305-6310
Cevc G (1988) *Ber Bunsen ges Phys Chem* **92**, 953-961
Cevc G, Watts A and Marsh D (1980) *FEBS Lett* **120**, 267-270
Cohen AM, Liu SC, Derick SH, Palek J, (1986) *Blood* **68**, 920-926
Cullis PR, Verkleij AJ and Ververgaert, PHJT (1978) *Biochim Biophys Acta* **513**, 11-20
Cullis PR, De Kruijff B, Hope MJ, Nayer R, Rietveld A and Verkleij AJ (1980) *Biochim Biophys Acta* **600**, 625-635
Dasseux J-L Faucon J-F, Lafleur M, Pezolet M and Dufourcq J (1984) *Biochim Biophys Acta* **510**, 75-86
Davis JH (1983) *Biochem Biophys Acta* **737**, 117-171
Deese AJ and Dratz EA (1986) Chapter 2 In: *Progess in Protein Lipid Interactions* (Watts A and De Pont JJHHM, eds) Elsevier, Amsterdam, Vol 2
De Kruijff B and Cullis, PR (1980) *Biochim Biophys Acta* **602**, 477-490
De Kruijff B, Cullis PR, Verkleij AJ, Hope MJ, Van Echteld CJA, Tarashi TF, Van Hoogevest P, Killian JA and Van der Steen, ATM, (1985a) Chapter 3 In: *Progress in Protein-Lipid IInteractions* (Watts A and De Pont JJHHM eds), Elsevier, Amsterdam, Vol 1
De Kruijff B, Cullis PR, Verkleij AH, Hope MJ, Van Echtel CJA and Tarashi, TF (1985b) In: *The Enzymes of Biological Membranes*, 2nd edn Vol 1, Plenum Press, New York, pp 131-204
Demel RA, London Y, Geurts van Kessel, WSM, Vossenberg, FGA and van Deenen LL (1973) *Biochim Biophys Acta*, **311**, 507-519
Dempsey CE and Watts A (1987) *Biochemistry* **26**, 5811-5816
Dempsey CE, Ryba NJP and Watts A (1986) *Biochemistry* **25**, 2180-2187
Dempsey CE, Cryer G and Watts A (1987) *FEBS Letts*, **218** 173-177
Dempsey CE, Bitbol M and Watts A (1989) *Biochemistry* (in press)
De Pont JJHHM, van Prooijen-van Eeden A and Bonting SL (1978) *Biochim Biophys Acta* **508**, 464-477
Devaux PF and Seigneuret M (1985) *Biochim Biophys Acta* **822**, 63-125
Devaux PF, Hoatson GL, Favre E, Fellmann P, Farren B, MacKay, AL and Bloom M (1986) *Biochemistry* **25**, 3804-3812
Dufourc EJ, Smith IPC and Dufourcq J (1986) *Biochemistry* **25**, 6448-6455
Dufourcq J, Faucon J-L, Fourche G, Dasseux J-L, LeMaire M and Gulik-Krzywicki T (1986) *Biochim Biophys Acta* **859**, 33-48
El Mashak EM and Tocanne JF (1980) *Biochim Biophys Acta* **729**, 44-52
Görrisen H and Marsh D (1989) (submitted)
Görrisen H, Marsh D, Rietveld A and de Kruijff B (1986) *Biochemistry* **25**, 2904-2910
Hakomori S-I (1984) *Trends in Biochem Sci* **9**, 453-456
Hartman W, Galla HJ and Sackmann E (1978) *Biochem Biophys Acta* **510**, 124-139
Hauser H, Pascher I, Pearson RH and Sundrell S (1981) *Biochim Biophys Acta* **650**, 21-51
Hippe S and Lüth H (1986) *J Theor Biol* **121**, 351-366
Horváth LI, Brophy PJ and Marsh D (1988) *Biochemistry* **27**, 353-366
HsuChen CC and Feingold DS (1973) *Biochemistry* **12**, 2105-2111
Israelachivili JN, Marcelja S and Horn RG (1980) *Q Rev Biophys* **13**, 121-200
Jain MK and Vaz WLC (1987) *Biochem Biophys Acta* **905**, 1-8
Jones AJS and Rumsby MG (1977) *J Neurochem* **25**, 565-572
Knowles PF, Watts A and Marsh D (1979) *Biochemistry* **18**, 4480-4487

Knowles PF, Watts A and Marsh D (1981) *Biochemistry* **20**, 5888-5894
Kuchinka E and Seelig J (1989) Biochemistry (in press)
Lee AG (1987) *J Bioenerg Biomemb* **19**, 581-603
Lavialle F, Lenin I and Molloy C (1980) *Biochim Biophys Acta* **600**, 62-71
Marchesi VT, (1983) *Blood* **61**, 1-11
Mateu L, Caron F, Luzzati V and Billecocq A (1978) *Biochim Biophys Acta*, **508**, 109-121
Marsh D (1986) In: *Progess in Protein Lipid Interactions* (Watts, A and De Pont, JJHHM, eds), Elsevier, Amsterdam, Vol 1, Chapter 4
Marsh D and Watts, A (1982) In: *Lipid-Protein Interactions* (Jost, PC and Griffith, OH, eds), Wiley-Interscience, New York, Vol 2, Chapter 2
Marsh D and Watts A (1988) Chapter 9 In: *Advances in Membrane Fluidity* (Aloia, RC, ed), Alan R Liss Inc, New York, Vol 2.
Mombers C, de Gier J, Demel RA, Van Deenen LLM, *Biochim Biophys Acta* **603**, 52-62
Moss J and Vaughan M (1979) *Ann Rev Biochem* **48**, 581-600
Ohshima H and Ohki S (1986) *Bioelect Bioenerg* **15**, 173-182
Ohshima H and Kondo T (1987) *J Coll Inter Sci* **116**, 305-311
Ohyashiki T, Taka M, Mohri T (1987) *Biochim Biophys Acta* **905**, 57-64
Palmer FB and Dawson RMC (1969) *Biochem J* **28**, 231-240
Powell GL, Knowles PF and Marsh D (1985) *Biochim Biophys Acta* **816**, 191-194
Quinn P and Dawson RMC (1969) *Biochem J* **115**, 65-75
Rand RP and Sengupta (1972) *Biochem Biophys Acta* **255**, 484-492
Ryba NJP, Dempsey CE and Watts A (1986) *Biochemistry* **25**, 4818-4825
Sanderman H (1986) Chapter 6 In: *Progess in Protein Lipid Interactions* (Watts A and De Pont JJHHM, eds), Elsevier, Amsterdam, Vol 2
Sankram MB, Brophy, PJ and Marsh, D (1989a) *Biochemistry* (submitted)
Sankram MB, Brophy PJ and Marsh D (1989b) *Biochemistry* (submitted)
Seelig J (1977) *Q Rev Biophys* **10**, 353-418
Seelig J and Seelig A (1885) *Biochim Biophys Acta* **815**, 153-158
Seelig J and Seelig A (1980) *Q Rev Biophys* **13**, 9 61
Seelig J, Seelig A and Tamm L (1982) Chapter 3 In: *Lipid-Protein Interactions* (Jost PC and Griffith OH, eds), Wiley-Interscience, New York, Vol 2
Seelig J, Macdonald PM and Scherer PG (1987) *Biochemistry* **26**, 7535-7541
Seelig A, Allegrini PR and Seelig J (1988) *Biochim Biophys Acta* **899**, 196-204
Sikorski AF, Michalak K, Bobrowska M, (1987) *Biochim Biophys Acta* **904**: 55-60
Sixl F and Galla HJ (1981)*Biochim Biophys Acta* **643**, 626-635
Sixl F and Watts A (1983) *Proc Natl Acad Sci USA* **80**, 1613-1615
Sixl F and Watts A (1982)*Biochemistry* **21**, 6446-6452
Sixl F and Watts A (1985) *Biochemistry* **24**, 7906-7910
Sixl F Brophy PJ and Watts A (1984) *Biochemistry* **23**, 2032-2039
Smith R (1977a) *Biochim Biophys Acta*, **491**, 581-590
Smith R (1977b) *Biochim Biophys Acta*, **470**, 170-184
Speicher DW and Marchesi VT, (1984) *Nature*, **311**, 177
Sternberg B, Gale P and Watts A (1989) *Biochim Biophys Acta* (in press)
Stollery JC and Vail W (1977) *Biochim Biophys Acta* **471**, 372-390
Tamm L and Seelig J (1983) *Biochemistry* **22**, 1474-1483
Vadas EB, Melançon D, Braun PE and Galley WC (1981) *Biochemistry*, **20**, 3110-3886
Van SP and Griffith OH (1975) *J Membr Biol* **20**, 155-170
Van Gorkom L, Horvath L, Sternberg B, Hemminga MA and Watts A (1989) (submitted)
Watts A (1981) *Nature*, **294**, 51
Watts A (1986) *Stud Biophys* **110**, 149-154
Watts A (1987a) In: *Membrane Receptors, Dynamics and Energetics* (Wirtz, KAW, ed), Plenum Press, New York, pp 329-339
Watts A (1987b) *Bull Mag Reson*, **9**, 66-70
Watts A (1987c) *J Bioenerg Biomemb* **19**, 625-653
Watts A (1989) *Studia Biophys* **127**, 29-36
Watts A and Poile TP (1986) *Biochim Biophys Acta* **861**, 368-372
Watts A, Harlos K, Maschke W and Marsh D (1978) *Biochim Biophys Acta* **510**, 63-74

Watts A Volotovski, ID and Marsh, D (1979) *Biochemistry*, **18**, 5006-5013
Watts A, Sixl F, Ryba NJP, Dempsey CE and Brophy PJ (1985) In: *Magnetic Resonance in Biology and Medicine* (Govil G, Khetrapal CL and Saran A, eds), Tata-McGraw Hill, New Delhi
Wolfs CJAM, Horvath LI, Marsh D, Watts A and Hemminga MA (1989) *Biochemistry*, (submitted)

The super-ordered fluid: A domain model of cell surface membranes

Michael Edidin
Department of Biology
The Johns Hopkins University
3400 N. Charles Street
Baltimore, MD 21218 USA

In the last twenty years models of phospholipid bilayer membranes have emphasized the dynamics of membrane components, the motion of molecules in the ordered fluid of the bilayer. Most commonly cited is the fluid mosaic model of Singer and Nicolson (1972). This model originated with studies of membrane composition, based on an earlier, static, picture of membrane architecture (for a historical review see Gennis, 1988). "Mosaic" indicates that the phospholipid bilayer is studded with intercalated proteins; "fluid" indicates that these proteins, and the lipids are free to move parallel or normal to the bilayer surface. Though the model is only weakly predictive, it does synthesize a body of experiments, and it remains the basis of much current thinking about cell surfaces.

In this paper I want to develop evidence for a model of cell surfaces, and, by extension, other membranes, which emphasizes their organization into discrete, bounded, regions, domains. This model arises from work on cell surface biophysics and biochemistry. Though it retains the idea of the phospholipid bilayer as an ordered (2-dimensional) fluid, the model emphasizes the heterogeneities created in this fluid, by immiscible lipids, and by molecular interactions of lipids and proteins both in the bilayer and in the aqueous milieu on either side. The scale of the domains is in the range of 100's to 1,000's of nm.

The existence of immiscible lipid phases is readily shown for synthetic bilayers (Small, 1986; Hui, 1987). These phases may be gel and fluid or may be co-existing fluid phases. Their demonstration led to the first of the models for membrane domains, the plate tectonic model of Jain and White (1977).

Patches of lipids below their phase transition are detectable in cell surface membranes at physiological temperatures (Brasitus and Schachter, 1980; Wolf et al., 1981a; Karnovsky et al., 1982; Bearer and Friend, 1980; Edidin and Sessions, 1984; Thompson and Tillack, 1985). Such patches should not be called phases, since phases form in a mixture at equilibrium, a condition not achieved by any living cell or organism. Instead we term these metastable features domains. A domain can be detected as a patch, or differentiated feature of a membrane if its lifetime is longer than the time scale on which it is probed, a scale of nsec for spectroscopic probes, a scale of minutes for observations in the fluorescence microscope. Lifetimes on the longer time scale are certainly likely for gel patches, from which molecules can only escape slowly (Thompson and Tillack, 1985).

The size scale of lipid domains, whether gel or fluid is variously estimated by different techniques and probes. Electron microscopy of membranes suggests that the immiscible lipid patches are in the range of 10's of nm in erythrocytes (Tillack et al., 1983) to µm in sperm acrosomal membrane and Plasmodium surface membranes (Bearer and Friend, 1980; Mikkelsen et al., 1988). Fluorescence microscopy of cells labeled with merocyanine 540, a fluorescent lipid probe that preferentially partitions into fluid lipids, suggests that there are µm-scale fluid lipid domains in erythrocyte membranes as well (Schlegel et al., 1986) which remain stable after fusion of erythrocytes with cultured fibroblasts.

Measurements of lateral diffusion of lipid probes also detect local organization of membrane lipids. The nature of the domains detected depends upon the probe used. Wolf and co-workers (1981a;1981b) used the cationic dye indocarbocyanine to probe membranes of sea urchin and mouse eggs. Lateral diffusion of the diI was a function of its alkyl chain length. Calibration of diffusion behavior of the diI's in liposomes of known composition, showed that longer-chain diI's partitioned preferentially into gel lipid domains (Klausner and Wolf, 1980). It also could be shown that fluorescence of the dye was quenched as it concentrated in a given domain. The size of these domains was undetermined, but since the photobleaching spots used are of the order of 1 μm radius, the domains are likely to have this characteristic linear dimension.

Another fluorescent lipid probe, NBD-phosphatidylcholine also reports lipid domains in cell membranes, but in this case it is the dependence of lateral diffusion of a single species of label on the scale of diffusion measurement that indicates the presence of domains (Yechiel and Edidin, 1987). The diffusion coefficient increase and the fraction of label free to diffuse decreases as the area of the spot measured increases in the range ~0.5 μm^2 to ~10 μm^2. Diffusion of labeled proteins also shows this trend. An example is shown in Figure 1.

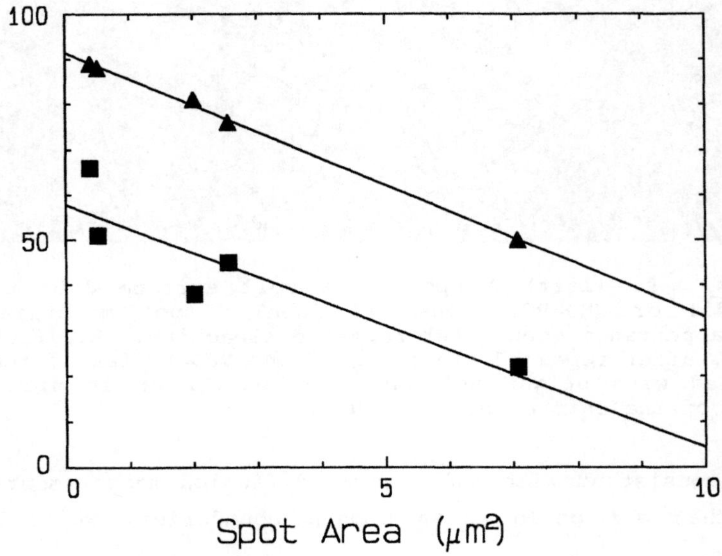

Figure 1. Dependence of the apparent mobility of a fluorescent lipid analog, NBD-PC and of a protein, labeled with fluorescein-Fab, on the area of the spot bleached in an FPR experiment.

Control measurements of diffusion of NBD-PC in large liposomes showed that the results on cells were not artifacts of the method for measuring lateral diffusion. In cells, the diI probes did not show a dependence of lateral diffusion on spot size. This controls for artifacts of cell surface geometry, for example microvilli or folds in the membrane. Given these controls, and considering the details of the data led us to conclude that the results with NBD-PC and proteins were due to the existence in membranes of large protein-rich domains, differing in composition and physical properties from the average for the entire membrane. The area of the domains must be similar to the areas investigated in our experiments, μm^2. Patches of fluorescence of this size are seen when one spot on an NBD-PC-labeled cell is repeatedly bleached, but are not seen on cells labeled with diI (Figure 2).

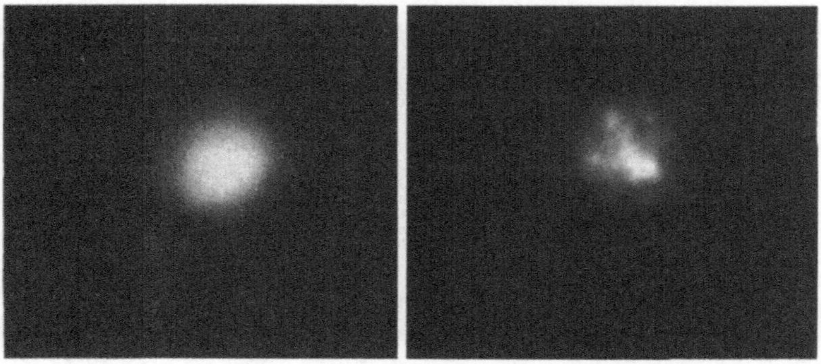

Figure 2. A. (left) A spot on a cell surface labeled with either diI or NBD-PC. The diI labeled spot maintains its uniform appearance even after repeated bleaching. B. (right) In contrast, after repeated bleaching of the NBD-PC label, the spot is mottled with bright and dark patches ~1 μm in diameter. Diameter of the entire spot is ~10 μm.

This is consistent with the lateral diffusion measurements that showed that diI could cross domain boundaries, while NBD-PC could not.

Other lateral diffusion measurments detect lipid domains, seen in terms of heterogeneity of diffusion coefficients, in plant protoplasts (Metcalf et al., 1986). Very large lipid domains are present in the Xenopus egg. This single cell is over 1 mm in diameter, covered by a single plasma membrane. Diffusion of fluorescent fatty acids differs between animal and vegetal halves of the egg. Fluorescence micrographs show a striking difference in fluorescence pattern in the two halves. Fluorescence of the animal half, where the diffusion of the probe is low, is patchy, while that of the vegetal half is continuous (Dictus et al., 1984). The authors conclude "Our data demonstrate that the plasma membrane of the Xenopus egg differs in its organization from other biologial membranes studied so far and make clear that this membrane is not of a fluid mosaic nature..."

Domains have been detected biochemically as well as biophysically. Lipase digestion differentiates lipid domains in erythrocyte (Shukla and Hanahan, 1982) and platelet (Wang et al., 1987) membranes. Patches of surface membrane differing in composition from the membrane average are detected by careful biochemical analysis of enveloped viruses budding from cells (Pessin and Glaser, 1981) and of lymphocyte plasma membrane fractions (Goppelt et al., 1986). Enzymes and receptors were enriched in particular fractions. Movement of receptors from one domain to another may be induced by ligand binding (Jesaitis et al., 1988).

This quite incomplete survey has been made to give some indication of the extent to which cell surface membranes exhibit lateral heterogeneity and patchiness. The survey results in a picture, rather than a working model. Indeed, we have several pictures, since each probe or approach to detecting domains gives us a different report about their composition and physical properties. Certainly the small clusters of glycolipids detected by electron microscopy could be contained within the larger domains implied by the diffusion of NBD-lipids. On the other hand, it is not clear if the results on diffusion of NBD-lipids, implying µm size protein-rich domains, and those on diI, also implying µm-size, protein-poor domains, report on the same or overlapping regions of the cell surface. Some alternative possibilities for the arrangement of membrane domains are shown in Figure 3.

Our picture or pictures also lack any indication of the way most surface membrane domains are created and maintained. Lipid-lipid immiscibility can, in principle, account for large or small lipid domains, but we do not know if these are formed

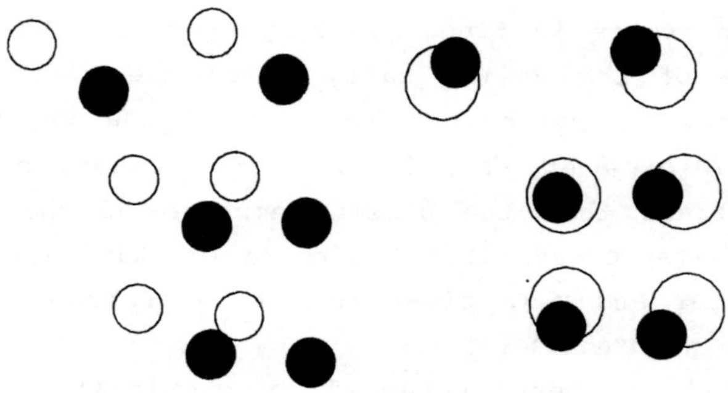

Figure 3. Alternative arrangements for the domains reported by the diffusion of two different lipid probes, the diIs and NBD-PC, and by the diffusion of labeled proteins. Left, domains detected by two of the probes, shown as open and closed circles, are spatially separated over the membrane. The organization of the spaces between domains is undefined in this drawing. Right, domains detected by one of the probes are partly or wholely within domains detected by a second probe. These relationships may change with time, but the time-scale for change is undefined.

in endomembranes and exported as a unit to the cell surface, or if they organize after lipids are exported to the surface. The former would seem likely in morphologically polarized epithelial cells. The lipid composition of the apical surface of these cells typically differs from the composition of their basal/lateral surfaces (see Brasitus and Dudeja, 1988 for a review of intestinal epithelial cell membranes) and presumably these lipids are sorted delivered to the surface as a unit. Inositol-lipid-linked proteins are also selectively targeted in such cells (cf. Lisanti et al., 1989).

Recent work probing synthetic bilayers made of a single phospholipid species shows that packing heterogeneities, are detectable with fluorescence probes. These regions are constantly forming and dissipating, but the authors suggest that their stability could be changed by membrane proteins (Davenport et al., 1989).

Membrane domains could also be organized by the interaction of a cytoskeleton with membrane lipids and/or proteins. Erythrocyte spectrin and the spectrin-like fodrins of nucleated are likely candidates for such interactions. The spectrin network of erythrocytes forms a mesh with a characteristic repeat distance based on the spectrin tetramer, 100-200 nm (Shen et al., 1986). The barrier to lateral mobility formed by this mesh, and hence apparent domain size detected, would depend upon the integrity of the mesh, the ratio of spectrin tetramers to spectrin dimers (Saxton, 1989). Recent work indicates that any sort of impermeable region, whether gel lipid or protein aggregate, will impede diffusion in a surface, and that diffusion may vary with distance scale in these circumstances, being largest on a small, Å scale, and reaching some steady-state value on distance scales that depend upon interactions with other molecules in the membrane (Abney et al., 1989). Other work using a percolation approach shows that a concentration of gel lipid above some threshold, will destroy the connectivity of the permeable regions of the membrane isolating small fluid domains (Vaz et al., 1989). In short, the characteristic size of a domain observed by different methods may not reflect the size of the structures organizing the domain, and different probes may behave differently for the same underlying membrane organization.

We have much more to learn about the ordering and spatial differentiation of cell surfaces. Membrane functions are likely to be affected by the patchiness of the cell surface. This patchiness can serve to keep proteins together in functional arrays. It could also separate molecules, so that they can never interact during their lifetime at the surface. Our understanding of such functional organization is sure to develop from the increasing recognition of the higher levels of ordering and complexity imposed on bilayer membranes.

References

Abney JR, Scalettar BA and Owicki JC (1989) Self diffuson of interacting membrane proteins. Biophys J 55:817-833

Bearer EL and Friend DS (1980) Anionic lipid domains: Correlation with functional topography in a mammalian cell membrane. Proc Natl Acad Sci USA 77:6601-6605

Brasitus TA and Dudeja PK (1988) Small and large intestinal plasma membrane: Structure and function in Lipid domains and the relationship to membrane function. (Aloia RC, Curtain CC and Gordon LM, eds) Allan R. Liss Inc, NY, pp 227-254

Brasitus TA and Schachter D (1980) Lipid dynamics and lipid-protein interactions in rat enterocyte basolateral and microvillus membranes. Biochemistry 19:2763-2769

Davenport L, Knutsopn JR and Brand L (1989) Fluorescence studies of membrane dynamics and heterogeneity in Subcellular Biochemistry. (Harris JR and Etemadi AH, eds) Plenum, NY, pp 145-188

Dictus WJAG, van Zoelen EJJ, Tetteroo PAT, Tertoolen LGJ, De Laat SW and Bluemink JG (1984) Lateral mobility of plasma membrane lipids in Xenopus eggs: Regional differences related to animal/vegetal polarity become extreme upon fertilization. Devel Biol 101:201-211

Edidin M and Sessions A (1984) Heterogeneity in the plasma membrane lipids of eukaryotic cells. Ann NY Acad Sci 414:8-18

Gennis RB (1988) Biomembranes Molecular structure and function. Springer-Verlag, New York, 1989

Goppelt M, Eichhorn R, Krebs G and Resch K (1986) Lipid composition of functional domains of T-lymphocyte plasma membrane. Biochim Biophys Acta 854:184-190

Hui SW (1987) Ultrastructural studies of the molecular assembly in biomembranes: Diversity and similarity. Curr Top Memb Transport 29:30-70

Jain MK and White HB (1977) Long range order in biomembranes. Adv Lipid Res 15:1-59

Jesaitis AJ, Bullock GM, Tulley JO and Allen RA (1988) lateral segregation of neutophil chemotactic receptors into actin-and fodrin-rich plasma membrane microdomains depleted in guanyl nucleotide regulatory proteins. J Cell Biol 107:921-928

Karnovsky MJ, Kleinfeld AM, Hoover RL and Klausner RD (1982) The concept of lipid domains in membranes. J Cell Biol 94:1-6

Klausner RD and Wolf DE (1980) Selectivity of fluorescent lipid analogues for lipid domains. Biochemistry 19:6199-6203

Lisanti MP, Sargiacomo M, Graeve L, Saltiel AR and Rodriguez-Boulan ER (1989) Polarized apical distribution of glycosyl-phosphatidylinositol-anchored proteins in a renal epithelial cell line. Proc Natl Acad Sci USA 85:9757-9761

Metcalf TN III, Wang JL and Schindler M (1986) Lateral diffusion of phospholipids in the plasma membrane of soybean protoplasts: Evidence for membrane lipid domains. Proc Natl Acad Sci USA 83:95-99

Mikkelsen RB, Kamber M, Wadwa KS, Lin PS and Schmidt-Ullrich R (1988) The role of lipids in Plasmodium falciparum invasion of erythrocytes. A coordinated biochemical and microscopic study. Proc Natl Acad Sci USA 85:5956-5960

Pessin JE and Glaser M (1981) Budding of Rous sarcoma virus and vesicular stomatitis virus from localized lipid regions in the plasma membrane of chicken embryo fibroblasts. J Biol Chem 255:9044-9050

Saxton M (1989) The spectrin network as a barrier to lateral diffusion in erythrocytes. Biophys J 55:21-28

Schlegel RA, Lumley-Sapanski K and Williamson P (1986) Insertion of lipid domains into plasma membranes by fusion with erythrocytes. Biochim Biophys Acta 846:234-241

Shen BW, Josephs R and Steck TL (1986) Ultrastructure of the intact skeleton of the human erythrocyte membrane. J Cell Biol 102:997-1006

Shukla SD and Hanahan DJ (1982) Identification of domains of phosphatidylcholine in human erythrocyte plasma membranes. J Biol Chem 257:2908-2911

Singer SJ and Nicolson GL (1971) The fluid mosaic model of the structure of cell membranes. Science 175:720-731

Small DM (1986) The Physical Chemistry of Lipids from Alkanes to Phospholipids in Handbook of Lipid Research (Hanahan D, ed) Plenum Press, NY

Thompson TE and Tillack TW (1985) Organization of glycosphingolipids in bilayers and plasma membranes of mammalian cells. Ann Rev Biophys Biophys Chem 14:361-386

Tillack TW, Allieta M, Moran RE and Young WE Jr (1983) Localization of globoside and Forssman glycolipids on erythrocyte membranes. Biochim Biophys Acta 733:15-24

Vaz WLC, Melo ECC, Thompson TE and Jovin TM (1989) Translational diffusion and fluid domain connectivity in mixed lipid bilayers. Biophys J 55:112a

Wang, CT, Tsai WJ, Chang SM, Shiao YJ and Yang CC (1987) Different susceptibilities of platelet phospholipids to various phospholipases and modifications induced by thrombin. Possible evidence of rearrangement of lipid domains. Biochim Biophys Acta 899:205-212

Wolf DE, Kinsey W, Lenarz W and Edidin M (1981a) Changes in the organization of the sea urchin plasma membrane upon fertilization: Indications from lateral diffusion rates of lipid-soluble fluorescent dyes. Devel Biol 85:195-198

Wolf DE, Edidin M and Handyside AH (1981b) Changes in the organization of the mouse egg plasma membrane upon fertilization and first cleavage: Indications fromthe lateral diffusion rates of fluorescent lipid analogs. Devel Biol 85:195-198

Yechiel E and Edidin M (1987) Micrometer-scale domains in fibroblast plasma membranes. J Cell Biol 105:755-760 [illegible] (19[illegible]) [illegible]

[illegible] Handbook of Thermal [illegible]

[illegible]

Membrane hydration

Klaus Gawrisch[a,+], V. Adrian Parsegian[a], and R. Peter Rand[b]

[a]DCRT, National Institutes of Health, Bethesda, Md. 20892, USA

[b]Department of Biological Sciences, Brock University, St. Catharines, Ontario, Canada, L2S3A1

[+]Permanent Address: Sektion Physik, Karl-Marx-Universität, Leipzig, 7010, G.D.R

Beginning from the idea that hydration of the bilayer surface is an essential part of maintaining its stability as a non-polar layer, seeing how hydration forces can reach out many Angstroms between surfaces, then realizing that hydration is an important factor in mechanical properties and even in lamellar vs. non-lamellar morphology, one realizes that interaction with water is essential to the natural life of lipids. Surface hydration is surely even more important in the behavior of the many saccharides and proteins that inhabit the membrane surface. Processes of cell membrane fusion, exocytosis and endocytosis all incur significant changes in accessibility to water; their energetics must be comparably affected. In this brief review we will not catalog the many specific membrane processes where one should be thinking about hydration. Rather, we will try to make clear the energetic and structural factors that should be kept in mind whenever one is thinking about membrane design, stability and function.

*****

Membrane "hydration" sounds almost paradoxical. Unique among cellular organelles, membranes are designed to be non-polar layers. As such they can divide aqueous compartments by controlling transport of solutes, and they can provide a locus for anchoring and organizing cellular activities on a two-dimensional surface. In order to exist as stable entities, non-polar membranes must make their own peace with their aqueous surroundings. It is well known that they do this by being constructed mainly of amphiphiles, substances with dual polar and non-polar parts, substances that can merge their non-polar moieties to form oily regions that are stabilized by polar groups at the interface with water. The water affinity of these surface-confined polar groups of lipids or associated molecules is the central theme of the present chapter.

Phospholipid Hydration Forces

Most of what is known about the physical force consequences of membrane hydration has been learned from direct measurements of forces between electrically neutral or charged phospholipid bilayers. This subject has been exhaustively reviewed recently (Rand & Parsegian 1989), so only the main points will be covered here. The central result of these measurements is that for separations up to about 20 Angstroms phospholipid bilayers repel with a force that varies exponentially with a decay constant of 1 to 3 Angstroms. Because this repulsion is seen even between bilayers with electrically neutral, zwitterionic polar groups immersed in distilled water, it has been associated with the work of removing water from the bilayer surface, a "hydration" (or perhaps better, "dehydration") force.

The strength of repulsion varies with the chemical identity of the phospholipid polar group. It matters whether the bilayer hydrocarbon chains are frozen or melted. The forces are remarkably insensitive to salt in the medium. Hydration forces between charged lipids are somewhat more difficult to detect since they must be distinguished from inevitable electrostatic double layer repulsion. Yet evidence from phospholipids (Cowley *et al.* 1978, Loosley-Millman *et al.* 1982) and, more strongly, from measurements between bilayers of synthetic amphiphiles shows that hydration forces operate to dominate the interaction of charged bilayers when approaching contact.

It has become the practice to codify hydration forces in terms of a coefficient P_o and an exponential decay constant λ to fit the force per unit area $P(d_w)$ between parallel bilayers as separation d_w with the expression

$$P(d_w) = P_o \exp(-d_w/\lambda).$$

Some representative values of P_o and λ are given in Table 1. The procedure for the determination of these parameters from osmotic stress measurements (Parsegian *et al.* 1986, Rand 1981, Parsegian *et al.* 1979) is elaborated elsewhere (Rand & Parsegian 1989). The separation d_w follows the mass-average convention that divides the repeat spacing of a multilayer into regions of pure water of thickness d_w and pure lipid of bilayer thickness d_l. There is some deformation of the bilayer that is ascertained by independently measured bilayer lateral compressibility (Evans & Needham 1987). Other definitions and determinations of separation give slightly different parameters (McIntosh & Simon 1986) but have virtually no effect on physically critical estimates of the work of bilayer dehydration. In Figure 1 we show typical plots of pressure vs separation for bilayers of several zwitterionic phospholipids.

The coefficient P_0 varies from 10^{10} to 10^{13} dynes/cm^2 (or erg/cm^3) while the decay parameter can be as small as about 1 Angstrom seen with phosphatidylethanolamines (PE) and as large as 2.5 Angstroms for phosphatidylcholines (PC). It has been argued recently that the cases of shorter range decay reflect a mixture of powerful attractive and repulsive hydration forces between bilayers of complementary polar structure (Rand *et al.* 1988); the longer range decay represents a pure repulsion such as is also seen between polysaccharides (Rau & Parsegian 1987) and between charged polyelectrolytes approaching contact (Rau *et al.* 1984). Forces measured between neutral lipids immobilized onto the crossed mica cylinders of a surface force apparatus (Marra & Israelachvili 1985, Israelachvili & Marra 1986) agree with those between multilayers once one takes account of curvature and membrane undulation (Horn *et al.* 1988).

Most puzzling is the extraordinary sensitivity of the measured forces to small changes in polar group structure. For example, a single methylation of the PE polar group will cause the lipid multilayer to take up 25% more water when under no stress. Similar sensitivity to methylation has been reported for changes in hydrocarbon chain melting temperature (Cevc 1987) and for the tendency to transform from a lamellar to an inverted hexagonal phase (Gruner *et al.* 1988).

Of greatest interest from a thermodynamic viewpoint is the work of bringing together two bilayers seen as a work of transfer of water from the space between membranes out to the bulk solution. Multiplying the pressure P by the molar volume of water (18 cm^3) and correcting for units (1 calorie = 4.18 x 10^7 ergs),

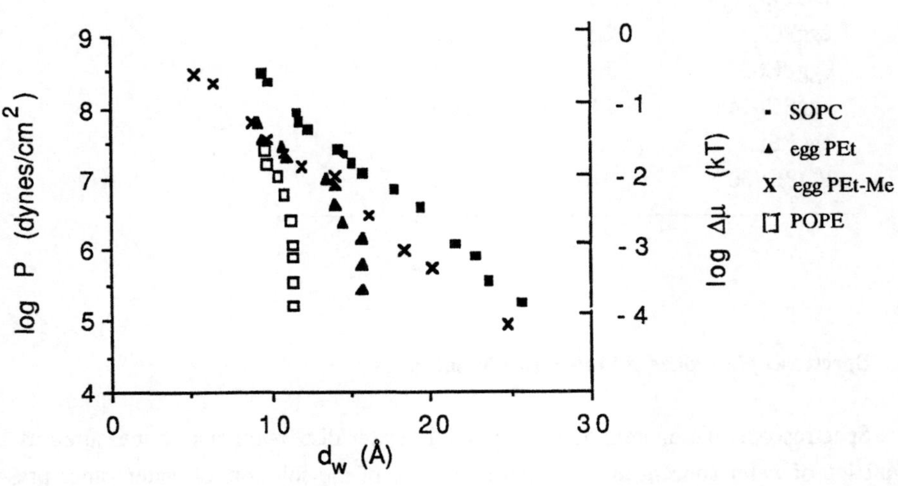

Fig. 1 Representative pressure vs. separation data for zwitterionic phospholipids. (Right axis - chemical potential per water molecule relative to bulk water in kT units.)

we find that even for physically significant pressures of 1 to 100 atmospheres (10^6 to 10^8 dyn/cm^2) the thermodynamic perturbation on water is only about .5 to 50 cal/mole, almost negligible compared to thermal energy of 600 cal/mole. This comparison shows why it is so difficult using probes of water itself to detect perturbations that are quite important in maintaining separate membrane surfaces. The reason why small perturbations are seen in force measurement lies in the size of the bilayer compared with the size of water molecules. To move a patch of membrane involves the displacement of hundreds or thousands of water molecules. The small ~cal/mole perturbations add up to give physically significant forces. They are virtually impossible to see in terms of a perturbation of one or a few water molecules that are typical of most probes. When perturbations of water by lipids reach thermally significant kcal/mole levels, they correspond to the extraordinary 1000 atmosphere pressures felt between bilayers approaching contact.

Table 1 Hydration force parameters for some zwitterionic phospholipids (compressibility-adjusted data). See Rand & Parsegian (1989) for details.

Lipid	λ (Å)	log P_0 (dyn/cm^2)
DPPC (25°C)	1.2	12.3
DPPC (50°C)	2.1	11.0
SOPC (30°C)	2.0	10.5
DOPC	2.1	10.6
eggPC	2.1	10.6
eggPEt-Me_2	1.8	10.4
eggPEt-Me	1.8	10.3
eggPEt	1.1	12.3
POPE (30°)	0.8	12.5

Spectroscopic Probes of Water near Membranes

Spectroscopic parameters of a phospholipid water dispersion can be measured as a function of water concentration, osmotic pressure of the solution, or water vapor pressure. For reasons already stated, one can expect to detect only the perturbations in a first (sometimes a second) water layer near to a membrane. Various spectroscopies provide information about the number of water molecules interacting with the surface, the kind of binding of waters to the lipids, and rates and symmetries of motions of waters at the surface. The division of water into perturbed and nonperturbed by categories is artificial in a dynamical and a structural sense. It is reasonable to assume that the perturbation in the water structure caused by a surface decays continually with the distance from that surface, as indicated by the force-distance curve measurements for bilayers.

Even comparison of experimental results obtained by different methods can be difficult. Each kind of a spectroscopy is sensitive to different types of structural changes and to motions on very different time scales. Some of the spectroscopical parameters are particularly sensitive to changes in structure, e.g. the quadrupole splitting of heavy water at a membrane surface. The measured values of that splitting are only 1% or less of the splitting of perfectly oriented water, indicating that the water molecules move more or less randomly at a membrane surface.

X-ray and neutron scattering, IR-spectroscopy

At very low water concentrations and low temperatures some phospholipids form monocrystals. The position of every atom in these crystals can be determined by X-ray diffraction (Hitchcock *et al.* 1974, Pearson & Pascher 1979, Hauser *et al.* 1980). The water molecules or other polar water substituents are located near the polar phosphate groups. This experimental result is verified by measurements of stretching modes of bonds of polar groups in IR-spectroscopy. At low water concentrations the signals of several bonds in the polar region of phospholipids and the water signals are shifted as a function of water concentration (Fookson & Wallach 1978, Wong & Mantsch 1988). Small angle X-ray and neutron scattering measurements indicate the uptake of water between phospholipid bilayers by increased repeat spacings (Luzzati & Husson 1962, Zaccai *et al.* 1975). Unlike NMR which provides a spatial average of the signals from repeating groups, X-ray and neutron diffraction give a distribution of locations of scattering centers - electrons with X-ray diffraction and nuclei of low atomic weight with neutron diffraction. The latter also has the possibility, in principle, of showing distribution of scattering particle velocities.

NMR

NMR methods are widely used for hydration studies. The signal of H_2O near a membrane surface is broadened or split by dipole-dipole interactions and/or diamagnetic susceptibility effects (Fleming *et al.* 1978). The correlation time of motions of water molecules at a membrane surface is slowed down. As a result the water molecules have changed spin-lattice relaxation times (Berendsen 1975).

NMR spectra are very sensitive to the rate of exchange processes. When a molecule switches between the two states A and B, which have different resonance frequencies ($\nu_A = \omega_A/2\pi$, $\nu_B = \omega_B/2\pi$) and the lifetimes τ_A and τ_B, NMR "sees" only a single resonance frequency, a time-averaged picture of the system, if $\tau_A, \tau_B << 1/(\omega_A - \omega_B)$. When $\tau_A, \tau_B >> 1/(\omega_A - \omega_B)$, each state of the system has its "own" resonance spectrum. The measured spectra indicate an apparent statistical distribution of stationary conformations or distribution of molecules in the system. The spin-lattice and spin-spin relaxation times of the resonance signals are related to the lifetimes of the different states.
The water at the membrane surface is in fast exchange with the other water between bilayers; measured parameters are proportional to the average lifetime of the water molecules at the membrane surface. The estimated number of water molecules in that first (or second) layer is usually interpreted as "bound" to the membrane. The experiment can be influenced by structural transitions in the membranes occurring as a function of water concentration.

The proton NMR signal of water is obscured by proton signals of the membrane constituents, especially at low water concentrations, while measurements of membrane hydration with ^2H_2O are not influenced by such effects. Finer & Darke (1974) measured quadrupolar splittings as a function of water concentration for several phospholipid water dispersions. The dependence of the splitting on the inverse number of waters per lipid can be interpreted in terms of layers of "bound" and "trapped" water between bilayers. The deuterium nucleus has in addition to its magnetic moment, an electric quadrupolar moment. Every electric field gradient at the ^{2}H-nucleus, not averaged out by fast motions, splits the ^{2}H-NMR signal into a doublet. The biggest electric field gradients are the inner field gradients of the water molecule itself. Because of fast, nearly isotropic, motions that gradient is widely averaged out. The measured effective electric field gradient acting on the ^{2}H nuclei is normal to the bilayer surface. If the water molecules are attached to the lipids by hydrogen bonds for a short time, the normal is the symmetry axis for the water motions. It is possible also, that the waters move isotropically and that the acting electric field gradient is of intermolecular origin. The potential jump of the membrane electric dipole potential at the boundary between the apolar lipid interior and the polar lipid headgroup region could provide electric field gradients strong enough to cause that splitting.

The T_1-relaxation mechanism of heavy water is dominated by quadrupolar relaxation. The relaxation times of water in a first hydration shell around phospholipid head groups are 5 times shorter than for bulk water (Borle & Seelig 1983). T_1-relaxation data of ^2H_2O are less sensitive to the presence of small amounts of paramagnetic ions in the membranes. Because of the differences in dynamic properties of hydrogen bonds in H_2O and ^2H_2O, deuteration could conceivably influence hydration. Up to now, however, there has been no indication that ^2H_2O changes structures of membranes and hydration properties.

Microscopy

The knowledge of the structural arrangement of lipid bilayers in multilamellar liposomes is important for data analysis of X-ray, neutron scattering and NMR experiments (Gawrisch *et al.* 1985). Usually during hydration liposomes of different sizes are formed spontaneously. Microscopy techniques, such as electron microscopy after negative staining or freeze etching and optical phase contrast microscopy, give rough estimates of changes which occur as a function of water concentration. Electron microscopy suffers sometimes from artifacts produced during sophisticated sample preparation. It has to be considered also, that the investigated lipid spot is only a very small part of the total sample and may be not representative. Nevertheless, new methods are beginning to give clearer pictures of lipid conformation (Zasadzinski *et al.* 1988).

Calorimetry of nonfreezable water

Calorimetry of multilayer preparations containing different amounts of water shows clearly that not all the water freezes at O°C or even at lower temperatures (Chapman *et al.* 1967, Ter-Minassian-Saraga & Madelmont 1982). While such data are occasionally used to estimate the extent of "perturbation" of water by membranes, they are in fact no more than qualitative indications. "Freezing" in the normally understood sense must recognize the extensive three dimensional structure of ice, impossible to realize in the confirms of a multilayer. The chemical potential of water, the key thermodynamic parameter in phase transition as in mass transfer, is a poorly understood average over an inhomogeneous region affected both by direct interactions of water with the surface and by the interaction between surfaces. The fact that water will often come out of a multilayer when it freezes (Grünert *et al.* 1984) further emphasizes the complexity of any correction of freezing, surface perturbation, and lipid morphology.

Adsorption isotherms

Adsorption isotherms and force distance curves are a more direct measure of the thermodynamic state of water between bilayers in multilayer phases. Adsorption isotherms are measured equilibrating weighted amounts of lipid samples with a definite water vapour pressure (Elworthy 1961, Jendrasiak & Mendible 1976). The uptake of water is determined with a balance. This methods works well up to relative humidities around 95%. At higher humidities a very high accuracy of temperature control for the whole equippment must be provided to get a correct adjustment of the humidity. The driving force for the water uptake decreases exponentially with increasing water concentration in the sample. The results may be influenced by capillary condensation of water in pores in the lipid film. That makes it difficult to reach equilibrium conditions at relative humidities near to 100%. There is a puzzling discrepancy between the amount of water sorbed from a 100% relative humidity vapour and from liquid water (Rand & Parsegian 1989).

Theoretical calculations

Theories of membrane hydration, despite many very intelligent efforts, are still unsatisfying. There has been a line of thinking proposed by Marcelja (Marcelja & Radic 1976, Gruen *et al.* 1984) that one can model the perturbation of water by a membrane surface in terms of a continuously varying order parameter whose physical identity -- a polarization, a disturbance of the hydrogen bond network, or other physical parameter -- is not specified. The idea has recently been elaborated by Cevc (1987) that there is a polarizing hydration potential associated with the lipid surface that might be identified with an electrostatic dipole potential (Simon *et al.* 1989). A neat way to look at surface perturbation has been suggested by Kornyshev (Dzhavakhidze *et al.* 1988) through the use of a "non-local dielectric" formalism. Computer simulations of molecular density or H-bond perturbation seem to be too insensitive to pick up the many very weak interactions that must be encompassed in an adequate theoretical model (Kjellander & Marcelja 1985a,b) although a model proposed quite recently suggests that many features can be explained in terms of perturbed H-bond networks (Attard & Batchelor 1988). Still, computer experiments might allow one to relate a particular intermolecular potential with a predicted perturbation of any order parameter. We believe that such computer simulations ultimately will be the only way to link models of water structure and its surface perturbation with the sum of the partial information provided by many different kinds of measurement.

The theory of membrane hydration, as well as of the hydration of macromolecules of the type that cover real cell surfaces, is still in need of serious effort.

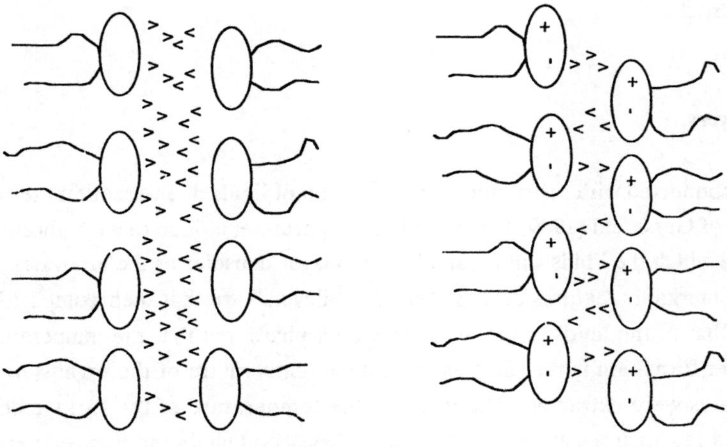

Fig. 2 Schematic representation of the polarizing perturbation of water between (left) purely repulsive bilayers or (right) bilayers experiencing both attractive and repulsive hydration forces (Rand *et al.* 1988).

Mechanical Properties, undulations enhancing hydration

Multilayer assemblies can take up water through mechanical undulations of lamellae that act to drive bilayers apart even though there is no interaction of water with the bilayer surface. Indeed the powerful idea of a steric undulatory repulsion, first developed by Helfrich (1978), has developed a strong following during the past few years. It is now clear that when forces between bilayers are weak enough to allow bilayer motion, the consequent steric forces of mutual collision will dominate the uptake of water in a lamellar lattice. What is becoming equally clear is that the entropically restrictive "collisions" do not occur as hard encounters between layers but rather are themselves mediated by long-range hydration or electrostatic double layer repulsion. There is a remarkable interplay of direct and configurational entropic forces whose consequence is a mutual enhancement of each.

This coupling of direct and steric forces has been demonstrated theoretically (Evans & Parsegian 1986) and experimentally between bilayer membranes (Horn *et al.* 1988) or parallel linear polyelectrolytes (Podgornik *et al.* 1989). Bilayers whose mean separation in excess water would formerly be thought to sit where there was a balance of van der Waals attractive and hydration or electrostatic repulsion are now recognized as sitting at greater separations that allow membrane undulations about the force minimum position (Evans & Parsegian 1986). Highly flexible layers might even break out of a direct force balance position altogether to "hydrate" to indefinite levels as envisaged in the early model of Helfrich.

Surface Bilayers

Possibly connected with distribution and hydration of lipids due to motion are recent observations of Gershfeld (1989) that there is a temperature, unique to each phospholipid mixture, at which the lipids will accumulate to bilayer densities at the air/water interface. Two remarkable features of these "surface bilayers" are that their composition is the same as that of the lipids in the underlying bulk phase and that the temperature of surface bilayer formation is the same as the growth temperature of the organisms from which the lipids were extracted. The fact that the temperature of the surface density maximum is a few degrees above the "main" hydrocarbon chain melting temperature (seen for those cases of pure artificial lipids where chain melting can be measured) suggests that there is a connection between loosening of the membrane and the expansion of bilayers from the multilamellar arrays of finite hydration in which they might otherwise remain.

Hydration of non-lamellar phases

Rather than pack into bimolecular leaflets, some membrane lipids and many pure synthetic phospholipids will form inverted hexagonal H_{II} phases, parallel tubes of water whose axes lie on a hexagonal net bounded by phospholipids whose hydrocarbon tails fill the intervening space. When the tails are too short to fill that space, these lipids are forced out of the hexagonal packing symmetry, a frustration that is apparent upon addition of filler alkanes (Gruner *et al.* 1986, Tate & Gruner 1987). In excess water and alkane, these lipids will form H_{II} structures whose spontaneous radii of curvature (Gruner 1985) represent a state of minimum energy. What is remarkable is that the enabling addition of alkane creates hexagonal phases that take up much more water than does the multi-lamellar structure that formed without alkane. One must recognize then yet another source of lipid hydration, this one driven by the tendency to form a surface of minimum energy.

In Sum:

Membrane hydration will have many functional and morphological consequences. Hydration forces can be measured to good accuracy, but various probes of water and theories of perturbation of water near surfaces give only hints of how these forces occur. There is a dual opportunity here: first, to sharpen our physical ideas about the origin of these forces; and second to keep in mind their enormous strength when thinking about the behaviour of cell membranes.

References

Attard, P., Batchelor, M.T.: A mechanism for the hydration force demonstrated in a model system. *Chem. Phys. Letters* **149**:206-211, 1988.

Berendsen, H.J.C.: Specific interactions of water with biopolymers. *in Water A Comprehensive Treatise, ed.* F. Franks, Vol. 5, p. 293-330, Plenum Press, N.Y. and London, 1975.

Borle, F., Seelig, J.: Hydration of *Escherichia coli* lipids. Deuterium T1 relaxation time studies of phosphatidylglycerol, phosphatidylethanolamine and phosphatidylcholine. *Biochim. Biophys. Acta* **735**:131-136, 1983.

Cevc, G.: How membrane chain melting properties are regulated by the polar surface of the lipid bilayer. *Biochemistry* **26**:6305-6310, 1987.

Chapman, D., Williams, R.M., Ladbrooke, B.D.: Physical studies of phospholipids. VI. Thermotropic and lyotropic mesomorphism of some 1,2-diacyl-phosphatidylcholines (lecithins). *Chem. Phys. Lipids* **1**:445-475, 1967.

Cowley, A.C., Fuller, N., Rand, R.P., Parsegian, V.A.: Measurement of repulsive forces between charged phospholipid bilayers. *Biochemistry* **17**:3163-3168, 1978.

Dzhavakhidze, P.G., Kornyshev, A.A., Levadny, V.G.: The structure of the interface in the solvent-mediated interaction of dipolar surfaces. *Il Nuovo Cimento* **10**:627-652, 1988.

Elworthy, P.H.: The adsorption of water vapour by lecithin and lysolecithin, and the hydration of lysolecithin micelles. *J. Chem. Soc.* 5385-5389, 1961.

Evans, E., Needham, D.: Physical properties of surfactant bilayer membranes: Thermal transitions, elasticity, rigidity, cohesion, and colloidal interactions. *J. Phys. Chem..* **91**:4219-4228, 1987.

Evans, E.A., Parsegian, V.A.: Thermal-mechanical fluctuations enhance repulsion between bimolecular layers. *Proc. Natl. Acad. Sci., USA* **83**:7132-7136, 1986.

Finer, E.G., Darke, A.: Phospholipid hydration studied by deuteron magnetic resonance spectroscopy. *Chem. Phys. Lipids* **12**:1-16, 1974.

Fleming, W.W., Fornes, R.E., Memory, J.D.: Evidence specifying the origin of the NMR doublet of water in oriented macro-molecular systems. *Macromolecules* **11**:1057-1058, 1978.

Fookson, J.E., Wallach, D.F.H.: Structural differences among phosphatidylcholine, phosphatidylethanolamine and mixed phosphatidylcholine phosphatidylethanolamine multi-layers: An infrared absorption study. *Archives of Biochemistry & Biophysics* **189**:195-204, 1978.

Gawrisch, K., Richter, W., Möps, A., Balgavy, P., Arnold, K., Klose, G.: The influence of the water concentration on the structure of egg yolk phospholipid/water dispersions. *Studia Biophysica* **108**:5-16, 1985.

Gershfeld, N.L.: Spontaneous assembly of a phospholipid bilayer as a critical phenomenon; influence of temperature, composition and physical state. *J. Phys. Chem..* 1989 (in press).

Gruen, D.W.R., Marcelja, S., Parsegian, V.A.: Water structure near the membrane surface. *in: Cell Surface Dynamics: Concepts and Models, ed.* Perelson, A.S., De Lisi, C., Wiegel, F.W.. Marcel Dekker, Inc., New York and Basel 1984, p. 59-91.

Gruner, S.M., Tate, M.W., Kirk, G.L., So, P.T.C., Turner, D.C., Keane, D.T., Tilcock, C.P.S., Cullis, P.R.: X-ray diffraction study of the polymorphic behavior of N-methylated dioleoylphosphatidylethanolamine. *Biochemistry* **27**:2853-2866, 1988.

Gruner, S.M., Parsegian, V.A., Rand, R.P.: Directly measured deformation energy of phospholipid H_{II} hexagonal phases. *Faraday Discuss. Chem. Soc.* **81**:29-37, 1986.

Gruner, S.M.: Intrinsic curvature hypothesis for bio-membrane lipid composition: A role for non-bilayer lipids. *Proc. Natl. Acad. Sci., USA* **82**:3665-3669, 1985.

Grünert, M., Börngen, L., Nimtz, G.: Structural phase transition due to a release of bound water in phospholipid bilayers at temperatures below 0°C. *Ber. Bunsenges. Phys. Chem.* **88**:608-612, 1984.

Hauser, H., Pascher, I., Sundell, S.: Conformation of phospholipids. Crystal structure of a lysophosphatidylcholine analogue. *J. Mol. Biol.* **137**:249-264, 1980.

Helfrich, W.: Steric interaction of fluid membranes in multilayer systems. *Z. Naturforsch.* **33a**:305-315, 1978.

Horn, R.G., Israelachvili, J.N., Marra, J., Parsegian, V.A., Rand, R.P.: Comparison of forces measured between phosphatidylcholine bilayers. *Biophys. J.* **54**:1185-1187, 1988.

Hitchcock, P.B., Mason, R., Thomas, K.M., Shipley, G.G.: Structural chemistry of 1,2 dilauroyl-DL-phosphatidylethanolamine: Molecular conformation and intermolecular packing of phospholipids. *Proc. Natl. Acad. Sci., USA* **71**:3036-3040, 1974.

Israelachvili, J., Marra, J.: Direct methods for measuring conformational water forces (hydration forces) between membrane and other surfaces. *Meth. Enzymol.* **127**:353-360, 1986.

Jendrasiak, G.L., Mendible, J.C.: The effect of the phase transition on the hydration and electrical conductivity of phospholipids. *Biochim. Biophys. Acta* **424**:133-148, 1976.

Kjellander, R., Marcelja, S.: Polarization of water between molecular surfaces: A molecular dynamics study. *Chemica Scripta* **25**:73-80, 1985a.

Kjellander, R., Marcelja, S.: Perturbation of hydrogen bonding in water near polar surfaces. *Chem. Phys. Lett.* **120**:393-396, 1985b.

Loosley-Millman, M., Rand, R.P., Parsegian, V.A.: Effects of monovalent ion binding and screening on measured electrostatic forces between charged phospholipid bilayers. *Biophys. J.* **40**:221-232, 1982.

Luzzati, V., Husson, F.: The structure of the liquid-crystalline phases of lipid-water systems. *J. Cell Biol.* **12**:207-219, 1962.

Marcelja, S., Radic, N.: Repulsion of interfaces due to boundary water. *Chem. Phys. Lett.* **42**:129-130, 1976.

Marra, J., Israelachvili, J.N.: Direct measurements of attractive, adhesive, and repulsive forces between phosphatidylcholine and phosphatidylethanolamine bilayers in aqueous electrolyte solutions. *Biochemistry*, **24**:4608-4618, 1985.

McIntosh, T.J., Simon, S.A.: Hydration force and bilayer deformation: A reevaluation. *Biochemistry* **25**:4058-4066, 1986.

Parsegian, V.A., Rand, R.P., Fuller, N.L., Rau, D.C.: Osmotic stress for the direct measurement of intermolecular forces. in: *Methods in Enzymology*. Vol. 127 - Biomembranes; Protons and Water: Structure and Translocation, ed. Packer, L., Chapter 29, page 400-416. Academic Press, New York, 1986.

Parsegian, V.A., Fuller, N.L., Rand, R.P.: Measured work of deformation and repulsion of lecithin bilayers. *Proc. Natl. Acad. Sci.* **76**:2750-2754, 1979.

Pearson, R.H., Pascher, I.: The molecular structure of lecithin dihydrate. *Nature* **281**:499-501, 1979.

Podgornik, R., Rau, D.C., Parsegian,V.A.: The action of interhelical forces on the organization of DNA double helices: Fluctuation-enhanced decay of electrostatic double layer and hydration forces. *Macromolecules* **22**:1780-1786, 1989.

Rand, R.P., Parsegian, V.A.: Hydration forces between phospholipid bilayers. *Biochim. Biophys. Acta*, 1989 (in press).

Rand, R.P., Fuller, N., Parsegian, V.A., Rau, D.C.: Variation of hydration forces between neutral phospholipid bilayers: Evidence for hydration attraction. *Biochemistry* **27**:7711-7722, 1988.

Rand, R.P.: Interacting phospholipid bilayers: Measured forces and induced structural changes. *Annu. Rev. Biophys. Bioeng.* **10**:277-314, 1981.

Rau, D.C., Parsegian, V.A.: Measurement of forces between xanthan polysaccharides. *Biophys. J.* **51**:503a, 1987.

Rau, D.C., Lee, B., Parsegian, V.A.: Measurement of the repulsive force between polyelectrolyte molecules in ionic solution: Hydration forces between parallel DNA double helices. *Proc. Natl. Acad. Sci., USA* **81**:2621-2625, 1984.

Simon, S.A., McIntosh, J., Magid, A.D.: Magnitude and range of hydration pressure between lecithin bilayers as a function of head group density. *J. Colloid. & Interface Sci.* 1989 (in press).

Tate, M.W., Gruner, S.M.: Lipid polymorphism of mixtures of dioleoylphosphatidylethanolamine and saturated and mono-unsaturated phosphatidylcholines of various chain length. *Biochemistry* **26**:231-236, 1987.

Ter-Minassian-Saraga, L., Madelmont, G.: Differential scanning calorimetry studies of hydration forces with phospholipid multilamellar systems. *J. Colloid & Interface Sci.* **85**:375-388, 1982.

Wong, P.T.T., Mantsch, H.H.: High-pressure infrared spectroscopic evidence of water binding sites in 1,2-diacyl phospholipids. *Chem. Phys. Lipids* **46**:213-224, 1988.

Zaccai, G., Blasie, J.K., Schoenborn, B.P.: Neutron diffraction studies on the location of water in lecithin bilayer model membranes. *Proc. Natl. Acad. Sci., USA* **72**: 376-380, 1975.

Zasadzinski, J.A.N., Schneir, J., Gurley, J., Elings, V., Hansma, K.: Scanning tunneling microscopy of freeze-fracture replicas of biomembranes. *Science* 239:1013-1015, 1988.

Cell surface electrostatics and electrokinetics

A. Voigt
Department of Surface-active Compounds
Central Institute of Organic Chemistry
Academy of Sciences of the G.D.R.
Rudower Chaussee 5
Berlin-Adlershof, 1199
German Democratic Republic

E. Donath
Department of Biophysics
Section of Biology
Humboldt-University of Berlin
Invalidenstraße 42
Berlin, 1040
German Democratic Republic

SUMMARY

Theoretical aspects of fixed charge distributions at cell surfaces are considered from the viewpoint of their discrete or continuous ("smeared") nature in surface layers of finite thickness. The method of Green functions provides an elegant means for a very compact presentation of the electric potential profile results under low potential conditions. The relationship between transmembrane and surface potentials is explained. Energetical aspects of and stress development in single and interacting charged layers are studied as well as a minimum electrostatic free energy model of the fixed charge distribution. In the second part the development of the electrokinetics of structured surfaces is outlined. The advantage of the application of irreversible thermodynamics is demonstrated with respect to generalization and compactness of the electrokinetic fundamental relationships. Analytical solutions for the minimum electrostatic free energy model are obtained. Results of particle electrophoresis in concentrated aqueous polymer solutions are discussed on the basis of depletion layer effects or related phenomena.

INTRODUCTION

The electrostatic structure of the cell surface constitutes a facet of the overall electrodynamic surface or interface properties in biological matter. In general, electrokinetic experiments on cells or cell systems in aqueous solutions give proof of the existence of fixed electric charges at the cell surface/aqueous solution interface. There exist numerous excellent experimental and theoretical studies on generation, structure and meaning of the electric double layer at the cell surface (cp. McLaughlin, 1977;1982;1989). Rather important biochemical and biophysical cell surface, model bilayer or monolayer parameters and phenomena are determined and influenced by the electrostatic surface structure:

pH (Matinyan et al., 1984), local binding and binding constants of ions (Sauve and Ohki, 1979; Loosley-Millman et al., 1982), metabolites, drugs and polymers (Norde and Lyklema, 1978;1978a; Marra et al., 1983), membrane phase state (Träuble and Eibl, 1974; Belaya and Feigel'man, 1984), surface charge clustering (Neumcke and Stampfli, 1984), special surface layer solubility and partition properties (Zimm, 1983; LeBret and Zimm, 1984; Yaroshchuk and Baran, 1987; Makino et al., 1987;1988), interface stability (Bisch et al., 1983; Gallez and Coakley, 1986), photosynthetic processes (Barber, 1980; Rubin and Barber, 1980), structure of the cell surface glycoprotein layers (segment density profiles, thickness), membrane transport (Neumcke and Läuger, 1970; Aytyan et al., 1977;1977a; Pastushenko and Chizmadshev, 1981; Arakelyan and Arakelyan, 1983), cell-cell interaction (e.g. vesicles cp. Ohsawa et al., 1981), cell adhesion (Maroudas, 1975; Donath and Gingell, 1983; Wolf and Gingell, 1983; Trommler et al., 1985; Belintsev, 1988).

Thus, to understand and to regulate the above mentioned structures and phenomena, reliable experimental techniques are applied:

electrokinetics (Donath and Pastushenko, 1979; Donath and Lerche, 1980; Donath and Voigt, 1986), fluorescence (Winiski et al., 1988; Langner et al., 1988), NMR (Winiski et al., 1986), ESR (Hartsel and Cafiso, 1986), X-ray diffraction (Loosley-Millman et al., 1982; Rau et al., 1984; Parsegian et al., 1985; Evans and Parsegian, 1986), neutron diffraction (Büldt et al., 1988), direct interbilayer force measurements (Marra and Israelachvili, 1985; Marra, 1986; Pashley et al., 1986; Israelachvili, 1987), adhesion to fluid interfaces (Gingell et al., 1977; Gingell and Todd, 1980; Todd and Gingell, 1980; Donath and Gingell, 1983), adhesion to a polarized solid electrode surface (Gingell and Fornés, 1975; 1976).

For an extensive discussion of the results obtained with these and other techniques cp. the reviews of McLaughlin (1977;1989).

The presented paper intends to outline the main steps of modeling and theoretical description of the surface electrostatic structure and energetics as well as of the theoretical basis of the electrokinetic experiments. It is known that the glycocalix of animal cells carries under physiological conditions a fixed surface charge of a density about 0,02 C/m^2 distributed inside a layer of a finite thickness (Parsegian and Ninham, 1970; Parsegian and Gingell, 1972;1973). There exists a well characterized situation for the red blood cell, where about $1{,}5 \cdot 10^7$ electric charges are distributed in a surface layer of a thickness about 60 Å and an area of about 140 μm^2, respectively (Donath, 1979; Donath and Pastushenko, 1979; Donath and Lerche, 1980).

DISCRETE NATURE OF FIXED CHARGES

Dependend on the experimental conditions (ionic strength, pH) the surface electrostatics has to be described in the framework of discrete or "smeared" fixed charge distributions (cp. McLaughlin, 1989). If the quotient of mean fixed charge distance inside the glycocalix $\bar{a}$ and the Debye length $1/\varkappa$ fulfills the relation $\bar{a} \cdot \varkappa > 1$ one has to use the discrete picture, otherwise the model approach basing on the assumption of a smoothly distributed fixed charge of (homogeneous) density ρ is valid.

Now, let us consider the discrete modeling in a system (Fig. 1) consisting of a dielectric interlayer of thickness a (dielectric permittivity ε_m) separating two electrolyte solutions of different compositions (reciprocal Debye lengths $\varkappa_1$, $\varkappa_2$; relative dielectric constants ε_s). The coordinate origin is in the center of the dielectric interlayer; the coordinate plane x,y is parallel to the interfaces. The z-axis is perpendicular to the interfaces and positively directed to the region 3. A point charge q is located near the interface in the electrolyte solution at position r_i with $r_i = x_i\vec{i} + y_i\vec{j} + z_i\vec{k}$

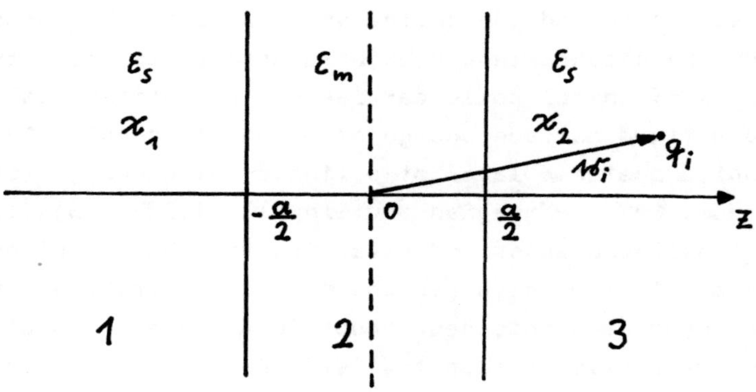

Fig. 1. Coordinate system and physical parameters for the calculation of discrete charge potentials. Explanations are given in the text.

The system of (linearized) Poisson-Boltzmann differential equations to be solved in regions 1 to 3 is (Δ - Laplace operator)

1 $\Delta\phi_1 = \chi_1^2\phi_1$ (1)

2 $\Delta\phi_2 = 0$ (2)

$$3 \quad \Delta\phi_3 = \chi_2^2\phi_3 - \frac{\delta(w-w_i)}{\varepsilon_0\varepsilon_s} \tag{3}$$

with ε_0 the absolute permittivity of vacuum and δ the Dirac's Delta-function. The boundary conditions are

$$\phi_1 = \phi_2 \qquad \text{for} \quad z = -a/2 \tag{4}$$

$$\varepsilon_s\frac{\partial\phi_1}{\partial z} = \varepsilon_m\frac{\partial\phi_2}{\partial z} \quad \text{for} \quad z = -a/2 \tag{5}$$

$$\phi_2 = \phi_3 \qquad \text{for} \quad z = a/2 \tag{6}$$

$$\varepsilon_m\frac{\partial\phi_2}{\partial z} = \varepsilon_s\frac{\partial\phi_3}{\partial z} \quad \text{for} \quad z = a/2 \tag{7}$$

To continue we apply a theoretical approach developed in a series of papers considering similar geometrical and physical problems (Stillinger, 1961; Levich and Yalamov, 1962;1962a; Nelson and McQuarrie, 1975; Aytyan et al., 1977; Enos and McQuarrie, 1981; Arakelyan and Arakelyan, 1983; Belaya and Feigel'man, 1984). The solution is found, involving Fourier-Bessel integral expressions, and has to be solved finally by numeric procedure. The resulting fundamental system of equations representing the solution for the potential problem in regions 1 to 3 is (J_0 is the zeroth order Bessel function)

$$\phi_1 = \frac{q}{4\pi\varepsilon_0\varepsilon_s}\int_0^\infty E(k)\exp\left(z\sqrt{K^2+\chi_1^2}\right)J_0(K|\vec{\rho}-\vec{\rho}_i|)\,dK \tag{8}$$

$$\phi_2 = \frac{q}{4\pi\varepsilon_0\varepsilon_m}\int_0^\infty [C(k)\exp(Kz) + D(k)\exp(-Kz)]\,J_0(K|\vec{\rho}-\vec{\rho}_i|)\,dK \tag{9}$$

$$\phi_2 = \frac{q}{4\pi\varepsilon_0\varepsilon_s}\int_0^\infty \left[\frac{K\exp\left(-|z-z_i|\sqrt{K^2+\chi_2^2}\right)}{\sqrt{K^2+\chi_2^2}} + V(k)\exp\left(-z\sqrt{K^2+\chi_2^2}\right)\right] J_0(K|\vec{\rho}-\vec{\rho}_i|)\,dK \tag{10}$$

with

$$|\vec{p}-\vec{p}_i| = +\sqrt{(x-x_i)^2+(y-y_i)^2} \tag{11}$$

The coefficients E(k), C(k), D(k) and V(k) are obtained from the respective algebraic system of equations provided equations (4) to (7) are substituted by equations (1) to (3). Ultimately, we find

$$E(k) = -\frac{2K^2\xi \exp(a\sqrt{K^2+\chi_1^2}/2)\exp(aK)\exp[(a/2-z_i)\sqrt{K^2+\chi_2^2}]([-]+[+])}{(\sqrt{K^2+\chi_1^2}-K\xi)(\sqrt{K^2+\chi_2^2}\cdot[-]-K\xi[+])} \tag{12}$$

$$C(k) = \frac{K\xi \exp[(a/2-z_i)\sqrt{K^2+\chi_2^2}]\exp(-aK/2)([-]-[+])}{\sqrt{K^2+\chi_2^2}[-]-K\xi[+]} \tag{13}$$

$$D(k) = \frac{K\exp[(a-z_i)\sqrt{K^2+\chi_2^2}]\exp(aK/2)([-]+[+])}{\sqrt{K^2+\chi_2^2}[-]-K\xi[+]} \tag{14}$$

$$V(k) = \frac{K\exp[(a-z_i)\sqrt{K^2+\chi_2^2}](\sqrt{K^2+\chi_2^2}[-]+K\xi[+])}{\sqrt{K^2+\chi_2^2}(\sqrt{K^2+\chi_2^2}[-]-K\xi[+])} \tag{15}$$

$$\xi = \frac{\varepsilon_m}{\varepsilon_s} \tag{16}$$

$$[-] = (\sqrt{K^2+\chi_1^2}-K\xi)\exp(-Ka)-(\sqrt{K^2+\chi_1^2}+K\xi)\exp(Ka) \tag{17}$$

$$[+] = (\sqrt{K^2+\chi_1^2}-K\xi)\exp(-Ka)+(\sqrt{K^2+\chi_1^2}+K\xi)\exp(Ka) \tag{18}$$

We will illustrate the numerical solution of equations (8) to (10) together with (11) to (18) (unpublished results). Assuming a very thick membrane ($a\rightarrow\infty$) with ε_m=2 and a point charge q at the position 5 Å separated from the membrane/electrolyte solution (100 mM/l) interface inside the electrolyte solution with ε_s =80

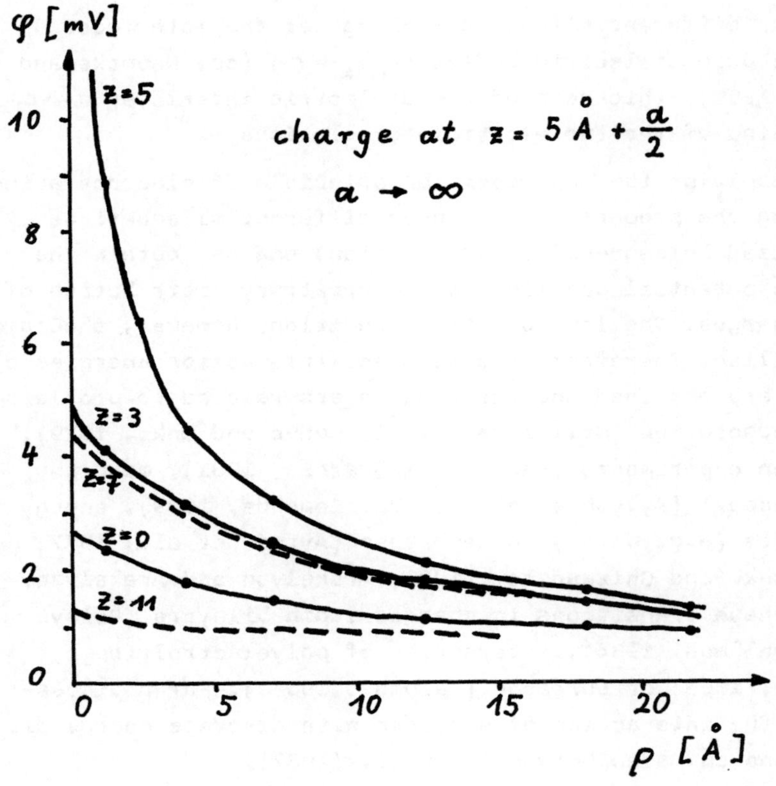

Fig. 2. Numerical example of electric potential profiles. An elementary charge is placed 5 Å apart from the membrane/electrolyte solution interface inside the electrolyte solution. Constants: ε_m=2; ε_s =80; ion concentration 100 mM/l; membrane thickness a→∞ . For given distances from the membrane z [Å] the potentials are plotted as a function of $\rho = \sqrt{(x-x_i)^2+(y-y_i)^2}$.

A more complete description of the discrete charge model and its consequences for colloid and membrane systems is given elsewhere. Here, we will mention only that also the case of a point charge inside the dielectric layer separating two different electrolyte solutions (with respect to dielectric permittivities and Debye reciprocal lengths) was solved (unpublished). Both solutions, for point charges inside and outside the dielectric layer, provide the known limit expressions for different situations: charge at the interfaces $z=\pm a/2$; pure dielectric bodies ($\varkappa_1, \varkappa_2 \to 0$) (cp. Neumcke and Läuger, 1969); thickness of the dielectric interlayer $a \to \infty$ (uncoupling of the two electrolyte solutions).

Now, applying the superposition principle of electrostatics and using the properties of linear differential equations (linearized Poisson-Boltzmann equation) one can obtain the electric potential profiles for an arbitrary distribution of fixed charges. The low potential condition, however, should be fulfilled. Therefrom, the self and interaction energies of charges are obtained and the results are related to problems of ion adsorption (Stillinger, 1961; Sauve and Ohki, 1979), titration experiments (Enos and McQuarrie, 1981), membrane ion transport (Aytyan et al., 1977a; Edmonds, 1989), energy of dipoles (e.g. water) in membranes (Aytyan et al., 1977; Pastushenko and Chizmadshev, 1981; Arakelyan and Arakelyan, 1983), phase transitions in charged lipid bilayers (Belaya and Feigel'man, 1984), interaction of polyelectrolytes (Yalamov, 1963) or surfaces (Yalamov, 1963a). For a discussion of the interaction of surfaces with discrete charge distributions cp. also Deryaguin et al. (1987).

Quite recently, we can observe a strong development of electrostatic interaction models considering discrete charge arrangements as statistical ensembles and applying correlation analysis and statistical physics of liquids (Yuchnovskij and Golovko, 1980; Kjellander, 1984; Kjellander and Marčelja, 1984; Attard et al., 1988;1988a;1988b; Attard and Mitchell, 1988). The results are promising and bridge partly the gap between Lifshitz-van der Waals theory of molecular forces and the electrostatic interaction theories as well as the gap between new experimental findings of structure-induced (e.g. hydration forces) interaction components and classical analysis of experimental results.

CONTINUOUSLY ("SMEARED") FIXED CHARGE DISTRIBUTIONS

If the quotient of mean fixed charge distance inside the glycocalix $\bar{a}$ and the Debye length $1/\varkappa$ fulfills the relation $\bar{a}\cdot\varkappa \lesssim 1$ the problem of an adequate description of the surface electrostatics can be solved in another and less difficult way. The fixed charges are considered as smeared or smoothly distributed over the cell surface layer and are represented by a space fixed charge density function ρ which depends only on the coordinate z perpendicular to the membrane surface. Plane charge distributions of density σ (C/m^2) at arbitrary positions inside the surface layer (especially at the membrane/glycocalix transition at z=0 and the glycocalix/bulk electrolyte solution transition at $z=\delta$). Representing the fixed charges in this way leads to a drastic reduction of mathematical efforts compared with the discrete charge case. The modeling following this picture started with Parsegian and Gingell (1972;1973). Thereafter, an extension of this approach was achieved and general expressions for the calculation of the electric potential profiles at single (Pastushenko and Donath, 1976a) and interacting (Voigt et al., 1982) cell surfaces were derived.

A. Representation of electric potential profiles using Green functions

Applying this method the potential profiles are represented by an integral expression. The only function, one has to know, is the fixed charge density profile. Because any set of boundary conditions generates its own Green function the mathematics in obtaining the final results is reduced to simple integration procedures (numeric or analytic). Of course, the method of Green functions is restricted to linear differential equations (linearized Poisson-Boltzmann equation).

Pastushenko and Donath (1976a) derived the following expression for the electric potential profile of a single charged layer

$$\psi(z) = \frac{\sigma}{\varepsilon_s \varepsilon_0 \chi} \exp(-\chi z) + \frac{\cosh(\chi z)}{\varepsilon_s \varepsilon_0 \chi} \cdot \int_0^{\delta} \rho_f(z') \exp(-\chi z') dz'$$
$$- \frac{1}{\varepsilon_s \varepsilon_0 \chi} \int_0^{z} \rho_f(z') \sinh[\chi(z - z')] dz' \qquad (19)$$

with σ a plane charge density at z=0 (membrane surface), ρ_f the density profile of fixed electric charges and δ the thickness of the surface layer (glycocalix). In case of a constant density of fixed charges $\rho_f = \rho =$ const. the following potential profile is obtained

$$\psi = \frac{\rho}{\varepsilon_s \varepsilon_0 \chi^2} [1 - \cosh(\chi z) \exp(-\chi \delta)] \qquad \text{for} \quad z \leq \delta$$

$$\psi = \frac{\rho}{\varepsilon_s \varepsilon_0 \chi^2} \{\cosh[\chi(z - \delta)] - \cosh(\chi z) \exp(-z\delta)\} \quad \text{for} \quad z \geq \delta \qquad (20)$$

At the same level of approximation there was derived the expression for the potential profile between two interacting cell surfaces (Voigt et al., 1982)

$$\psi(z) = \int_{-d/2}^{d/2} G(z, z') \rho_f(z') dz' \qquad (21)$$

with the Green function G(z,z')

$$G(z, z') = \frac{1}{\varepsilon_s \varepsilon_0 \chi \sinh(\chi d)} \times \begin{cases} \cosh[\chi(z' + d/2)] \cosh[\chi(z - d/2)] & \text{for} \quad z > z' \\ \cosh[\chi(z' - d/2)] \cosh[\chi(z + d/2)] & \text{for} \quad z < z' \end{cases} \qquad (22)$$

and d the separation distance between the interacting membrane surfaces.

In the same way we derived the potential expressions for single surface layers of spherical and cylindrical geometry. For the profile inside a sphere of radius r_0 and a surface layer of thickness δ between $r_o - \delta$ and r_o we obtained (unpublished)

$$\psi(r) = -\frac{1}{\varepsilon_s \varepsilon_0 r} \int_0^{r_0} G(r,r') r' \rho(r') dr' \tag{23}$$

with the Green function $G(r,r')$

$$G(r,r') = \frac{1}{\chi \sinh(\chi r_0) - r_0 \chi^2 \cosh(\chi r_0)} \times \begin{cases} \{r_0 \chi \cosh[\chi(r'-r_0)] + \sinh[\chi(r'-r_0)]\} \cdot \sinh(\chi r) & \text{for} \quad r<r' \\ \{r_0 \chi \cosh[\chi(r-r_0)] + \sinh[\chi(r'-r_0)]\} \cdot \sinh(\chi r') & \text{for} \quad r>r' \end{cases} \tag{24}$$

For a cylinder of radius r_o and a surface layer of thickness δ between $r_o - \delta$ and r_o we derived the expressions

$$\psi(r) = -\frac{1}{\varepsilon \varepsilon_0} \int_0^{r_0} G(r,r') r' \rho(r') dr' \tag{25}$$

with the Green function $G(r,r')$

$$G = (r,r') = \begin{cases} \left[\frac{K_1(\chi r_0)}{I_1(\chi r_0)} I_0(\chi r') + K_0(\chi r') \right] I_0(\chi r) & \text{for} \quad r<r' \\ I_0(\chi r') \left[\frac{K_1(\chi r_0)}{I_1(\chi r_0)} I_0(\chi r) + K_0(\chi r) \right] & \text{for} \quad r>r' \end{cases} \tag{26}$$

with $K_1(\ )$, $I_1(\ )$ and $I_o(\ )$ modified Bessel functions of first and zeroth order.

B. Surface and Transmembrane potentials

We will make a few remarks on the relationship between surface and transmembrane potentials. Does a transmembrane potential influence and change the surface potential? If a transmembrane potential acts on a membrane separating electrolyte solutions of different composition we obtain the following potential expressions for the regions 1 to 3 (cp. Fig. 1) with U the transmembrane potential difference

$$\Phi_1 = \frac{U}{2} - \frac{U\chi_2 \exp[\chi_1(z + a/2)]}{\chi_1\left[1 + \chi_2/\chi_1\left(1 + \frac{a\chi_2\varepsilon_s}{\varepsilon_m}\right)\right]} \tag{27}$$

$$\Phi_2 = -\frac{U\varepsilon_s\chi_2 \cdot z}{\varepsilon_m\left[1 + \chi_2/\chi_1\left(1 + \frac{a\chi_2\varepsilon_s}{\varepsilon_m}\right)\right]} - \frac{U\left(1 + \frac{a\chi_2\varepsilon_s}{2\varepsilon_m}\right)}{1 + \chi_2/\chi_1\left(1 + \frac{a\chi_2\varepsilon_s}{\varepsilon_m}\right)} + \frac{U}{2} \tag{28}$$

$$\Phi_3 = -\frac{U}{2} + \frac{U\exp[-\chi_2(z - a/2)]}{1 + \frac{\chi_2}{\chi_1}\left(1 + \frac{a\chi_2\varepsilon_s}{\varepsilon_m}\right)} \tag{29}$$

In Fig. 3 is schematically illustrated the situation. The origin of the potential difference is unimportant (from outside applied; asymmetry in the system). In general, one observes a slight change in the surface potentials and an asymmetry of the intramembrane potential profile. The surface potential change related to the undisturbed surface potential ψ_0 is given by

$$\frac{\Delta\psi}{\psi_0} = \frac{1}{1 + (\chi_1/\chi_2) + (\varepsilon_s/\varepsilon_m)\chi_1 a} \tag{30}$$

which is simplified in the symmetric case to

$$\frac{\Delta\psi}{\psi_0} = \frac{1}{2 + (\varepsilon_s/\varepsilon_m)\chi_1 a} \tag{31}$$

with the capacities of the membrane and the diffuse double layer, C_m and C_d, respectively

$$C_m = \varepsilon_0\varepsilon_m/a \tag{32}$$

$$C_d = \varepsilon_0\varepsilon_s\chi_1 \tag{33}$$

An inspection of equations (30) to (33) provides an impression on the experimental conditions under which the change in surface potential has to be taken into account.

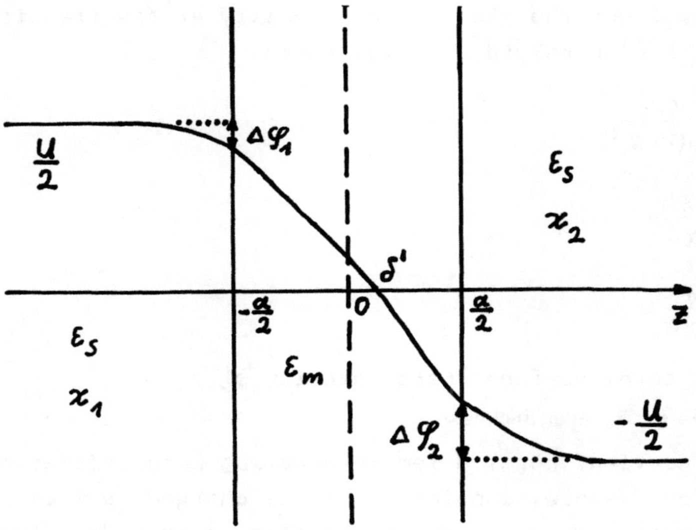

Fig. 3. Influence of transmembrane potential on surface potential. A membrane (dielectric permittivity ε_m, thickness a) separates two electrolyte solutions of different composition (dielectric permittivity ε_s, reciprocal Debye lengths $\varkappa_1$ and $\varkappa_2$). The transmembrane potential difference is U. Asymmetry of the intramembrane potential profile is expressed in δ' and the changes in surface potentials are expressed as $\Delta\varphi_i$

C. Minimum electrostatic free energy model

If the fixed charges inside the surface layer could rearrange as to minimize the free electrostatic energy a modified approach has to be used (Donath and Voigt, 1983). A minimum is realized for single and interacting surface layers in the case of constant potential inside the layers. To obtain this result the total charge density of the surface layer σ_0 (C/m^2) must split into a constant space charge density part inside the surface layer and a plane charge density part at the transition surface layer/bulk electrolyte solution. For a single layer the respective expressions for the (constant) surface potential ψ, the (constant) space charge density of the layer ρ and the plane charge density at the transition layer/bulk electrolyte solution σ are

$$\psi = \frac{\sigma_0}{\varepsilon_s \varepsilon_0 (1 + \chi \delta)} \tag{34}$$

$$\rho = \frac{\sigma_0 \chi}{1 + \chi \delta} \tag{35}$$

$$\sigma = \frac{\sigma_0}{1 + \chi \delta} \tag{36}$$

with the total surface charge density σ_0

$$\sigma_0 = \rho \delta + \sigma \quad \delta\text{: layer thickness} \tag{37}$$

We will mention, here, a few interesting pecularities of the model. For example, consider a single charged surface layer of thickness δ with an additional plane charge density σ_m at the membrane surface (z=0). If the ionic strength is low, all the fixed charges, able to redistribute, are concentrated at the transition region surface layer/bulk electrolyte solution (z=δ). Thus, the fixed charges are distributed in the manner of two plane charge densities. Now, raising the ionic strenght above a critical value ($\varkappa_{crit}$ is the respective value of the reciprocal Debye length) part of the fixed charges at z=δ is redistributed to build up a layer of constant space charge density ρ

$$\rho = \chi \sigma_m / \sinh(\chi \delta_1) \tag{38}$$

between z=δ and z=$\delta_1 < \delta$ with δ_1 obtainable from

$$\chi \delta_1 = \sinh^{-1}\left\{\frac{\sigma_m}{\sigma_0}[1 + (\delta - \delta_1)\chi]\right\} \tag{39}$$

The critical ionic strength can be derived from the expression

$$\chi_{crit} = \frac{1}{\delta}\sinh^{-1}\left(\frac{\sigma_m}{\sigma_0}\right) \tag{40}$$

Before closing this point let us stress the formal character of this minimum electrostatic free energy model. Really, there are many constraints for a free redistribution of fixed charges. Nevertheless, the model suggests to the possible reactions the electric field can promote inside the surface layers. The model, though being a mathematical result, provides us with rules for thinking.

D. Energetics of single and interacting charged layers

The fundamental aspects of the energetics of Gouy-Chapman double layers are considered in the outstanding monography of Verwey and Overbeek (1948). Transfering their ideas to charged single and interacting surface layers we obtained the following expression for the electrostatic free energy E

$$E = \int_z \int_\rho \psi \, d\rho \, dz \tag{41}$$

which can be simplified considering the linear relationship between potential and charge density reflected in the linearized Poisson-Boltzmann equation (Voigt et al., 1982)

$$E = \frac{1}{2} \int_z \psi(z)\rho(z) dz \tag{42}$$

Applying this result to a single surface layer of thickness δ and constant space density of fixed charges ρ one gets

$$E_\rho = \frac{\rho^2}{4\varepsilon_s\varepsilon_0\chi^2}\left[2\delta - \frac{1}{\chi} + \frac{1}{\chi}\exp(-2\chi\delta)\right] \tag{43}$$

Comparing this expression with that for a simple Gouy-Chapman double layer

$$E_\sigma = \frac{\sigma^2}{2\varepsilon_s\varepsilon_0\chi} \tag{44}$$

the ratio of both results in

$$E_\rho/E_\sigma = \frac{1}{2}\left\{\frac{2}{\chi\delta} + \frac{1}{(\chi\delta)^2}[\exp(-2\chi\delta) - 1]\right\} \tag{45}$$

As expected the ratio depends only on the thickness and the milieu parameters aggregated in the Debye length $1/\chi$.
In case of the minimum electrostatic free energy model the result for a single surface layer is (Donath and Voigt, 1983)

$$E_{min} = \frac{\sigma_0^2}{2\varepsilon_s\varepsilon_0\chi(1+\chi\delta)} \tag{46}$$

To obtain explicit expressions for the electrostatic interaction energy of surface layers one has to solve

$$E_{int} = E_{12} - E_1 - E_2 \tag{47}$$

with E_1 and E_2 the energies of the (isolated from each other) single surface layers and E_{12} the energy of the system with interaction.

For a cell surface (glycocalix of thickness δ and constant density of fixed charges ρ) interacting with a (plane) dielectric body (plane surface charge density σ) in aqueous electrolyte solution the following expression results

$$E_{int}(d) = \frac{\rho}{2\varepsilon_s\varepsilon_0\chi^2}\left\{\frac{2\sinh(\chi\delta)\cdot\sigma}{\sinh(\chi\delta)} - \frac{\rho\sinh(\chi\delta)\sinh[\chi(d-\delta)]}{\chi\sinh(\chi d)}\right\}$$
$$-\frac{\rho^2}{4\varepsilon_s\varepsilon_0\chi^3}[\exp(-2\chi\delta)-1] + \frac{\sigma^2}{2\varepsilon_s\varepsilon_0\chi}[\coth(\chi\delta)-1] \tag{48}$$

Finally, in closing this section of demonstrating the calculation procedure for surface layer energy expressions, we will present the expression for the interaction of two identical cell surfaces according to the rules of the minimum electrostatic free energy model

$$E_{int}(d) = \frac{\sigma_0^2}{\varepsilon_s\varepsilon_0\chi} \frac{1-\tanh[\chi(d/2-\delta)]}{(1+\chi\delta)\{\chi\delta + \tanh[\chi(d/2-\delta)]\}} \tag{49}$$

E. Stress development in single and interacting charged surface layers

The presence of charged surface layers modifies also the picture of acting pressures and stresses in systems with and without interaction compared with the Gouy-Chapman double layer system where the thickness of the fixed charge layer is zero. The first integral of the Poisson-Boltzmann equation considering the surface layer is

$$\pi^\infty = \text{const.} = \pi(z) - \frac{\varepsilon_s\varepsilon_0}{2}\left(\frac{d\psi}{dz}\right)^2 - \int_{\delta+0}^{z} \rho'(z)\frac{d\psi}{dz}dz \tag{50}$$

with π^∞ the osmotic pressure in the bulk solution, $\pi(z)$ the

osmotic pressure at position z. The second term on the r.h.s. is the Maxwell stress and the third term is the feature of the surface layer - a stress induced by the electric field on the fixed charges. The consideration of this last term in the force balance is new (Donath and Voigt, 1988).
If the interaction of two surface layers is calculated one arrives at the following expression

$$\pi_e = \pi(0,d) - \frac{\varepsilon_s \varepsilon_0}{2}\left[\frac{d}{dz}\psi(0,d)\right]^2 - \int_{\delta+0}^{0} \rho^1 \frac{d\psi}{dz}dz - \pi^\infty \tag{51}$$

π_e is the electrostatic component of the disjoining pressure (Deryaguin et al., 1987). The terms on the r.h.s. of equation (51) have to be estimated at one of the membrane/surface transition planes assuming a separation distance d of the interacting membranes. For penetrating surface layers the expression is also valid, of course. In Fig. 4 the overall situation is illustrated. Besides the van der Waals attraction and the osmotic part of repulsive force which act mainly on the membrane surface itself, the macromolecular stress is transfered to the membrane via the charged layer polyelectrolytes. Consequences are discussed elsewhere (Donath and Voigt, 1988).

F. Nonlinear Poisson-Boltzmann equation and the advantage of the minimum free energy model

If the electric potential values exceed $\approx$25 mV the above (sections A to D) derived expressions assuming the applicability of the linearized Poisson-Boltzmann equation become invalid. Unfortunately, we have not been able to find an explicit solution for an arbitrary surface charge distribution, ρ_f (z) in case of the nonlinear equation. Thus, a numerical algorithm had to be developed (Donath and Voigt, 1986) in solving the Poisson-Boltzmann equation

$$\psi''(z) = -\frac{1}{\varepsilon\varepsilon_0}\left[\rho_f(z) + e_0 \sum_i n_{i0} z_i \exp(-e_0 z_i \psi(z)/kT)\right] \tag{52}$$

with the boundary conditions

$$\psi'|_{z=0} = \frac{\sigma_m}{\varepsilon_s \varepsilon_0} \tag{53}$$

$$\psi'|_{z=\delta-0} = \left\{\frac{2kT}{\varepsilon_s \varepsilon_0} \sum_i n_{io} [\exp(-z_i e_0 \psi/\delta/kT) - 1]\right\}^{1/2} + \frac{\sigma}{\varepsilon_s \varepsilon_0} \tag{54}$$

inside the surface layer. T is the absolute temperature, k the Boltzmann constant. e_o, z_i and n_{io} are the elementary charge, electrovalency of ionic species i and the bulk electrolyte concentration of ionic species i, respectively.

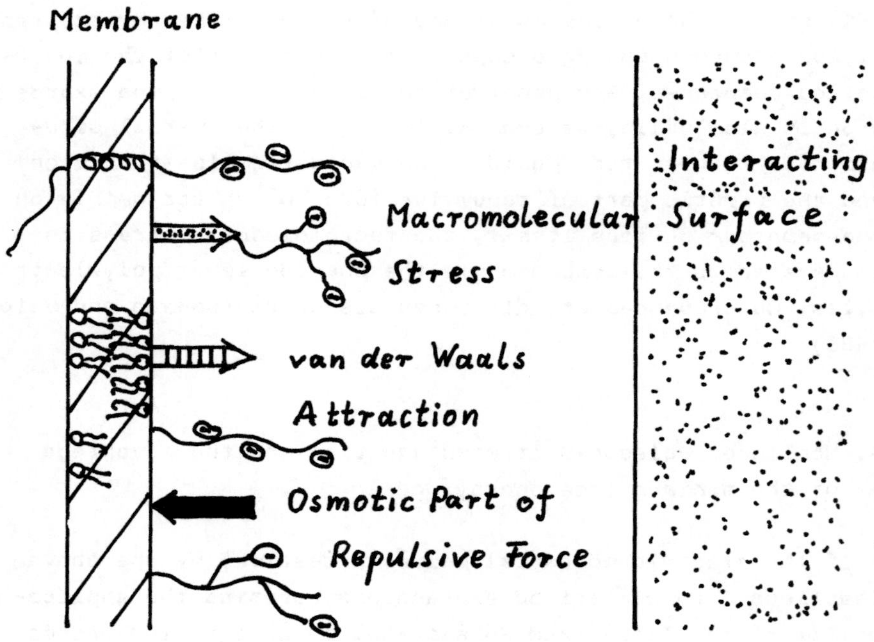

Fig. 4. Illustration of the different forces components action sites in an interaction situation.

The solution algorithm proceeds in the following way: Assume, there is a function $\psi^*(z)$ which satisfies the boundary conditions (53) and (54) and is sufficiently close to the solution of equation (52). Now, the electric potential and its derivatives are expanded in a Taylor series to calculate the coefficients of expansion $\psi^*(z)$. Equation (52) is transfered this way into a linear differential equation with respect to ψ (z)

$$\psi''(z) = \rho(z)\psi(z) = f(z) \tag{55}$$

with

$$\rho(z) = \frac{e_0^2}{\varepsilon_s \varepsilon_0 kT} \sum_i n_{i0} z_i^2 \exp[-e_0 z_i \psi^*(z)/kT] \tag{56}$$

and

$$f(z) = -\frac{1}{\varepsilon_s \varepsilon_0}\left[\rho(z) + e_0 \sum_i n_{i0} z_i \exp\{-z_i e_0 \psi^*(z)/kT\}\right] - \frac{e_0^2}{\varepsilon_s \varepsilon_0 kT} \sum_i n_{i0} z_i^2 \exp[-e_0 z_i \psi^*(z)/kT] \cdot \psi^*(z) \tag{57}$$

This equation has to be solved by means of a differential method. One can cope with this problem in a rather straightforward way by very few iteration cycles, if a zeroth approximation is constructed by means of the explicit solution of the linear problem.
The algorithm, of course, works for both cases, with and without interaction of the surface layers.

There is one exceptional case of practical relevance, the above presented minimum electrostatic free energy model, for which an analytically closed expression could be derived for a single surface layer (Donath and Voigt, 1986). For a z' : z' valent electrolyte solution follows the expression of the potential inside the surface layer (implicit)

$$\chi = \frac{1}{2\delta \cosh(z' e_0 \psi_0/2kT)}\left\{\left[1 + \frac{2e_0 \delta z \sigma_0}{\varepsilon_s \varepsilon_0 kT \tanh(z' e_0 \psi_0/2kT)}\right]^{1/2} - 1\right\} \tag{58}$$

and of the decay outside the layer

$$\chi = \delta + \chi^{-1} \ln\left\{\frac{\tanh(z' e_0 \psi_0/4kT)}{\tanh[z' e_0 \psi(z)/4kT]}\right\} \tag{59}$$

G. Miscellaneous aspects

We will cite a few further developments of surface electrostatics.

Taking into account the partition of ions between bulk and surface phases Ohshima and coworkers derived analytical expressions for different surface phenomena (Ohshima et al., 1987; Makino et al., 1987;1988). Effects induced by an adsorption layer of nonionogenic polymers with respect to a changed dissolving power of the water inside the layer were considered by Dukhin et al. (1984) and Yaroshchuk and Baran (1987).

Modeling of potential distributions in case of a nonuniform charge arrangement is done by Ohshima et al. (1986).

The relation between Donnan potential and surface potential applied to charged surface layers at membranes was investigated by Ohshima and Ohki (1985).

If the electrolyte concentration is very low the procedure to describe the electrostatic particle interaction has to be done by means of the methods of statistical physics. A colloid-colloid pair potential function is derived and interpreted (Beresford-Smith et al., 1985).

By Gross (1988) was considered what happens, if fixed charges are redistributed over the cell surface under the action of an applied field. Membrane potential changes were observed.

If a redistribution of fixed electric charges between the interaction region of particles and a region free of interaction (e.g. meniscus in films; cell membrane; inner leaflet of the bilayer) takes place, the electrostatic interaction pressure is decreased compared with the case without interaction (Aytyan et al., 1981). An inspection of this rather general case shows, immediately, the general meaning. Constant potential or constant charge approximations are, really, limit cases, even under equilibrium conditions. A forthcoming unifying approach should properly take into account the equilibration processes between all parts of the system.

A model of membrane shape changes induced by electrostatic effects on the spontaneous curvature was developed by Kozlov et al. (1985) and Lerche et al. (1987;cp. Kozlov and Markin, 1987).

Bivas and Petrov (1981) assumed that mechanical and electrical membrane properties are strongly related to each other by a flexoelectricity behaviour.

The problem of modeling electrostatics of membranes with cylindrical pores was successfully solved by Rice and Horne (1985) and Jordan (1982).

There is no dought, that there exist strong parallel developments of surface electrostatics in colloid chemistry and biophysics. That's why, the progress reflected in different monographs of colloid chemistry (Verwey and Overbeek, 1948; Dukhin, 1975; Dukhin and Deryaguin, 1976; Hunter, 1981; Deryaguin, 1986; Deryaguin et al., 1987; Israelachvili, 1987) promotes the understanding of the respective problems in biophysics.

ELECTROKINETICS OF STRUCTURED SURFACES

A. Reminiscence - towards a surface layer approach

In this part of the paper we will summarize a few developments and problems in electrokinetics of structured surfaces.

There is a discrepancy between the observed electrophoretic mobility of human erythrocytes and the mobility calculated using Smoluchowski's formula and biochemical data. It was discussed in several papers (Vassar et al., 1969; Parsegian and Ninham, 1970; Levine et al., 1983). Another problem in applying the Smoluchowski equation to erythrocytes is the observed dependency of zero-mobility on the ion concentration (Heard and Seaman, 1961; Nordt et al., 1978; Donath and Steidel, 1980). There is also observed an increase in the calculated charge density with increasing ionic strength. Both observations are in contradiction to the Gouy-Chapman double layer model at a smooth surface. Similar inconsistencies have been observed with other surfaces too.

There exist several attempts to cope with these contradictions. Lyklema and Overbeek (1961) discussed the influence of a changed viscosity near the interface. Hunter (1960) took into account a nonconstant surface charge by considering effects of the finite size of counterions. Heard and Seaman (1960) interpreted the increase in charge density in terms of an Langmuir-adsorption process of coions. This approach was criticized by Haydon (1961) because of the lack of specificity. Haydon (1961a) stressed that a permeable surface would increase the calculated surface charge density. He discussed the dispersion of ionogenic groups in large regions of non-ionogenic surface material. The same author published a qualitative treatment of the porosity of the surface region.

As early as 1970 by Parsegian and Ninham (1970; cp. also Parsegian and Gingell, 1972;1973) was introduced the concept of a "fuzzy" surface layer of finite thickness. However, they did not consider the hydrodynamic problem of the surface layers in electrokinetic experiments. A solution to the problem connecting both the electrostatic and the hydrodynamic parts was proposed by Pastushenko and Donath (1976;1976a). It uses the linearized Poisson-Boltzmann equation for the determination of the electric potential profile. Recently, Jones (1979), Wunderlich (1982) and Levine et al. (1983) obtained similar theoretical results. The linear approach was successfully applied to cell electrophoretic measurements (Donath and Lerche, 1980; Levine et al., 1983; Dolgaya and Mironov, 1984). Quite recently, McDaniel et al. (1984) presented a set of experimental investigations providing strong evidence for the validity of the model.

Frequently, in cell biology and colloid chemistry one has to perform the experiments in aqueous solutions under low ionic strength conditions. In this case, moderate surface charge densities result in high surface potential values and the linear models become invalid and insufficient.

Solutions of the problem considering the nonlinearized Poisson-Boltzmann equation were obtained by several authors in different ways (numeric or analytic) during the last years (Sharp and Brooks, 1985; Donath and Voigt, 1986;1986a; Snabre et al., 1986). First applications are available now.

B. Electrokinetics of structured surfaces from the viewpoint of irreversible thermodynamics

The application of irreversible thermodynamics to the electrokinetics of structured surfaces seems to be fruitful and promising. It becomes possible, for the case of convenience, to separate the hydrodynamic and electrostatic parts of the problem. This is advantageous with respect to the mathematical and numerical effort in obtaining the fundamental relationships for the different electrokinetic phenomena.

The corresponding phenomenological equations are (de Groot, 1959; Dukhin and Deryaguin, 1976)

$$I = L_{11}\Delta\phi + L_{12}\Delta P \qquad (60)$$

$$J = L_{21}\Delta\phi + L_{22}\Delta P \qquad (61)$$

with I and J, electric current and volume flow. $\Delta\varphi$ and ΔP are the driving forces electric potential difference and pressure difference. The coefficients L_{11} and L_{22} are the electrical and hydrodynamical (hydraulic) conductivities and the coefficients $L_{12}=L_{21}$ are streaming current and electroosmotic volume flow normalized to the respective driving forces.

Now, any electrokinetic phenomenon can be adequately described by a combination of the coefficients L_{ij} (de Groot, 1959). Thus, all,one has to do is the derivation of the three coefficients L_{11}, L_{22} and L_{12}. We will summarize the main results for the plane geometry case and for nonoverlapping double layers.

L_{11} is obtained for a capillary with rectangular cross section of width c, height d and length l from the expression

$$L_{11}=\frac{2\cdot c}{L}K_s+\frac{A}{L}K \tag{62}$$

with the surface conductivity $K_s=K_m+K_c$, as a sum of a migrative (K_m) and a convective (K_c) part, K is the bulk electrolyte solution conductivity (tabulated or easy accessible to experimental determination), A is the cross section area d•c. For the linear models and for the (nonlinear) minimum electrostatic free energy model analytical solutions were obtained (Donath and Voigt, 1986;1986a). The final expressions for the minimum model are

$$K_m=z'e_0n_0\sum_i u_i\left\{\left[\exp\left(-\frac{z_i'e_0\psi_0}{kT}\right)-1\right]\delta+\frac{2}{\chi}\left[\exp\left(-\frac{z_i'e_0\psi_0}{2kT}\right)-1\right]\right\} \tag{63}$$

u_i: ionic mobilities

$$K_c=\frac{8kTn_0\varepsilon_s\varepsilon_0\delta}{\eta}\sinh^2\left(\frac{z'e_0\psi_0}{2kT}\right)\left\{\frac{\tanh(a\delta)}{a\delta}\left[1-\frac{\chi^2}{a^2}\cosh^2\left(\frac{z'e_0\psi_0}{2kT}\right)\right]\right.$$
$$+\frac{\chi^2}{a^2}\cosh\left(\frac{z'e_0\psi_0}{2kT}\right)\left[\cosh\left(\frac{z'e_0\psi_0}{2kT}\right)+2\frac{\cosh(a\delta)-1}{\chi\delta\cosh(a\delta)}\right]$$
$$\left.+\frac{1}{\chi\delta[\cosh(z'e_0\psi_0/2kT)+1]}\right\}\quad \text{a: layer friction coefficient} \tag{64}$$

Thus, we can give an explicit expression for L_{11}.

The hydraulic conductivity L_{22} is derived from

$$L_{22}=\frac{L}{\Delta P}\int_0^d V_h(z)dz \tag{65}$$

with $v_h(z)$ the hydrodynamic flow velocity profile generated by the pressure difference ΔP ($\Delta\varphi=0$). The final solution expression is

$$L_{22}=\frac{2}{\eta}\left\{\frac{\frac{\sinh(a\delta)}{a}+\frac{d}{2}-\delta}{a^2\cosh(a\delta)}-\frac{d}{2a^2}-\frac{d^3}{24}-\frac{\delta^3}{6}+\frac{d\delta^2}{4}+C\left(\frac{d}{2}-\delta\right)\right\}$$

with

$$C=\frac{\delta-\frac{d}{2}-\frac{\sinh(a\delta)}{a}}{a}\tanh(a\delta)-\frac{d^2}{2}+\frac{d\delta}{2}+\frac{\cosh(a\delta)}{a^2}-\frac{1}{a^2} \tag{66}$$

For L_{12} we obtained a very general expression. It contains all possible and imaginable cases of the linear model with smeared electric fixed charges

$$L_{12} = L_{21} = \frac{A}{L\eta\chi}\frac{1}{1-(a^2/\chi^2)}\left\{\frac{I_{sa}\cdot\chi + I_{ex}\cdot a}{a\cosh(a\delta)} - \frac{a}{\chi}\exp(-\chi\delta)I_{c\chi}\right.$$

$$\left.\left[\tanh(a\delta) + \frac{a}{\chi}\right]\right\} \quad (67)$$

η:viscosity

with the integral expressions

$$I_{sa} = \int_0^{\delta} \rho_f(z)\sinh(az)\,dz \quad (68)$$

$$I_{e\chi} = \int_0^{\delta} \rho_f(z)\exp(-\chi z)\,dz \quad (69)$$

$$I_{c\chi} = \int_0^{\delta} \rho_f(z)\cosh(\chi z)\,dz \quad (70)$$

For the minimum electrostatic free energy model we obtained an analytically closed expression. This facilitates the comparison with the experiments.

$$L_{12} = L_{21} = \frac{A}{L\eta}\left\{\varepsilon_s\varepsilon_0\psi_0 + \sinh\left(\frac{z'e_0\psi_0}{2kT}\right)\left[(8kTn_0\varepsilon_s\varepsilon_0)^{1/2}\frac{\tanh(a\delta)}{a}\right.\right.$$

$$\left.\left. + \frac{4z'e_0n_0\cosh(z'e_0\psi_0/2kT)}{a}\left(1 - \frac{1}{\cosh(a\delta)}\right)\right]\right\} \quad (71)$$

First applications of the presented approach were submitted by Donath and Voigt (1986;1986a).

C. Hydrodynamics in particle electrophoresis

Recently, there were published several papers investigating the relationship between surface layer structure and surface layer hydrodynamics under the conditions of particle elec-

trophoresis (Cohen Stuart et al., 1984a; Goff and Luner, 1984; Yaroshchuk, 1985; Holt and Dalgleish, 1986). Another set of papers deals with electrophoresis in concentrated polymer solutions (Bäumler and Donath, 1987; Pratsch and Donath, 1988; Donath et al., 1989).
New experimental techniques were successfully applied to unfavoured conditions for classical electrokinetics: concentrated colloid systems (Marlow and Fairhurst, 1988), invisible small latex particles (Goff and Luner, 1984).
A theoretical approach for electrophoresis of particles with nonuniform Zeta-potential was developed (Anderson, 1985).

We will stress, here, only two aspects. An elegant approach to electrophoresis of particles with surface layer was presented by Yaroshchuk (1985). The hydrodynamic part of the problem is remarkably simplified. Assuming a not very low density of immobile surface layer material the effect of viscous forces may be neglected in favour of the frictional forces and the mathematical expenditure was reduced drastically.
The other aspect relates to the above mentioned papers of Donath and coworkers on particle electrophoresis in concentrated neutral polymer (Dextran, PEG) solutions. Assuming a polymer depletion layer in the transition region of the cell surface coat and the bulk solution, the thickness of which is of the order of magnitude of the polymer molecule radius, an excellent fit to the experimental data was obtained. The result of the depletion layer is a decrease of the viscosity of the near-surface solution compared with the bulk phase. Taking not into account this effect will lead to a remarkable overestimation of the Zeta-potential.
First theoretical and experimental approaches to particle electrophoresis in concentrated polymer (e.g. polyelectrolyte) solutions are very promising for an improved understanding of particle electrophoresis itself. Thus, a Renaissance in the application of electrophoretic techniques seems to be possible because of an extension of the investigatable systems. Nevertheless, a lot of interesting aspects of particle electrophoresis could not be discussed here.

REFERENCES

Anderson JL (1985) Effect of Nonuniform Zeta Potential on Particle Movement in Electric Fields. J Colloid Interface Sci 105:45-54
Arakelyan VB, Arakelyan SB (1983) Ėnergetičeskij profil′ dipol′noj molekuli v tonkoj membrane. Biol Žurn Armenii 36: 755-779
Attard P, Kjellander R, Mitchell DJ, Jönsson B (1988) Electrostatic fluctuation interactions between neutral surfaces with adsorbed, mobile ions or dipoles. J Chem Phys 89:1664-1680
Attard P, Mitchell DJ (1988) The forces between surfaces of mobile, orientable dipoles. The method of reflection coefficients. J Chem Phys 88:4391-4396
Attard P, Mitchell DJ, Ninham BW (1988a) The attractive forces between polar liquid bilayers. Biophys J 53:457-460
Attard P, Mitchell DJ, Ninham BW (1988b) Beyond Poisson-Boltzmann images and correlations in the electric double layer. I. Counterions only. J Chem Phys 88:4987-4996
Aytyan SKh, Belaya ML, Chizmadshev YuA (1981) Membrane interaction with regard to redistribution of charged components. Biofizika 26:701-708
Aytyan SKh, Dukhin SS, Chizmadshev YuA (1977) Ėffekty sil izobraženija pri peremeščanii zarjada v membranach. Ėlektrochimija 13:779-783
Aytyan SKh, Dukhin SS, Pastushenko VF, Chizmadshev YuA (1977a) O svjazi "vorotnych tokov" s vnutrimembrannym smeščeniem zarjada. Dokl Akad Nauk SSSR 233:1215-1218
Barber J (1980) Membrane surface charges and potentials in relation to photosynthesis. BBA 594:253-308
Bäumler H, Donath E (1987) Does Dextran Indeed Significantly Increases The Surface Potential Of Human Red Blood Cells? studia biophysica 120:113-122
Belaya ML, Feigel′man MV (1984) Melting temperature of charged lipids lattice on the membrane surface. Biol Membr 1: 438-440
Belintsev BV (1988) Cell adhesion physical factors governing the cell-substrate interaction. Biol Membr 5:1100-1107
Beresford-Smith B, Chan DYC, Mitchell DJ (1985) The Electrostatic Interaction in Colloidal Systems with Low Added Electrolyte. J Colloid Interface Sci 105:216-234
Bisch PM, Steinchen A, Sanfeld A (1983) Coupled Longitudinal and Transversal Wave Motion at Fluid Interfaces with Electrical Double Layers. J Colloid Interface Sci 95:561-572
Bivas I, Petrov AG (1981) Flexoelectric and Steric Interactions between Two Bilayer Lipid Membranes Resulting from their Curvature Fluctuations. J theor Biol 88:459-483
Büldt G, Dencher NA, Dresselhaus D, Hentschel MP, Mischel M, Papadopoulos G, Plöhn H-J, Zaccai G (1988) Membrane Structure and Charge Distribution Related to Ion Transport as Studied by Neutron Diffraction. studia biophysica 127: 223-230
Cohen Stuart MA, Waajen FHWH, Cosgrove T, Vincent B, Crowley TL (1984) Hydrodynamic Thickness of Adsorbed Polymer Layers. Macromolecules 17:1825-1830

Cohen Stuart MA, Waajen FHWH, Dukhin SS (1984a) Electrokinetic effects of adsorbed neutral polymers. Coll & Polym Sci 262:423-426
de Groot SR (1959) Thermodynamics of Irreversible Processes. North-Holland Publ Comp Amsterdam
Deryaguin BV (1986) Teorija ustojčivosti kolloidov i tonkich plenok. Nauka Moscow
Deryaguin BV, Churaev NV, Muller VM (1987) Poverchnostnye sily. Nauka Moscow
Dolgaya EV, Mironov SL (1984) Issledovanie svojstv poverchnosti nejronov spinal'nych gangliev krys s ispol'zovaniem metoda mikroėlektroforeza. Nejrofiziologija 16:176-182
Donath E (1979) Zur Charakterisierung der elektrostatischen Eigenschaften von Zelloberflächen mit vorgelagerter Glycocalix. Thesis (A)
Donath E (1984) Zur Biophysikalischen Charakterisierung von Membranwechselwirkungsprozessen unter besonderer Berücksichtigung der Elektrostatik als Voraussetzung von Fusionsprozessen. Thesis (B)
Donath E, Gingell D (1983) A sharp cell surface conformational transition at low ionic strength changes the nature of the adhesion of enzyme-treated red blood cells to hydrocarbon. J Cell Sci 63:113-124
Donath E, Lerche D (1980) Electrostatic and structural properties of the surface of human erythrocytes. I. Cell-electrophoretic studies following neuraminidase treatment. Bioelectrochem Bioenerg 7:41-53
Donath E, Pastushenko V (1979) Electrophoretical study of cell surface properties. The influence of the surface coat on the electric potential distribution and on general electrokinetic properties of animal cells. Bioelectrochem Bioenerg 6:543-554
Donath E, Pratsch L, Bäumler H-J, Voigt A, Taeger M (1989) Macromolecule depletion at membranes. studia biophysica 130:117-122
Donath E, Steidel R (1980) Electrostatic and structural properties of the surface of human erythrocytes. 3. Cell electrophoretical studies following addition of human serum albumin. Acta Biol Med Germ 39:207-215
Donath E, Voigt A (1983) Charge Distribution Within Cell Surface Coats of Single and Interacting Surfaces - a Minimum Free Electrostatic Energy Approach. Conclusions for Electrophoretic Mobility Measurements. J theor Biol 101:569-584
Donath E, Voigt A (1986) Streaming Current and Streaming Potential on Structured Surfaces. J Colloid Interface Sci 109:122-139
Donath E, Voigt A (1986a) Electrophoretic Mobility of Human Erythrocytes. On the Applicability of the Charged Layer Model. Biophys J 49:493-499
Donath E, Voigt A (1988) Effect of thick fixed-charge layers on electrostatic interaction - a theoretical approach. Coll & Polym Sci 266:1024-1030
Dukhin SS (1975) Ėlektroprovodnost' i ėlektrokinetičeskie svojstva dispersnich sistem. Naukova Dumka Kiev
Dukhin SS, Deryaguin BV (1976) Ėlektroforez. Nauka Moscow

Dukhin SS, Yaroshchuk AE, Deryaguin BV (1984) On the role of the electrostatic factor in the stabilization of dispersions protected by the adsorption layers of polymers. Kolloidn Ž 46:225-230

Edmonds DT (1989) A Kinetic Role for Ionizable Residues in Membrane Channel Proteins. Eur Biophys J in press

Enos BE, McQuarrie DA (1981) The Effect of Discrete Charges on the Electrical Properties of Membranes. II. J theor Biol 93:499-522

Eriksson JC, Ljunggren NS (1988) Molecular calculations for interacting surfactant monolayers. Progr Colloid Polymer Sci 76:188-202

Evans EA, Parsegian VA (1986) Thermal-mechanical fluctuations enhance repulsion between bimolecular layers. Proc Natl Acad Sci USA 83:7132-7136

Gallez D, Coakley WT (1986) Interfacial Instability At Cell Membranes. Prog Biophys molec Biol 48:155-199

Gingell D, Fornés JA (1975) Demonstration of intermolecular forces in cell adhesion using a new electrochemical technique. Nature 256:210-211

Gingell D, Fornés JA (1976) Interaction of red blood cells with a polarized electrode. Evidence of long range intermolecular forces. Biophys J 16:1131-1153

Gingell D, Todd I (1975) Adhesion of red blood cells to charged interfaces between immiscible liquids. A new method. J Cell Sci 18:227-239

Gingell D, Todd I (1980) Red blood cell adhesion. II. Interferometric examination of the interaction with hydrocarbon oil and glass. J Cell Sci 41:135-149

Gingell D, Todd I, Parsegian VA (1977) Long-range attraction between red cells and a hydrocarbon surface. Nature 268: 767-769

Goff JR, Luner P (1984) Measurement of Colloid Mobility by Laser Doppler Electrophoresis: The Effect of Salt Concentration on Particle Mobility. J Colloid Interface Sci 99: 468-483

Gross D (1988) Electromobile surface charge alters membrane potential changes induced by applied electric field. Biophys J 54:879-884

Hartsel SC, Cafiso DS (1986) A Test of Discreteness-of-Charge Effects in Phospholipid Vesicles: Measurements Using Paramagnetic Amphiphiles. Biochemistry 25:8214-8219

Haydon DA (1961) The surface charge of cells and some other small particles as indicated by electrophoresis. I. The Zeta potential - surface charge relationship. Biochim Biophys Acta 50:450-457

Haydon DA (1961a) The surface charge of cells and some other small particles as indicated by electrophoresis. II. The interpretation of the electrophoretic charge. Biochim Biophys Acta 50:457-462

Heard DH, Seaman GVF (1960) The influence of pH and ionic strength on the electrokinetic stability of human erythrocyte membrane. J Gen Physiol 43:635-654

Heard DH, Seaman GVF (1961) The action of lower aldehydes on the human erythrocytes. Biochim Biophys Acta 53:366-374

Heinrich R, Gaestel M, Glaser R (1982) The Electric Potential

Profile Across the Erythrocyte Membrane. J theor Biol 96: 211-231
Holt C, Dalgleish DG (1986) Electrophoretic and Hydrodynamic Properties of Bovine Casein Micelles Interpreted in Terms of Particles with an Outer Hairy Layer. J Colloid Interface Sci 114:513-524
Hunter RJ (1960) Constancy of the surface charge density of human erythrocytes at different ionic strength. Arch Biochem Biophys 88:308-312
Hunter RJ (1981) Zeta Potential in Colloid Science. Principles and Applications. Academic Press New York London
Israelachvili JN (1987) Intermolecular and Surface Forces. With Applications to Colloidal and Biological Systems. Academic Press London San Diego New York Boston Sydney Tokyo Toronto
Jones IS (1979) A theory of electrophoresis of large colloid particles with adsorbed polyelectrolyte. J Colloid Interface Sci 68:451-461
Jordan PC (1982) Electrostatic Modeling of Ion Pores Energy Barriers and Electric Field Profiles. Biophys J 39:157-164
Kjellander R (1984) On the Image-Charge Model for the Hydration Force. J Chem Soc Faraday Trans II 80:1323-1348
Kjellander R, Marčelja S (1984) Correlation and image charge effects in electric double layers. Chem Phys Lett 112:49-53
Konev SV, Kaler GV (1988) Transmembrane electric potential as a regulator of functional activities of biomembranes. Biofizika 33:1018-1022
Kozlov MM, Lerche D, Markin VS (1985) Free energy and spontaneous curvature of charged membrane with glycocalix. Biol Membr 2:806-812
Kozlov MM, Markin VS (1987) Model of Red Blood Cell Membrane Skeleton: Electrical and Mechanical Properties. J theor Biol 129:439-452
Langner M, Winiski A, Eisenberg M, McLaughlin A, McLaughlin S (1988) in Ledeen RW, Hogan EL, Tettamanti G, Yates AJ (eds) New Trends in Ganglioside Research; Neurochemical and Neuroregenerative Aspects. Liviana Press Padova
LeBret M, Zimm BH (1984) Distribution of Counterions Around a Cylindrical Polyelectrolyte and Manning's Condensation Theory. Biopolymers 23:287-312
Lerche D, Kozlov MM, Markin VS (1987) Electrostatic free energy and spontaneous curvature of spherical charged layered membrane. Biorheology 24:23-34
Levich VG, Yalamov YuI (1962) K teorii rastvorov poliėlektrolitov. I. Malye stepeni ionizacii. Ž fiz chim 36:1096-1102
Levich VG, Yalamov YuI (1962a) Nekotorye voprosy teorii poliėlektrolitov pri malych stepenjach ionizacii. Dokl Akad Nauk 142:851-854
Levine S, Bell GM, Pethica BA (1964) Counter-Ion Penetration in Ionized Monolayers at a Dielectric Interface. J Chem Phys 40:2304-2309
Levine S, Levine M, Sharp KA, Brooks DE (1983) Theory of the electrokinetic behavior of human erythrocytes. Biophys J 42:127-135

Loosley-Millman ME, Rand RP, Parsegian VA (1982) Effects of monovalent ion binding and screening on measured electrostatic forces between charged phospholipid bilayers. Biophys J 40:221-232
Lyklema J, Overbeek JThG (1961) On the interpretation of electrokinetic potentials. J Colloid Sci 16:501-512
Makino K, Ohshima H, Kondo T (1987) Electrostatic interaction of ion-penetrated membranes. Effects of ionic solubility. Coll & Polym Sci 265:911-915
Makino K, Ohshima H, Kondo T (1988) Electrostatic Interaction of Two Dissimilar Membranes: Effects of Ionic Solubility, Permittivity and Thickness of Surface Charge Layers. Colloids Surfaces 33:153-166
Marlow BJ, Fairhurst D, Pendse HP (1988) Colloid Vibration Potential and the Electrokinetic Characterization of Concentrated Colloids. Langmuir 4:611-626
Maroudas N (1975) Adhesion and Spreading of Cells on Charged Surfaces. J theor Biol 49:417-424
Marra J (1986) Direct Measurement of the Interaction between Phosphatidylglycerol Bilayers in Aqueous Electrolyte Solutions. Biophys J 50:815-825
Marra J, Israelachvili JN (1985) Direct measurements of forces between phosphatidylcholine and phosphatidylethanolamine bilayers in aqueous electrolyte solutions. Biochemistry 24:4608-4618
Marra J, van der Schee HA, Fleer GJ, Lyklema J (1983) Polyelectrolyte Adsorption from Saline Solutions. in Ottewill RH, Rochester CH, Smith AL (eds) Adsorption from Solution. Academic Press London
Matinyan NS, Ershler IA, Abidor IG (1984) Proton Equilibrium on the Bilayer Lipid Membrane Surfaces. Biol Membr 1:254-267
McDaniel RV, McLaughlin A, Winiski AP, Eisenberg M, McLaughlin S (1984) Bilayer membranes containing the ganglioside G_{MI}: models for electrostatic potentials adjacent to biological membranes. Biochemistry 23:4618-4623
McLaughlin S (1977) Electrostatic Potentials at Membrane-Solution Interfaces. Current Topics in Membrane and Transport 9:71-144
McLaughlin S (1982) Divalent Cations, Electrostatic Potentials, Bilayer Membranes. in Martonosi AN (ed) Membranes and Transport 1:51-55. Plenum Press New York London
McLaughlin S (1989) The Electrostatic Properties of Membranes. Ann Rev Biophys Biophys Chem 18 in press
Nelson AP, McQuarrie DA (1975) The Effect of Discrete Charges on the Electrical Properties of Membranes. I. J theor Biol 55:13-27
Neumcke B, Läuger P (1969) Nonlinear electrical effects in lipid bilayer membranes. II. Integration of the generalized Nernst-Planck equations. Biophys J 9:1160-1170
Neumcke B, Läuger P (1970) Space Charge-Limited Conductance in Lipid Bilayer Membranes. J Membrane Biol 3:54-66
Neumcke B, Stampfli R (1984) Heterogeneity of external surface charges near sodium channels in the nodal membrane of frog nerve. Pflügers Arch 401:125-131
Norde W, Lyklema J (1978) The Adsorption of Human Plasma

Albumin and Bovine Pancreas Ribonuclease at Negatively Charged Polystyrene Surfaces. III. Electrophoresis. J Colloid Interface Sci 66:277-284
Norde W, Lyklema J (1978a) The Adsorption of Human Plasma Albumin and Bovine Pancreas Ribonuclease at Negatively Charged Polystyrene Surfaces. IV. The Charge Distribution in the Adsorbed State. J Colloid Interface Sci 66:285-294
Nordt FJ, Knox RJ, Seaman GVF (1978) Red cell aging. II. Anomalous electrophoretic properties of neuraminidase treated human erythrocytes. J Cell Physiol 97:207-220
Ohsawa K, Ohshima H, Ohki S (1981) Surface potential and surface charge density at the cerebralcortex synaptic vesicle and stability of vesicle suspensions. BBA 648:206-214
Ohshima H, Kondo T (1987) Electrophoretic mobility and Donnan potential of a large colloidal particle with a surface charge layer. J Colloid Interface Sci 116:305-311
Ohshima H, Makino K, Kondo T (1986) Potential Distribution across a Membrane with Surface Charge Layers: Effects of Nonuniform Charge Distribution. J Colloid Interface Sci 113:369-374
Ohshima H, Makino K, Kondo T (1987) Electrostatic interaction of two parallel plates with surface charge layers. J Colloid Interface Sci 116:196-199
Ohshima H, Ohki S (1985) Donnan potential and surface potential of a charged membrane. Biophys J 47:673-678
Parsegian VA (1973) Long-range physical forces in the biological milieu. Ann Rev Biophys Bioengen 2:221-253
Parsegian VA, Gingell D (1972) Some Features of Physical Forces Between Biological Cell Membranes. J Adhesion 4:283-306
Parsegian VA, Gingell D (1973) A Physical Force Model of Biological Membrane Interactions. in Lee LH (ed) Recent Advances in Adhesion. Gordon & Breach London. 153-192
Parsegian VA, Ninham BW (1970) A major artifact in cell surface charge measurements by electrophoresis due to the presence of cell peripheral material. manuscript
Parsegian VA, Rand RP, Rau DC (1985) Lessons from direct measurement of forces between biomolecules. Exxon Symp on Physics of Complex Liquids. J.Wiley
Pashley RM, McGuiggan PM, Ninham BW, Brady J, Evans DF (1986) Direct measurements of surface forces between bilayers of double-chained quaternary ammonium acetate and bromide surfactants. J Phys Chem 90:1637-1642
Pastushenko VF, Chizmadshev YuA (1981) Energetic Profile of Dipole Molecules in Membranes. Biofizika 26:458-463
Pastushenko V, Donath E (1976) On the electrophoretic mobility of cells coated with charged glycoprotein layer. studia biophysica 56:7-8 and 2B2-B7
Pastushenko V, Donath E (1976a) Potential distribution near the cell coated with a charged glycoprotein layer. studia biophysica 56:9-10 and 2B10-C2
Pethica BA, Hall DG (1982) Electric Field Effects on Membranes. J Colloid Interface Sci 85:41-51
Pratsch L, Donath E (1988) Poly-(ethyleneglycol) depletion layers on human red blood cell surfaces measured by electrophoresis. studia biophysica 123:101-108

Rau DC, Lee B, Parsegian VA (1984) Measurement of the repulsive force between polyelectrolyte molecules in ionic solution: Hydration forces between parallel DNA double helices. Proc Natl Acad Sci USA 81:2621-2625
Rice RE, Horne FH (1985) Analytical Solutions to the Linearized Poisson-Boltzmann Equation in Cylindrical Coordinates for Different Ionic-Strength Distributions. J Colloid Interface Sci 105:172-182
Rubin BT, Barber J (1980) The role of membrane surface charge in the control of photosynthetic processes and the involvement of electrostatic screening. BBA 592:87-102
Sauve R, Ohki S (1979) Interactions of Divalent Cations with Negatively Charged Membrane Surfaces. I. Discrete Charge Potential. J theor Biol 81:157-179
Scheutjens JMHM, Fleer GJ (1981) Some implications of recent polymer adsorption theory. in Tadros ThF (ed) The Effect of Polymers on Dispersion Properties. Academic Press London New York Paris San Diego San Francisco Sao Paulo Sydney Tokyo Toronto. 145-168
Schumacher GA, van de Ven TGM (1987) Brownian Motion of charged Particles surrounded by Electric Double Layers. Faraday Disc Chem Soc 83:75-85
Sharp KA, Brooks DE (1985) Calculation of the electrophoretic mobility of a particle bearing bound polyelectrolyte using the nonlinear Poisson-Boltzmann equation. Biophys J 47: 563-566
Snabre P, Mills P, Thiam AB (1986) Erythrocyte electrokinetic behaviour under low ionic strength conditions. Coll & Polym Sci 264:103-109
Stillinger FH (1961) Interfacial Solutions of the Poisson-Boltzmann-Equation. J Chem Phys 35:1584-1589
Todd I, Gingell D (1980) Red blood cell adhesion. I. Determination of the ionic conditions for adhesion to an oil-water interface. J Cell Sci 41:125-133
Träuble H, Eibl H (1974) Electrostatic Effects on Lipid Phase Transitions: Membrane Structure and Ionic Environment. Proc Natl Acad Sci USA 71:214-219
Trommler A, Gingell D, Wolf H (1985) Red Blood Cells Experience Repulsion But Make Molecular Adhesions With Glass. Biophys J 48:835-841
van de Ven TGM (1988) On the Role of Ion Size in Coagulation. J Colloid Interface Sci 124:138-145
van Dulm P, Norde W, Lyklema J (1981) Ion Participation in Protein Adsorption at Solid Surfaces. J Colloid Interface Sci 82:77-82
Vassar PS, Kendall MJ, Seaman GVF (1969) Electrophoresis of human leukocytes. Arch Biochem Biophys 135:350-355
Verwey EJV, Overbeek JThG (1948) Theory of the stability of lyophobic colloids. The interaction of sol particles having an electric double layer. Elsevier Amsterdam
Voigt A, Donath E, Heinrich R (1982) Free energy of the electrostatic interaction of cells with adjacent charged glycoprotein layer: a theoretical approach. J theor Biol 98: 269-282
Voigt A, Donath E, Richter L, Lauckner S, Becker R, Neumann G (1982) Blood components - surface interactions. Conclu-

sions for surface design. Proc "International Symposium on Kinetic Modeling in Artificial Organs" ISAO. Rostock-Warnemünde. 415-419

Winiski AP, Eisenberg M, Langner M, McLaughlin S (1988) Fluorescence Probes of Electrostatic Potential 1 nm from the Membrane Surface. Biochemistry 27:386-392

Winiski AP, McLaughlin AC, McDaniel RV, Eisenberg M, McLaughlin S (1986) An Experimental Test of the Discreteness-of-Charge Effect in Positive and Negative Lipid Bilayers. Biochemistry 25:8206-8214

Wolf H, Gingell D (1983) Conformational response of the glycocalix to ionic strength and interaction with modified glass surfaces: Study of live red cells by interferometry. J Cell Sci 63:101-112

Wunderlich RW (1982) The effects of surface structure on the electrophoretic mobilities of large particles. J Colloid Interface Sci 88:385-397

Yalamov YuI (1963) Teorija ėlektrostatičeskogo vzaimodejstvija makroionov s diskretnym poverchnostnym zarjadom v rastvore ėlektrolita. Ž fiz chim 37:1123-1126

Yalamov YuI (1963a) Teorija ėlektričeskogo vzaimodejstvija ploskich poverchnostej s odinakovym diskretnym poverchnostnym zarjadom, razdelennych rastvorom ėlektrolita. Ž fiz chim 37:1393-1397

Yaroshchuk AE (1985) On the taking into account of the contribution of the hydrodynamic permeability of the adsorption layers of nonionogen surfactants to the electroosmotic slip. Kolloidn Ž 47:369-375

Yaroshchuk AE, Baran AA (1987) Electrokinetic Manifestations of the adsorption (dipole) potential jump. Kolloidn Ž 49:972-980

Yuchnovskij IR, Golovko MF (1980) Statističeskaja teorija klassičeskich ravnovesnych sistem. Naukova Dumka Kiev

Zimm BH (1983) Counter-Ion Condensation and System Dimensionality. J Biomolecular Structure and Dynamics 1:461-471

Interfacial stability

W.T. Coakley
School of Pure and Appl. Biol.
Univ. of Wales Coll. of Cardiff
Museum Place
PO Box 915 Cardiff CF1 3TL
United Kingdom

D. Gallez
Service de Chim. Phys.
Universite Libre de Bruxelles
Campus Plaine, CP 231
1050 Brussels
Belgium

Abstract. Aspects of the behaviour of cell plasma membranes, either singly or as pairs in close apposition, resemble the behaviour of thin fluid films of low interfacial tension. Modification of a low interfacial tension thin film or of its environment can destabilise the film. Such destabilisation leads either to a transition from one state to another stable state of different morphology or to breakup of the film to give droplets. Theoretical treatments of the dependence of the stability of thin films on surface charge density, environmental ionic strength, electrical stress in the film, viscous and mechanical properties of the film and presence of cross-linking polymers are presented. The results of experimental studies of destabilization of erythrocyte membranes are reviewed to give (i) growth of spatially periodic disturbances on the membrane leading to vesicle formation or (ii) spatially periodic contact between cells exposed to a variety of adhesive polymers, or to electric fields. Some quantitative agreement between theory and experiment is obtained. Examples of interfacial instability involving cell surfaces in a number of in vivo situations are discussed.

1. INTRODUCTION.

The development of general theoretical treatments of the stability of low interfacial tension interfaces (Miller and Scriven 1970) particularly those of high curvature (Murphy 1966) has been motivated, in part, by the belief that some understanding of biological membrane behaviour would arise from consideration of such stability. Later studies (Jain et al 1978; Sanfeld et al 1979, Dimitrov 1982; Gallez 1983) were explicitly aimed at consideration of lipid bilayer, cell membrane, and cell surface stability (for review see Dimitrov and Jain 1984, Gallez and Coakley 1986).

With regard to in vitro experimentation the membrane vesicle formation which is associated with surface wave growth accompanying the thermal fragmentation of erythrocytes provided an experimental system in which some predictions of instability theory could be examined on single cells (Coakley and Deeley, 1980). The observation of discrete spatially periodic contacts in polycation induced adhesion of erythrocytes at room temperature (Coakley et al. 1985) provided a system in which to examine interfacial instability on surfaces of adjacent cells.

In vivo situations where interfacial instability appears to be involved in cell surface morphology change in isolated cells include (i) vesiculation of cell processes such as pseudopodia and microvilli on exposure of cells to cytoskeleton depolymerising regimes and (ii) wave growth on localized regions of amoeba during induced pinocytosis. Instability of closely apposed membranes appears to play a role in (i) the sperm acrosome reaction, (ii) focal contact formation during amoeboid movement, (iii) cell adhesion induction by some polymers and (iv) the production of local prefusion contacts in chemically and physically induced cell fusion.

2. THIN FILM MODEL AND INSTABILITY PRINCIPLES.

In this section we will briefly develop the thin film model used for the two cases involved (see Fig. 1). In Case a the system consists of three bulk phases: the thin film (phase 2; identified with the membrane (mainly the bilayer part) of width h about 10nm, and the two adjacent electrolytic solutions (phases 1 and 3), separated by two surfaces. The surfaces of the film are transition zones, modelled as two dimensional surface phases with appropriate physicochemical properties like surface tension, surface coverage, surface elasticity, surface charge. In contrast, slowly varying inhomogeneities like the long range part of the van der Waals interactions or the diffuse part of the electrical double layer are treated as body forces. The balance of these forces is the following: long range attractive van der Waals forces arise because of the small thickness of the film; electrical interactions give rise to compressive electrical forces inside the membrane and to electrical free energy contributions of the double layer; in addition, steric repulsive forces appear at very small thickness (as for black films) due to the overlap of the hydrocarbon chains of the lipids.

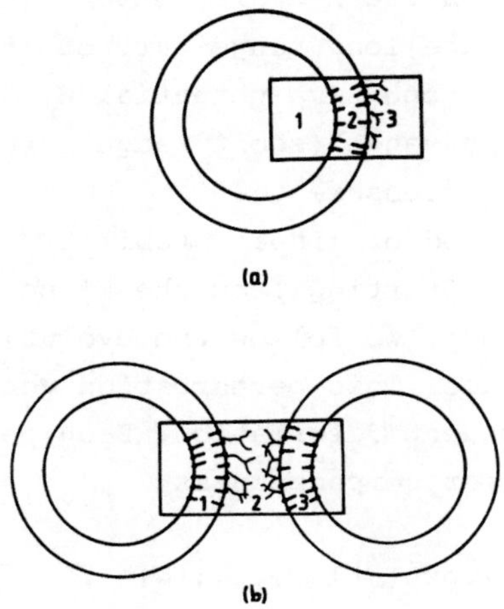

Fig. 1. Case a; the membrane itself modelled as a thin film; Case b; The aqueous film between the two membranes is the thin film.

In case b, the system also consists of three bulk phases but now the thin film (phase 2) is an aqueous film and phases 1 and 3 are the adjacent membranes. The balance of forces is different: we still have van der Waals attractive forces, the electric double layers are now overlapping inside the film, and new short range forces appear due to hydration of the surfaces (for two interacting cells these forces may be repulsive or attractive due to the complex nature of the biological surface; for a discussion see Gallez and Coakley, 1986; Israelachvili and McGuiggan, 1988).

For both cases the state of motion of the fluid in each bulk phase is given by the Navier-Stokes equation (Felderhof 1968):

$$\rho d\bar{v}/dt = -\bar{\nabla}p + \mu\nabla^2\bar{v} + \bar{F}_e + \bar{F}_w + \bar{F}_r \quad \ldots\ldots\ldots\ldots (1)$$

together with the incompressibility condition $\bar{\nabla}\circ\bar{v} = 0$ ..... (2) and with

$$\bar{F}_e = \nabla\circ\bar{\bar{T}}, \quad \bar{F}_w = -\rho\bar{\nabla}W_w, \quad \bar{F}_r = -\rho\bar{\nabla}W_r \quad \ldots\ldots\ldots\ldots (3)$$

here ρ and μ are the uniform mass density and viscosity respectively, $\bar{v}$ is the velocity of the phase under consideration and p represents the hydrostatic pressure. The electrical force $\bar{F}_e$, the van der Waals force $\bar{F}_w$ and the repulsive force $\bar{F}_r$ are introduced as body forces in (1). $\bar{\bar{T}}$ is the Maxwell tensor. The attractive potential W_w arises from the long range part of the van der Waals interactions while the repulsive potential W_r describes steric interactions at medium range (short range forces are included here in the hydrostatic pressure p).

Let us now sketch briefly the method of linear stability analysis applied to the film system. Starting from the plane reference state for the system at rest, we follow the evolution of a small perturbation of the surface. This perturbation generates motion in the fluid and subsequent deformation. Each perturbed quantity is expressed in Fourier components as

$$\delta G(r,t) = \delta G(z).\exp(ik_x x + ik_y y).\exp(\omega t) \quad ......(4)$$

where $k = (k_x^2 + k_y^2)^{\frac{1}{2}}$ is the wavenumber and $\omega = \omega_r + \omega_i$ is the complex frequency of the perturbation. Such expressions for the perturbed quantities are introduced into the linearized Eqn. 1 and the integration constants of the general solution in the bulk phases are determined from the appropriate boundary conditions (see Bisch et al 1983).

A general surface mass and surface charge balance is used, involving surface convection, surface diffusion and adsorption-desorption from the bulk. The general solution for the z-component of the velocity field is decomposed into a part $v^a(z)$ symmetric upon changing the sign of the spatial coordinate z, and into an antisymmetric part $v^b(z)$. The derivatives of these fields with z are $Dv^a(z)$ and $Dv^b(z)$ respectively. The secular determinant (leading to the dispersion relation) then has the general form of Eqn. 5 where v_s corresponds to interface displacements perpendicular to the surface and Dv_s to those parallel to the surface. The form of the secular determinant implies a bending mode (BE) with a superposition of v_s^a and Dv_s^a fields, and a squeezing mode (SQ) with superposition of v_s^b and Dv_s^b fields (see Fig. 2)

For asymmetrical systems the two modes are coupled by coupling terms CO while for symmetrical systems they are decoupled (CO = 0). For case a, the bending mode describes the deformation of the membrane while the squeezing mode describes

$$\det \begin{array}{|cc|cc|} \multicolumn{1}{c}{v_s^a} & \multicolumn{1}{c}{Dv_s^a} & \multicolumn{1}{c}{v_s^b} & \multicolumn{1}{c}{Dv_s^b} \\ \hline \multicolumn{2}{|c|}{BE} & \multicolumn{2}{c|}{CO} \\ \hline \multicolumn{2}{|c|}{CO} & \multicolumn{2}{c|}{SQ} \\ \hline \end{array} = 0 \quad \ldots\ldots(5)$$

the thickness variation of the membrane leading to rupture. For case b, we will consider essentially the squeezing mode which leads to periodic film rupture and pattern formation in the intercellular layer.

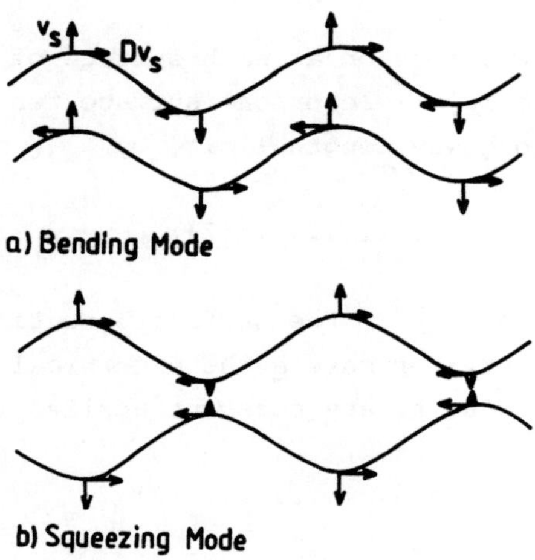

Fig.2. For the bending mode the two interfaces move transversely in phase. For the squeezing mode the two interfaces are 180° out of phase.

The exact forms of the dispersion equation $\omega = f(k)$ for these two modes in different situations are discussed elsewhere (e.g. Gallez 1983). Suffice it to say here that in general, for a root ω_r real positive, the system is unstable and the perturbation will grow, as can be seen by considering the exponential growth in Eq.4. For a root ω_r negative, the system is stable (exponential decay). For $\omega_r = 0$ and $\omega_i \neq 0$ the system oscillates. Among the possible wavelengths there is a marginal one k_o^{-1} for which $\omega_r = \omega_i = 0$. It is also possible to calculate a dominant wavelength k_m^{-1} for which $d\omega/dk = 0$. The corresponding ω_m is the fastest rate of growth of the perturbation, i.e. this dominant wavelength is the most probable wavelength. For the SQ mode, the inverse ω_m^{-1} corresponds to the time of rupture.

Let us first discuss the general stability criteria obtained by the model. The marginal stability case (the state of zero growth of the perturbations) gives us already indications about the interplay of the different physico-chemical parameters and defines the marginal wavelength. For both cases considered here waves having wavelengths longer than this marginal wavelength are unstable (justifying the choice of the long wavelength limit here as $kh \ll 1$).

For general symmetrical systems (same phase 1 and 3, same surface properties), the marginal stability criterion for the uncoupled BE mode reads (Gallez, 1983)

$$\sigma_t = 2\sigma_s + \sigma^o_{dl} + \sigma^i_{dl} + \sigma_m + \sigma_w + \sigma_r > 0 \quad \ldots\ldots (6)$$

The pure surface tension σ_s is the tension at each surface of the film. The outer and inner double layer electrical surface tensions are given, for case a, in the long wavelength limit, by

$$\sigma^{o,i}_{dl} = -\varepsilon\kappa\,(\phi^{o,i}_s)^2/(8\pi) \quad \ldots\ldots\ldots\ldots (7)$$

where κ is the inverse Debye length $\phi^{o,i}_s$ the surface potentials and ε the bulk dielectric constant. For case a the electrical contribution to the surface tension due to any constant applied field in phase 2 reads

$$\sigma_m = -\varepsilon_m\,(\phi_m)^2/(4\pi\, h) \quad \ldots\ldots\ldots\ldots (8)$$

where ϕ_m is the transmembrane potential (i.e. the potential difference between bulk phases 1 and 3) and ε_m the membrane dielectric constant. Both electrical terms are negative and hence reduce the total surface tension. For systems with low interfacial tension, as is the case for biological membranes, any change in the environment which renders the electrical tension more negative will thus tend to destabilise the BE mode.

The van der Waals attractive contribution to the surface tension for the long wavelength limit is given, where A is the Hamaker constant, by:

$$\sigma_w = -A/(4\pi\, h^2) \quad \ldots\ldots\ldots\ldots (9)$$

The repulsive contribution σ_r to the surface tension is positive and decreases exponentially with h and is given by Eqn.10 where P and λ are constants:

$$\sigma_r = P \exp(-h/\lambda) \quad \ldots\ldots\ldots(10)$$

The stability criterion for the uncoupled SQ mode reads

$$k^2\sigma_t - 2\, d\Pi_t/dh > 0;\ \text{where } \Pi_t = \Pi_e + \Pi_w + \Pi_r \quad \ldots\ldots (11)$$

Π_t is the total disjoining pressure due to electrical (Π_e), van der Waals (Π_w) and repulsive (Π_r) disjoining pressures. These terms are given exactly by Gallez (1983) and Gallez et al (1984) for cases a and b respectively.

Inspection of the stability criteria for the BE and SQ modes shows that both take into account the total tension σ_t but that for thickness variations (SQ mode) there is a second component due to the free energy of interaction between the surfaces and this is predominant at long wavelengths (small k).

These stability criteria will be applied to the observed instabilities on biological membranes, for case a (instability of the membrane itself, leading to deformation or rupture) and for case b (instability of the aqueous film between membranes leading to periodic pattern formation between aggregating cells or between cells making prefusion contact).

In general, the applied constraints are either chemical or electrical. For case b a detailed analysis of the role of chemical agents (like polycations) has been given previously (Gallez and Coakley, 1986). Essentially it has been proposed that the molecular cross-linking due to the polycations can be described by an attractive disjoining pressure (Π_A) incorporated in Eqn. 11. This additional term lowers the total disjoining pressure, allowing the squeezing mode to be destabilised.

The linear stability analysis provides estimates of the dominant wavelength (i.e. the wavelength which will give rise to the observed pattern) and of the fastest rate of growth of the instability (i.e. the time of formation of the pattern). Fig. 3 (Gallez and Coakley 1986) gives the rate of growth of the squeezing mode for two cells aggregated with polylysine. The dominant wavelength, λ_{MAX} = 0.63μm (for h = 20nm) is of the same order of magnitude as the average spacing between contact regions of the periodic pattern (0.8μm). The rate of growth can be estimated to be of the order of $30s^{-1}$ for an intercellular viscosity of 1 poise (due to

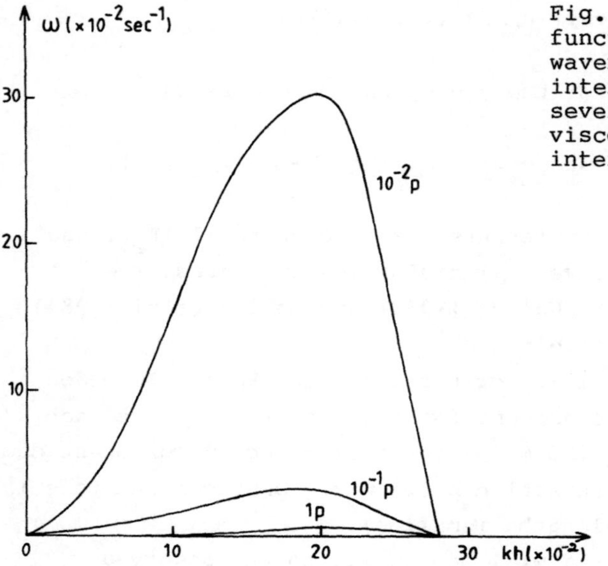

Fig. 3. Rate of growth as a function of the dimensionless wavenumber kh for two interacting erythrocytes, for several values of the viscosity of the intercellular layer.

glycocalyx molecules). This corresponds to a pattern development time of 0.1s which is, appropriately, less than the recorded time (0.3s; Tilley et al 1987) for cell-cell adhesive mutual engulment.

Let us now discuss the role of electrical constraints, in particular the situation where exposure of adjacent cells to high electric fields would effect the stability criteria discussed above. As a particular example the sequence of events during electrofusion of cells (Arnold and Zimmermann 1984) is: (i) The cells are first brought into close membrane contact by exposure to an alternating, non-uniform field (dielectrophoresis) of comparatively low magnitude; (ii) a short electric field pulse of high amplitude is then applied to achieve reversible electrical breakdown in the membrane contact zone. We will consider the effect of this short pulse on membrane stability.

For a cell of radius 'a' in an approximately uniform electric field E the surface potential due to interfacial polarisation is (see Arnold and Zimmermann 1984 or Neumann 1988)

$$\phi_s = -1.5Ea\cos\theta \quad \text{.........} \quad (12)$$

where θ is the angle between the radius through the position on the cell surface and E. The surface potential due to polarisation will be different along the cell surface with a maximum absolute value at the poles of the cell (in the field direction).

The actual value of the membrane potential results now from contributions from the diffusion potential ϕ_d of the cell (inside negative relative to the outside potential) from the potential drop due to the interfacial polarisation ϕ_s (Eq. 12) and from the contribution of the surface charges $\phi_s^{o,i}$. At the poles, it reads

$$\phi_m = \Delta\phi_d - 1.5Ea + \phi_s^{o,i} \quad .... \ (13)$$

This value of the membrane potential is used to evaluate the electric surface tension (Eq. 8). As the interfacial polarisation is dominant in Eq. 13, if the electric field pulse is increased above a threshold value the electric membrane potential will increase and the electrical surface tension term will become more negative.

In view of this potential profile, application of a field pulse will have two main effects on membrane stability: (a) first the total surface tension of the membrane will be lowered by the application of the field pulse and the bending mode of the membrane will be destabilised (criterion (6)) (b) second, as the surface potentials of two cells in close contact are now of opposite sign, the cells will attract each other. The variation of the disjoining pressure will change sign (criterion (11)) and the squeezing mode for the aqueous film will be destabilised.

Fig.4 shows an estimate of the balance of forces during electrofusion , for a value of the field pulse of 2.5 kV/cm. The attractive electrical forces arise because of the opposite surface potentials. At an intercellular distance of about 15nm (Stenger and Hui 1986) which seems to be the equilibrium approach distance for cells treated with pronase, it has been considered that the hydration forces (due mainly to the hydration of the lipids) begin to be operative and that the attractive electrical and van der Waals forces are of the same order of magnitude. Prior to the application of the field pulse the hydration forces will then be sufficient to keep the surfaces at an equilibrium distance (which seems to be the case after close contact by dielectrophoresis) but the application of the electric field strength of a value necessary for the membrane fusion is predicted to always

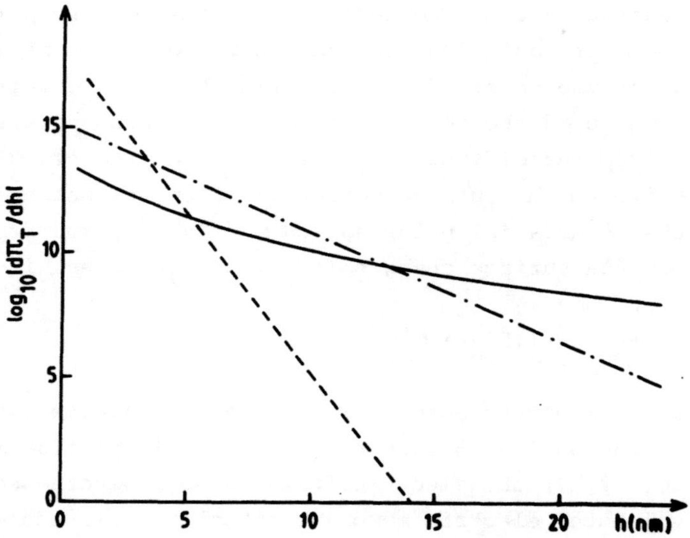

Fig.4; Variation of the various contributions to $d\Pi_t$ /dh as a function of the cell-cell distance h. — van der Waals attractive contribution, -- hydration repulsion, -·- electrical attraction during the field pulse.

destabilise the aqueous layer between the cells. After the field pulse, tight membrane-membrane contact is established 'in patches' (Sugar et al 1987): these authors observed that water was trapped in between these patches but remained continuous with the extracellular fluid which in the framework of our model corresponds to the development of an unstable squeezing mode. Experimental evidence of regular contact points between erythrocytes submitted to high external fields will be discussed in the following section.

3. EXPERIMENTAL

3 (i).The erythrocyte model system: Human erythrocytes lack the conventional cytoskeleton of nucleated cells. The membrane is stabilised by a 'membrane skeleton' composed of the actin binding protein spectrin, short filaments of actin, and connecting proteins which link the stabilising infrastructure to Band 3 protein and probably to the cytoplasmic aspect of Glycophorin C (Bennett 1989). The submembraneous spectrin skeleton forms a (usually)

hexagonal (180nm span) structure anchored to the membrane associated short actin filaments (Fowler 1986).

3 (ii). Interfacial instability on single heated erythrocytes: Erythrocytes fragment into spheres at 50°C i.e. close to the 49.6° C thermal denaturation transition in spectrin (Brandts et al. 1977). Cine-microscopy showed that there were only two fundamental pathways by which cells fragmented on heating through 49°C. The first mode involves the development of a surface wave on the rim of a cell and the pinching of vesicles from the crests of the growing surface waves (Fig.5; Coakley and Deeley 1980).

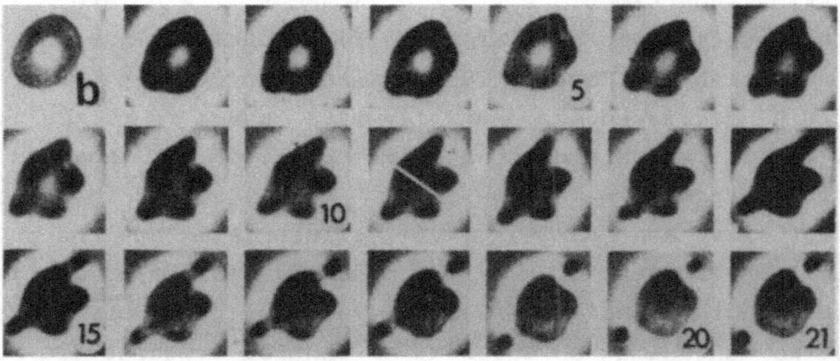

FIG. 5. Cine micrograph sequence (16 frames/s) of wavegrowth on the rim of an erythrocyte heated through the denaturation temperature of spectrin at 0.33 K/s. A clear axis of symmetry for wavegrowth is shown on the eleventh detail of the sequence. Opposite ends of the cell become unstable synchronously to give two externalized membrane vesicles (Coakley and Deeley 1980).

The second form of response to heating involved a rounding up of the cell at the instant of spectrin denaturation to trap one of the cell dimples within the rounded cell as a disc enclosing some of the extracellular aqueous phase. The curved rim of the internalized dimple was unstable and developed a surface wave which sent vesicles into the cell (Deeley and Coakley 1983). Typically, less than three per cent of freshly drawn and washed erythrocytes underwent the membrane internalization morphological change. This percentage increased markedly on cell surface charge depletion (Coakley and Deeley 1980, Doulah and Coakley 1984), membrane hyperpolarization (Doulah and Coakley, 1984), or exposure to cationic drugs (Coakley et al 1983).

In typical experiments cells were exposed to a number of different extracellular phases or cell surface treatments where one parameter was varied. For each experimental regime the number of waves developing on the cell rim at the spectrin denaturation temperature was scored for many cells. W, the average number of waves per cell, was calculated for each experimental treatment. W for control cells heated in phosphate buffered saline was about 7.0. Very few fragmenting cells developed less than four waves on the cell rim during thermal fragmentation. Hence, under conditions which reduced W, the value of W tended to an asymptotic value, W_a close to 4. It was possible, essentially independently, to alter the electrical contributions σ_{dl} (Eq.7) and σ_b (a bilayer electrical contribution similar to Eq.8) to the total surface tension σ_t (Doulah and Coakley 1984, Gallez and Coakley 1986). The value of $W-W_a$ decreased exponentially with change in σ_t (Gallez and Coakley, 1986). The rate constants from plots of $W-W_a$ against changes in σ_t achieved by change in σ_{dl} or σ_b were comparable in magnitude (Gallez and Coakley, 1986), illustrating how the physicochemical environment of a cell can modify vesicle production.

The wavelength associated with a W value of 7 for the average number of waves growing on the rims of control erythrocytes heated in saline (Coakley and Deeley 1980) is 3.4μm. The theoretical prediction for the dominant wavelenght of a bending wave on a plane film having a bending elasticity modulus B is given (Gallez and Coakley, 1986) by

$$\lambda_m = 4\pi(-B/3\sigma_t)^{\frac{1}{2}} \;\ldots\ldots\; (14).$$

The bending elasticity modulus for erythrocyte membrane is $1.8.\ 10^{-}$ J (Evans 1983). This value provides an upper limit for the value of B at the instant of thermal denaturation of the membrane skeleton. Eqn.14 then gives, for $\lambda_m = 3.4\mu m$, that -10^{-3} mN/m $< \sigma_t <$ 0 at the instant of wave growth initiation. Surface wave growth occurs only on the cell rim during thermal fragmentation. The confinement of vesicle production to the rim of heated erythrocytes (and to the rim of internalized dimples) has an analogy in Golgi vesicle production in vivo which (electron microscopy shows; Krystic 1979) is confined to the rim of the stack. Surface wave growth on the rim of amoeba surface membrane is part of the induced pinocytosis process in amoeba (Klein and Stockem 1979). The emphasis on the cell rim in the fragmentation studies and in the in vivo Golgi situation (suggesting a relationship between rim curvature, wavelength and final vesicle size) suggests that while theory based on infinite planes has helped predict how variables will affect wavelength and wave growth rate the analysis of the stability of membrane discs requires new theoretical development as a specifically biophysical problem. It is of interest here that erythrocytes do not inevitably fragment at the thermal denaturation temperature. For instance 70% of cells exposed to low concentrations of tetracaine remain disc-like at the spectrin denaturation temperature but internalise membrane (as a single vesicle at the centre of the cell dimple) as temperature is further increased (Coakley et al 1983).

3.(ii) Interfacial instability on cylindrical cell membrane processes: Interfacial instability is involved in the vesiculation of cylindrical plasma membrane processes when cytoskeletal proteins are depolymerised or denatured. Studies of vesiculation in a model system have been carried out with cylinders drawn from heated erythrocytes. Cells on a heated microscope stage were subjected to flow stress at the instant of spectrin denaturation (Crum et al 1979). A cell body with intact membrane was torn away from the glass surface leaving a membrane bound toroid, connected to the cell body by a membrane tether. Vesiculation of the glass-attached toroid occured by surface wave growth at a rate of about $30s^{-1}$. A membrane 'tether' retaining contact between the glass and the cell body in the suspending phase also consisted of a row of vesicles. Stable cylindrical tethers can be drawn from erythrocytes subjected to a slowly increasing flow stress at room temperature but tethers produced by rapidly applying stress are beaded (Hochmuth et al 1973). It appears that the sudden application of stress markedly reduces the elastic properties of the membrane.

Goldin et al (1969) showed that a viscoelastic fluid which has some inherent structure can remain stable only if the elastic modulus E exceeds a critical value. Irregularly beaded tethers, similar in appearance to threads of viscoelastic fluid are observed (Deeley et al 1979) when erythrocytes are stressed hydrodynamically following heating to temperatures close to 50°C. It has been predicted, using a viscoelasticity rheology, that reduction of the elastic modulus of the membrane will favour instability of the BE mode for plane membranes (Steinchen et al 1982). A theoretical treatment of the stability of membrane bound liquid cylinders remains to be carried out.

Regular beading of long filamentous cell processes has been reported for cells other than erythrocytes. Cytokeleton depolymerizing high pressures lead to vesiculation of the axopodia of the marine protozoan Actinophyrs sol (Tilney et al 1966) and the pseudopodia of Amoeba proteus (Marsland 1964). The long cylindrical processes attaching cytochalasin B treated cultured cells to a substratum bead when exposed to the microtubule depolymerising drug colcemid (Croop and Holtzer 1975). During fertilization of the Hydroides egg tubular interdigitation of sperm and egg membranes preceeds breakdown of the tubules into vesicles (Colwin and Colwin 1967). Exposure of epithelial cell brush borders to the micromolar calcium concentrations required to activate the actin filament severing protein, villin (Burgess and Prum 1982) leads to disassembly of the microfilament cores of the cylindrical microvilli and to vesiculation. Microvillus vesiculation can also preceed close contact of bacteria to the surface of cultured human intestinal mucosa (Knutton 1987). These recent observations offer some insights into the molecular basis of destabilization since Ca^2 ionophore treatment of microvilli causes a similar effect, presumably by Ca^{2+} activation of gelsolin.

Cinemicroscopy observations of growing mycoplasma (wall-less procaryotes) show that long filamentous forms can develop in rapidly growing cultures (Bredt 1973). The filaments breakdown into coccoid forms over a 2 min period. The nature of the contractile force which results in the coccoid form is not resolved; actin filament depolymerising cytochalasin B inhibits growth in only a limited range of mycoplasma (Razin 1985). Membrane lipids and hopanoids can influence mycoplasma morphology (Ourisson et al 1987).

3.(iii) Squeezing waves on erythrocytes:

3.(iii,a) Chemical constraints: Molecules capable of inducing adhesion of erythrocytes include polycations (Coakley et al 1985), polysaccharides (Skalak et al 1981), lectins (Grieg and Brooks 1979) and immunoglobulins.

Electron micrographs of the seam of contact between erythrocytes exposed to some polymers e.g. dextran (Skalak et al, Coakley et al. 1988) show the uniform separation expected of parallel cell surfaces. However recent studies of the contact seams of polylysine treated cells have shown a spatial periodicity of discrete contact regions separated by an average lateral distance of about 0.8 µm (Coakley et al 1985, Hewison et al 1988). Examples of the periodic contact are seen for polylysine treated cells in the transmission electron micrographs of Fig. 6a. The scanning electron micrograph of Fig. 6b shows a clump which broke during preparation for microscopy to reveal the contact region. It is clear from Figs 6a and 6b that the undulation pattern is confined to a region of the cell surface which forms contact. The freeze fracture electron micrograph of Fig. 6c, where cells were fixed by cryofixation, also shows a regularity of cell-cell contacts.

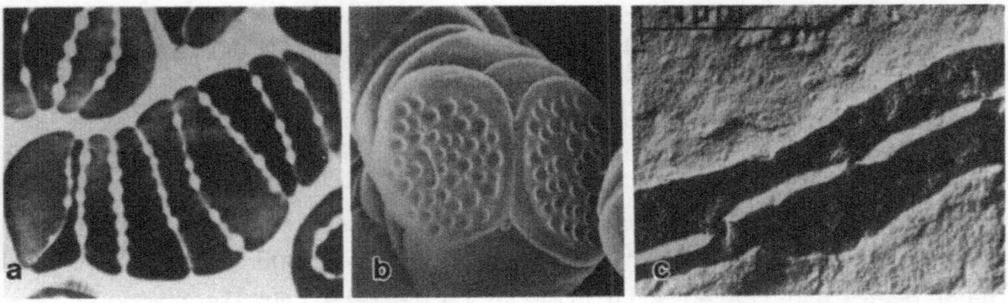

Fig. 6. Spatial periodicity of cell contacts in erythrocytes exposed to 20µg/ml polylysine of molecular weight 14,000; (a) Transmission electron micrograph (b) Scanning electron micrograph of broken clump showing exposed contact faces; (c) freeze-fracture electron micrograph of cryo-fixed erythrocytes: (a,b Coakley et al 1985; (c) Hewison et al 1988).

The threshold polycation concentration for cell adhesion is that at which the polycation begins to modify the electrophoretic mobility of the cell (Katchalsky et al. 1959, Coakley et al. 1985), suggesting that protrusion of polycation from the glycocalyx is the condition necessary for adhesion. Gallez and Coakley (1986) proposed that the molecular cross-linking can be described by an additional attractive disjoining pressure incorporated in Eqn. 11 and found reasonable agreement between the predicted and observed wavelengths and disturbance growth rates for cells being drawn together by crosslinking macromolecules (Fig. 3).

Erythrocytes exposed to the lectin WGA initially formed contact at the cell rim and then, over the following minutes contact spread by formation of discrete local contacts (Fig. 7a,b; Darmani et al. 1989) in a manner which contrasted with the rapid mutual cell engulfment step which was a feature of dextran, polylysine and polyethylene glycol induced cell contact (van Oss and Coakley 1988).

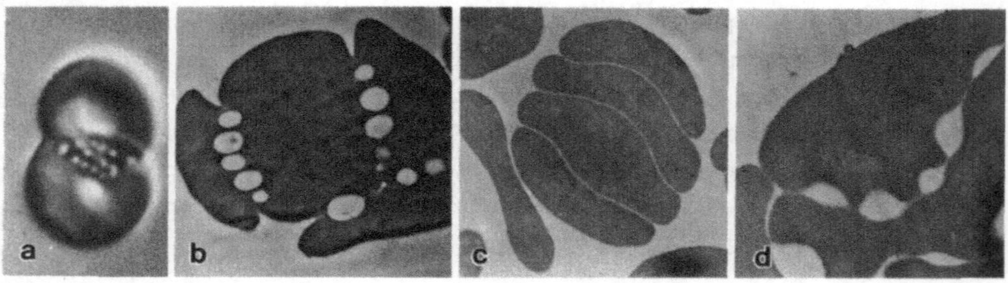

Fig. 7. (a) Light and (b) electron micrographs of erythrocytes following exposure to WGA (Darmani et al 1989): (c) Uniform separation of erythrocytes in 2% dextran T500 showing uniform contact separation and (d) spatially periodic contacts in pronase treated erythrocytes exposed to 4% dextran T500 (Coakley et al 1988).

The contact between dextran treated cells is of uniform separation (Fig. 8c; Coakley et al 1988). However when erythrocytes in dextran have been pretreated with pronase to deplete the glycocalyx the contact between cells becomes spatially periodic (Fig. 8d). Among the erythrocyte systems examined to date (Table 1) cells exposed to dextran following pretreatment by pronase or by combinations of pronase and neuraminidase are the only systems where the distance between contacts depends strongly on experimental conditions i.e. it decreases with increasing enzyme pretreatment (Darmani, personal communication). This observation is qualitatively consistent with the theoretical predictions in that increased cell-cell attraction (on reducing the repulsive glycocalyx and electrostatic contributions to the stability criterion 11) will, for constant dextran concentration, initially shift the contact from a stable equilibrium contact to a spatially discrete periodic one and for more drastic enzyme treatments move the system to shorter dominant wavelengths. The dextran adhesion system offers a promising system in which to compare experimental results with theoretical predictions.

TABLE 1. Contact seams of erythrocytes exposed to a variety of polymers:

Polymer:	Contact	Refs:
Polylysine, 14 - 225 kDa	Periodic (≈0.8μm)	(a)
Lectin (WGA)	Periodic (≈0.8μm)	(b)
Dextran: MW 75,000; 4%w/v	Uniform	(c)
Dextran, MW 450,000; 2%w/v	Uniform	(d)
Dextran; MW 450,000; 2%w/v & 4% w/v, pronase treated).	Periodic (2.5-0.8μm)	(d,e)

Refs: (a) Coakley et al 1985, Hewison et al 1988; (b) Darmani et al 1989; (c) Skalak et al 1981 (d) Coakley et al 1988); (e) Coakley et al 1988; Darmani, personal communication.

3(iii,b). Electrical constraints: As discussed in the preceeding section, the application of a field pulse will have two main effects on the membrane stability of fusing cells: First, the total surface tension of the membrane will be lowered by the application of the field pulse and the bending mode of the membrane will be destabilised (criterion 6). It has been shown (Zimmermann 1982) that the application of an electric pulse to single cells can cause crenation. Hewison (1988) found crenation of pronase pretreated cells in L1 fusion medium (0.25mM Mg-acetate, 0.1 mM Ca-acetate, 0.36 mM KH_2PO_4 and 0.6 mM K_2HPO_4, 280 mM inositol) exposed to a 10μs 2.5kV/cm electric pulse. The model predicts that crenation should be observed in regions of the membrane which are perpendicular to the electric field. Second, as the surface potentials of two cells in close contact are now of opposite sign, the cells attract each other, and the squeezing mode for the aqueous film will be destabilised (criterion 11). Indeed chains of erythrocytes exposed in L1 medium to electrofusion conditions (4.5kV/cm, 20μs) and subsequently fixed with glutaraldehyde had a regular distribution of contact points in the contact region between adjacent cells which can be attributed to the development of a squeezing wave (Hewison 1988). In addition, when plant protoplasts in contact are exposed to a fusogenic high voltage pulse (Zimmermann 1982) small vesicles are produced at the cell junction.

3(iv). Local contact formation in living systems: Spatially discrete contact formation is widely observed in cell-cell and membrane-membrane interactions. Local contact of the apposing faces of the acrosome outer membrane and the plasma membrane of boar spermatozoa during the acrosome reaction leads to the production of a sheet of essentially equally sized hybrid vesicles (implying regularly spaced (0.25µm) contact/fusion sites) consisting of hemispheres of both membranes (Russel et al 1979). Vesicle formation in sea urchin sperm membrane is a very rapid process as it is one of five distinct reaction stages seen in sperm fixed within 1s of initiation of the acrosome reaction (Dan, 1967).

Coakley and Gallez (1989) have recently discussed, in the context of the interfacial instability approach, the dependence, on ionic strength and on cell-glass separation, of the spacing of 'focal contacts' made by amoeba locomoting on glass. Spreading of tissue culture cells also involves formation of spatially periodic contacts (Segel et al 1983). The new dot contacts (0.1-0.2 µm diameter) which arise during fibroblast spreading are not initially associated with large actin bundles but some of the contacts later mature and become associated with the microfilament bundles characteristic of developed focal contacts (Vasiliev 1985, 1987). In phagocytosis, another form of contact spreading, electron micrographs of yeast ingestion by cultured human endothelial cells (Rotrosen et al 1985) show a regular contact separation distance.

The suggestion above associating the formation of periodic membrane contact with surface wave development requires that the cell boundary involved should be of low surface tension and not be very stiff. In addition to the rapid undulations which occur in the amoeboid ventral surface and at the margins of spreading mammalian cells it is known that erythrocyte thickness undergoes transient variations (erythrocyte flicker) of amplitude up to 200 nm. The damped surface wave associated with such erythrocyte fluctuations has a wavelength of about 0.6µm (Fricke and Sackmann 1984).

4. CONCLUSIONS:

The occurence of spatial regularity in vesicle formation in a range of systems focuses attention on the stability criterion for a 'bending' mode and suggests questions e.g the stability of membrane enclosed discs and cylinders which might usefully be examined. Such treatments provide insights to, for instance, the biophysical influence of lipids in stabilising filamentous growth of mycoplasma filaments. The relevance to biologial processes (e.g. acrosome reaction, cell locomotion, cell adhesion) of a stability criterion for a 'squeezing' mode, which involves membrane interaction terms, will be clarified as the understanding of areas such as (i) the magnitude and range of surface interaction forces in aqueous systems and (ii) the properties of the various adhesive proteins continues to develop. It is a truism to say that biological processes can only operate within the limitations set by the physicochemical properties of the systems involved. An understanding of this physical chemistry will allow informed intervention to bring about processes such as cell adhesion or cell fusion.

5. ACKNOWLEDGEMENTS.

DG would like to thank the Belgian Government (ARC) for support for this research. WTC and DG are grateful to the EEC for a Stimulation Action Grant (Ref.: 85200151 U.K. 108UJUI).

6. REFERENCES.

Arnold WM, Zimmermann U (1984) Electric-field induced fusion and rotation of cells. Biol Membranes 5:389-454

Bennett V (1989). The spectrin-actin junction of erythrocyte membrane skeletons. Biochim biophys Acta 988:107-121

Bisch PM, Wendel H, Gallez D (1983) Linear hydrodynamics of viscous thin films. General theory. J Coll Interf Sci 92:105-120

Brandts JF, Erickson L, Lysko, K, Schwartz AT, Taverna RD (1987) Calorimetric studies of the structural transitions of the human erythrocyte membrane. The involvement of spectrin in the A transition. Biochemistry 16:3450-3454

Bredt W, Heunert HH, Hofling KH, Milthaler B (1973) Microcinematographic studies of Mycoplasma hominis cells. J Bacteriol 113:1223-1227
Burgess DR, Prum BE (1982) Re-evaluation of brush border motility: Calcium induces core filament solation and microvillar vesiculation J. Cell Biol 94:97-107
Coakley WT, Bater AJ, Deeley JOT (1978) Vesicle production on heated and stressed erythrocytes. Biochim biophys Acta 512:318-330
Coakley WT, Darmani H, Irwin SA, Robson K and Gallez D (1988) Spatially periodic cell-cell contacts in an erythrocyte model system. Studia Biophys 127:69-74
Coakley WT, Deeley JOT (1980) Effects of ionic strength, serum protein and surface charge on membrane movements and vesicle production in heated erythrocytes. Biochim biophys Acta 602:355-375
Coakley WT, Gallez D (1989) Membrane-membrane contact; involvement of interfacial instability in the generation of discrete contacts. Bioscience Reports, submitted.
Coakley WT, Hewison LA, Tilley D (1985) Interfacial instability and the agglutination of erythrocytes by polylysine. Eur Biophys J 13:123-130
Coakley WT, Nwafor A, Deeley JOT (1983) Tetracaine modifies the fragmentation mode of heated human erythrocytes and can induce heated cell fusion. Biochim biophys Acta 727:293-302
Colwin LH, Colwin AL (1967) Membrane fusion in relation to sperm-egg association. In; Fertilization, vol.1 (eds CB Metz, A Monroy) Academic Press, NY pp. 295-368.
Croop J and Holtzer H (1975) Response of myogenic and fibrogenic cells to cytochalsin B and to colcemid. J Cell Biol 65:271-285
Crum LA, Coakley WT, Deeley JOT (1979) Instability development on heated human erythrocytes. Biochim biophys Acta 554:76-89
Dan JC (1967) Acrosome reaction and lysins: In; Fertilization, vol.1 (eds CB Metz, A Monroy) Academic Press NY pp. 237-294
Darmani H, Coakley WT, Hann AC, Brain A (1989) Spreading of wheat germ agglutinin induced erythrocyte contact area: Evidence for formation of local spatially discrete contacts. Cell Biophys, in press.
Deeley JOT, Crum LA and Coakley WT (1979) The influence of temperature and incubation time on deformability of human erythrocytes. Biochim biophys Acta 554:90-101
Deeley JOT, Coakley WT (1983) Interfacial instability and membrane internalization in human erythrocytes heated in the presence of serum albumin. Biochim biophys Acta 727:293-302
Dimitrov DS (1982). Instability of thin liquid films between membranes. Colloid & Polymer Sci 260:1137-1144
Dimitrov DS, Jain RK (1984) Membrane stability. Biochim biophys Acta 779:437-468
Doulah FA, Coakley WT (1984) Intrinsic electric fields and membrane bending. J Biol Phys 12:44-51
Evans EA (1983) Bending elastic modulus of red blood cell membrane derived from buckling instability in micropipette aspiration tests. Biophys J 43:27-30
Felderhof BV (1968) Dynamics of free liquid films. J Chem Phys 49:44-51
Fowler VM (1986) Cytoskeleton: New views of the red cell network. Nature 322:777-778

Fricke K, Sackmann E (1984) Variation of frequency spectrum of the erythrocyte flickering caused by aging, osmolarity, temperature and pathological changes. Biochim biophys Acta 803:145-152
Gallez D (1983) Repulsive stabilization in black lipid membranes. Biophys Chem 18:165-179
Gallez D, Coakley WT (1986) Interfacial instability at cell membranes. Prog Biophys molec Biol 48:155-199
Gallez D, Prevost M, Sanfeld A (1984) Repulsive hydration forces between charged lipidic bilayers. A linear stability analysis. Colloids Surf 10:123-131
Grieg RG, Brooks DE (1979) Shear-induced Concanavalin A agglutination of human erythrocytes. Nature 282:738-739
Goldin M, Yerushalmi J, Pfeffer R, Shinnar R (1969) Breakup of a laminar jet of viscoelastic fluid. J Fluid Mech 38:689-711
Hewison LA (1988) Spatial periodicity of cell-cell contact: An interfacial instability approach. Ph D Thesis, Univ of Wales.
Hewison LA, Coakley WT, Meyer HW (1988) Spatially periodic discrete contact regions in polylysine-induced erythrocyte-yeast adhesion. Cell Biophys 13:151-157
Hochmuth RM, Mohandas N, Blackshear PL (1973). Measurement of the elastic modulus for red cell membrane using a fluid mechanical technique. Biophys. J. 13:747-762
Israelachvili JN, McGuiggan PM (1988) Forces between surfaces in liquids. Science 24:795-800
Jain RK, Maldarelli C and Ruckenstein E (1978) Onset of microvilli in normal and neoplastic cells. AIChE Symposium Series, Biorheology 74:120-124
Katchalsky A, Danon D, Nevo A, de Vries A (1959) Interactions of basic polyelectrolytes with the red blood cell 2. Agglutination of red blood cells by polymeric bases. Biochim biophys Acta 33:120-138
Klein HP, Stockem W (1979) Pinocytosis and locomotion of amoebas. Dynamics and motive force generation during induced pinocytosis in A. proteus. Cell Tissue Res 197:263-279
Knutton S, Lloyd DR, McNeish A (1987) Adhesion of enteropathogenic Escherichia coli to human intestinal enterocytes and to cultured human intestinal mucosa. Infect Immunity 55:69-77
Krstic RV (1979) Ultrastructure of the mammalian cell, Springer Verlag, New York
Marsland D (1964) Primitive Motile Systems. In: Cell Biology (Allen RD and Kamiya N eds) pp 173-185, Academic Press, NY.
Miller CA, Scriven LE (1970). Interfacial instability due to electrical forces in double layers. J Colloid Interfac Sci 33:360-370
Murphy CL (1966). Thermodynamics of low tension and highly curved interfaces. Ph D Thesis, University of Minnesota, Minneapolis, University Microfilms, Zerox, Ann Arbor.
Neumann E (1988) The electroporation hysteresis. Ferroelectrics 77:1-9
Ournisson G, Rohmer M, Porolla K (1987) Prokaryotic hopanoids and other polyterpenoid sterol surrogates. Ann Rev Microbiol 41:301-333
Rotrosen D, Edwards JE Jr, Gibson TR, Moore JC, Cohen AH, Green I (1985) Adheherence of Candida to cultured vascular endothelial cells: Mechanisms of attachment and endothelial penetration. J. Infect Dis 152:1264-1274
Razin S (1985) Molecular biology and genetics of mycoplasma

(mollicutes). Microbiol Rev 49:419-454
Russell L, Peterson R, Freud M (1979) Direct evidence for formation of hybrid vesicles by fusion of plasma and outer acrosomal membranes in boar spermatozoa. J Exp Zool 208:41-56
Sanfeld A, Steinchen A, Hennenberg M, Bisch PM, van Lamsweerde-Gallez D, Dalle-Vedove W (1979). In; Lecture Notes in Physics 105:168-204
Segel LA, Volk T, Geiger B (1983) On spatial periodicity in the formation of cell adhesions to a substrate. Cell Biophys 5:95-104
Skalak R, Zarda PR, Jan K-M, Chien S (1981) Mechanics of rouleau formation Biophys J 35:771-781
Steinchen A, Gallez D, Sanfeld A (1982) A viscoelastic approach to the hydrodynamic stability of membranes. J Coll Interf Sci 85:5-15
Stenger DA, Hui SW (1986) Kinetics of ultrastructural changes during electrically induced fusion of human erythrocytes. J Mem Biol 93:43-53
Sugar I, Forster W, Neumann E (1987) Model of cell electrofusion. Membrane electroporation, pore coalescence and percolation. Biophys Chem 26:321-335
Tilley D, Coakley WT, Gould RK, Payne SE and Hewison LA (1987) Real time observations of polylysine, dextran and polyethylene glycol induced mutual adhesion of erythrocytes held in suspension in an ultrasonic standing wave field. Eur Biophys J 14:499-507
Tilney LG, Hiramoto Y, Marsland D (1966) Studies on the microtubules in Helizoa. III. A pressure analysis on the role of the structures in the formation and maintenance of the axopodia of Actinosphaerium nucleofilum. J Cell Biol 29:77-95
van Oss CJ, Coakley WT (1988) Mechanisms of successive modes of erythrocyte stability and instability in the presence of various polymers. Cell Biophys 13:141-150
Vasiliev JM (1985) Spreading of non-transformed and transformed cells. Biochim biophys Acta 780:21-65
Vasiliev JM (1987) Actin cortex and microtubular system in morphogenesis: cooperation and competition. In; Cell Behaviour: Shape, Adhesion and Motility (Heaysman JEM, Middleton CA, Watt FM, eds) J Cell Sci; Supplement 8.
Zimmermann U (1982) Electric field-mediated fusion and related electrical phenomena. Biochim biophys Acta 694:227-277

Surface free energy contribution to cell interactions

C.J. van Oss

Departments of Microbiology and Chemical Engineering,
State University of New York at Buffalo
Buffalo, NY 14214, U.S.A.

INTRODUCTION

The quantitative distinction between, and the measurement of, apolar and polar surface tension components and parameters of liquids and solids has become possible in the late 1980s. Using the Lifshitz approach, it could be demonstrated that the surface tension components associated with the three electrodynamic (van der Waals) interactions, i.e., the dispersion (London), induction (Debye) and orientation (Keesom) forces, follow the same equations and should be treated in the same manner (Chaudhury, 1984; van Oss et al., 1988a). These are the apolar interactions which, together, are designated as Lifshitz-van der Waals (LW) interactions. Once this matter was clarified, the entirely distinct polar, i.e., electron-acceptor/electron-donor, or Lewis acid-base (AB) interactions could be studied as separate phenomena that follow a different set of equations (van Oss et al., 1987a; 1988a,b).

It soon became evident that the surfaces of many common water-soluble polymers (e.g., polyethylene glycol) and biopolymers (e.g., polysaccharides, hydrated proteins, sialoglycoproteins) are very strong hydrogen-acceptors, and very feeble, to zero, hydrogen-donors. The surfaces of all such materials with a very asymmetrical polarity have been designated as monopolar surfaces (van Oss et al., 1987a). Compounds with strongly monopolar surfaces (which usually are Lewis acids), will repel each other, when immersed or dissolved in water, even in the absence of any electrical surface charge. That monopolar repulsion, which is instigated by short-range hydrogen-bonds, is transformed into a much longer-range repulsive interaction by means of the orientation of water molecules of hydration (van Oss and Good, 1988a), an effect which is familiar to many as hydration pressure (Gruen et al., 1984; Parsegian et al., 1985). Monopolar repulsion of polymers and biopolymers in aqueous media is at the origin of:

the pronounced solubility of polysaccharides, nucleic acids and hydrated proteins (van Oss et al., 1986; 1987a), the high osmotic pressure and its virtual molecular weight indepence in the cases of polyethylene glycols and dextrans (van Oss et al., 1989a), the phase separation (coacervation) in aqueous solutions of polymer mixtures (van Oss et al., 1987b), the stabilization of particles by polymers, and of cells by biopolymers of the glycocalix (van Oss, 1989b), and the protection against phagocytic ingestion of pathogenic bacteria which are enveloped in slimy capsules of hydrated polysaccharide or protein (van Oss, 1978; 1986).

On the other hand, polymers, particles or cells with less strongly polar, as well as with apolar surfaces will, when immersed in water, attract each other through hydrophobic interactions. By means of contact angle measurements with a number of different apolar and polar liquids, the quantitative degree of monopolar repulsion, or of hydrophobic attraction can be determined for any given system, with fair to excellent accuracy for all short-range interactions. From these, longer range interaction energies can then be obtained by means of the decay-with-distance functions pertaining to each of the different types of interaction. Thus energy balance diagrams are generated, which comprise both the DLVO and the non-DLVO, polar forces. It will be seen that the polar (AB) interactions play, quantitatively, the most important role in the energy balance of cell suspensions. Lifshitz-van der Waals (LW) and electrostatic (EL) forces are somewhat less important, and osmotic (OS) interactions sensu stricto, usually are negligible.

THEORY

Principles Applicable to both Apolar and Polar Interfacial Interactions

The free energy of cohesion of material, 1, is:

$$\Delta G_{11} = -2\gamma_1 \qquad [1]$$

where γ_1 is the surface tension of material 1, which can be a solid or a liquid.

The free energy of adhesion between materials 1 and 2, in vacuo, is, according to Dupré (1869):

$$\Delta G_{12} = \gamma_{12} - \gamma_1 - \gamma_2 \qquad [2]$$

The free energy of interaction between materials 1 and 2, immersed in liquid 3, is (cf. Dupré's equation [2]):

$$\Delta G_{132} = \gamma_{12} - \gamma_{13} - \gamma_{23} \qquad [3]$$

and the interaction between two particles or molecules of material 1, immersed in a liquid 2 then is:

$$\Delta G_{121} = -2\gamma_{12} \qquad [4]$$

Apolar, or Lifshitz-van der Waals (LW) Interactions

The apolar component of the interfacial tension between substances 1 and 2 is expressed as (Girifalco and Good, 1957; Fowkes, 1963; 1967; Good and Elbing, 1970):

$$\gamma_{12}^{LW} = (\sqrt{\gamma_1^{LW}} - \sqrt{\gamma_2^{LW}})^2 \quad [5]$$

From this, ΔG_{12}^{LW}, ΔG_{132}^{LW} and Δ_{121}^{LW} can be derived, using eqs. [2-4], while ΔG_{11}^{LW} follows directly from eq. [1]. In general:

$$\Delta G^{LW} = -A/12\pi\ell^2 \quad [6]$$

for semi-infinite parallel flat plates, situated a distance ℓ apart. Eq. [6] is valid for all situations described by subscripts 11, 12, 132, 131, where the subscripts pertain to ΔG as well as to A. A_{11} is the Hamaker constant of material 1, where material 1 is either in the liquid or solid state. Whereas the Hamaker constant initially was reserved exclusively to van der Waals London, or dispersion forces, we now comprise all three classes of electrodynamic forces into the Hamaker constant (Chaudhury, 1984). These, at least in the unretarded mode, decay with distance at the same rate. Thus, using eq. [1], eq. [6] can be rewritten as:

$$2\gamma_i^{LW} = A_{ii}/12\pi\ell_o^2 \quad [7]$$

in all cases where $\ell = \ell_o$, and ℓ_o is the minimum equilibrium distance of closest approach between molecules or particles of substance i. It was found empirically that for virtually all materials in the condensed state (van Oss et al., 1988a), $\ell_o \approx 1.57Å \pm 0.09Å$, so that Hamaker constants can be obtained directly from measurements of a substance's apolar surface tension component γ_i^{LW}. By using eqs. [2-5], in conjunction with eq. [6], Hamaker constants of the types: A_{12}, A_{132} and A_{131} can also be readily obtained, once γ_1^{LW}, γ_2^{LW} and if necessary γ_3^{LW} are known.

The decay of ΔG^{LW}, for semi-infinite parallel flat plates, with distance thus is:

$$\Delta G_\ell^{LW} = \Delta G_{\ell_o}^{LW} \frac{\ell_o^2}{\ell^2} \quad [8]$$

for non-retarded van der Waals interactions, i.e., for $\ell \leq 100Å$ ($\ell_o \approx 1.57Å$). For two equal spheres of radius R:

$$\Delta G_\ell^{LW} = \Delta G_{\ell_o}^{LW} \frac{\ell_o}{\ell} \pi R \quad [9]$$

and for a sphere of radius R and a flat plate:

$$\Delta G_\ell^{LW} = 2\Delta G_{\ell_o}^{LW} \frac{\ell_o}{\ell} \pi R \quad [10]$$

In eqs. [9-11], $\Delta G_{\ell_o}^{LW}$ is found from eqs. [2-5], as a function of surface tensions γ_i^{LW} and/or interfacial tensions γ_{ij}^{LW}. Thus, eqs. [8-10] may be used for finding the values of ΔG_{11}^{LW}, ΔG_{12}^{LW}, ΔG_{132}^{LW} and ΔG_{121}^{LW}, for ℓ from ≈1.6Å to

≈100Å. For values of ℓ significantly higher than ≈100Å, ℓ and ℓ_o in eqs. [9-10] become ℓ^2 and ℓ_o^2; ℓ^2 and ℓ_o^2 become ℓ^3 and ℓ_o^3 (eq. [8]) to account for retardation in dispersion (van der Waals-London) interactions, which in most cases represent the major portion of LW interactions, in particular in media such as saline water. In eqs. [8-10], ΔG^{LW} pertains to the flat parallel plate configuration.

Polar, or Hydrogen-bonding (Brønsted, or Lewis acid-base) (AB) Interactions

The apolar and the polar surface tension components are additive (van Oss et al., 1988b):

$$\gamma = \gamma^{LW} + \gamma^{AB} \quad [11]$$

However the polar interactions are fundamentally asymmetrical (van Oss et al., 1988a;b), so that it is necessary to express the polar (AB) interfacial free energy between substances 1 and 2 in terms of the products of their electron-acceptor (γ^+) and electron-donor (γ^-) surface tension parameters as:

$$\Delta G_{12}^{AB} = -2\left(\sqrt{\gamma_1^+\gamma_2^-} + \sqrt{\gamma_1^-\gamma_2^+}\right) \quad [12]$$

so that:

$$\Delta G_{11}^{AB} = -4\sqrt{\gamma_1^+\gamma_2^-} \quad [13]$$

In view of eq. [2], the polar component of the surface tension of a substance i then can be defined as:

$$\gamma_i^{AB} = 2\sqrt{\gamma_i^+\gamma_i^-} \quad [14]$$

Thus, the two polar surface tension parameters of γ^{AB} are, clearly, not additive (van Oss et al., 1988b). From the Dupré equation [2] written as:

$$\gamma_{12}^{AB} = \Delta G_{12}^{AB} + \gamma_1^{AB} + \gamma_2^{AB} \quad [2A]$$

and eqs. [2] and [14], one finds for the polar component of the interfacial tension:

$$\gamma_{12}^{AB} = 2\left(\sqrt{\gamma_1^+\gamma_1^-} + \sqrt{\gamma_2^+\gamma_2^-} - \sqrt{\gamma_1^+\gamma_2^-} + \sqrt{\gamma_1^-\gamma_2^+}\right) \quad [15]$$

which may also be written as:

$$\gamma_{12}^{AB} = 2\left(\sqrt{\gamma_1^+} - \sqrt{\gamma_2^+}\right)\left(\sqrt{\gamma_1^-} - \sqrt{\gamma_2^-}\right) \quad [15A]$$

Thus, when:

$\gamma_1^+ > \gamma_2^+$ and $\gamma_1^- < \gamma_2^-$ or when: $\gamma_1^+ < \gamma_2^+$ and $\gamma_1^- > \gamma_2^-$

γ_{12}^{AB} will be negative. (It should be noted that γ_{12}^{LW} can never become negative: eq. [5]). Combining eqs. [5] and [15], the total interfacial tension can then be expressed as:

$$\gamma_{12} = \left(\sqrt{\gamma_1^{LW}} - \sqrt{\gamma_2^{LW}}\right)^2 + 2\left(\sqrt{\gamma_1^+\gamma_1^-} + \sqrt{\gamma_2^+\gamma_2^-} - \sqrt{\gamma_1^+\gamma_2^-} - \sqrt{\gamma_1^-\gamma_2^+}\right) \quad [16]$$

As eq. [15] can readily become more negative than eq. [5] is positive, especially in aqueous media, the total interfacial tension γ_{12} can easily become negative (van Oss et al., 1987a). An exceedingly important consequence of this is that when γ_{12} is negative, ΔG_{121} is positive (eq. [4]). When the interfacial free energy between two cells, particles, or macromolecules, 1, immersed in liquid, 2, is positive, this implies a repulsion between the cells, particles, or macromolecules, immersed in that liquid. Thus, in the total absence of any electric surface potential, cells, particles, or macromolecules, with polar (and especially with monopolar) surface properties, immersed in water, can repel each other.

For ΔG_{132}, eq. [3] can be used, using the three interfacial terms, γ_{ij}, as defined in eq. [16]. ΔG_{132}^{LW}, as well as ΔG_{132}^{AB} can be positive, or negative, which therefore also applies to ΔG_{132}. When ΔG_{132} is positive, this means that a repulsion occurs between two different kinds of cells, particles or macromolecules, immersed in a liquid.

When ΔG_{132} is negative, an attraction exists between these two different kinds of cells, particles or macromolecules, immersed in a liquid. A negative value of either ΔG_{121} or ΔG_{132} quantitatively describes the hydrophobic energy of attraction between the same cells, particles, or macromolecules, 1, or different cells, particles or macromolecules, 1 and 2, immersed in a liquid.

Even though hydrogen bonds necessarily interact only at very short distances, which are of the order of 1 or 2Å, polar repulsions (positive ΔG^{AB} values, causing ΔG overall to be positive), as well as polar ("hydrophobic") attractions (negative ΔG values) still are operative at considerable distances away from the interacting interfaces, in aqueous media (Christenson, 1988). For repulsions ($\Delta G^{AB} > 0$), this action at a distance is mediated by "hydration pressure", which is caused by the orientation of water molecules of hydration, which is strongest at close proximity to the polar surface, the monopolarity of which is the primary cause of that orientation (van Oss and Good, 1988). The degree of orientation of the water molecules of hydration decays exponentially with the distance , but with a very sensitive force balance the effects of this "hydration pressure" may still be measurable at $\approx$ 100 nm or more (Israelachvili, 1985).

Whilst for repulsive polar forces in water ($\Delta G^{AB} > 0$) the propagation of the repulsion is effected by the orientation (i.e., an increase in order, concomitant with a decrease in density) of water molecules of hydration (which repel each other), attractive ("hydrophobic") polar forces in water ($\Delta G^{AB} < 0$) are propagated by means of the polar cohesive attraction between the surrounding water molecules (van Oss and Good, 1988b). The hydrophobic attraction appears to be accompanied by an increase in density among the water molecules closest to the apolar (hydrophobic) surface (cf. Derjaguin et al., 1987). The decay of ΔG^{AB} (where ΔG^{AB} may be either positive or negative), for parallel flat plates, with distance ℓ is:

$$\Delta G_{\ell}^{AB} = \Delta G_{\ell_o}^{AB} \exp\left(\frac{\ell_o - \ell}{\lambda}\right) \qquad [17]$$

where λ is the decay length of water. The value of λ has been estimated to be as low as ≈ 0.19 (Chan et al., 1979) to ≈ 0.26 nm (Parsegian et al., 1979), which is close to the radius of gyration of single, non-hydrogen-bonded water molecules, to as high as 13 nm, from empirical determinations (Christensson, 1988; Rabinovich and Derjaguin, 1988). In all probability, λ for normally hydrogen-bonded water at ≈ 20°C has a value between 0.2 and 1 nm (van Oss, 1989b). For the polar part of the interaction between two identical spheres of radius R (when ΔG^{AB} is determined in the flat parallel slab position):

$$\Delta G_{\ell}^{AB} = \Delta G_{\ell_o}^{AB} \pi R \lambda \exp\left(\frac{\ell_o - \ell}{\lambda}\right) \qquad [17a]$$

while for a sphere of radius R and a flat plate (and for two croned cylinders), the value of ΔG^{AB} is twice as high as the value for two identical spheres (eq. [18]). The value of ℓ_o in polar interactions has not been established with anywhere near the accuracy with which the ℓ_o value of apolar (LW) interactions was ascertained (van Oss et al., 1988a), but ℓ_o^{AB} is unlikely to be very different from the value of ℓ_o^{LW}, i.e., 1.5 to 1.6Å. As in LW interactions, ΔG^{AB} is found from eqs. [1-4], as a function of surface tension γ_i^{AB} (eq. [14]) and/or interfacial tension γ_{ij} (eq. [15]). Thus, eqs. [17] and [18] may be used for finding the values of ΔG_{11}^{AB}, ΔG_{12}^{AB}, ΔG_{132}^{AB} and ΔG_{121}^{AB}, for all values of ℓ. In eqs. [17-17a]. ΔG^{AB} pertains to the flat parallel plate configuration.

Osmotic (OS) Interactions

It is useful to ascertain whether, and to which extent, an osmotic interaction energy ΔG^{OS} needs to be taken into account in the total interaction energy ΔG, already comprising ΔG^{LW}, ΔG^{AB} and ΔG^{EL}. The osmotic pressure Π of a solution of a polymer, 1, in an aqueous medium, 2, (and thus also the osmotic pressure exerted by the glycoprotein strands of the glycocalix at a cell surface) may be described as:

$$\Pi = c_1 RT \left(\frac{1}{M_{n_1}} + B_2 c_1 / d^2 + 1/3 \, B_3 \, c_1^2/d_1^3 + \ldots\ldots.. \right) \qquad [18]$$

where:

$$B_2 = \left(\frac{n_1}{M_{n_1}}\right)^2 V_2 \, (0.5 - \chi_{12}) \qquad [19]$$

where c_1 is the fractional concentration of polymer 1, M_{n_1} its number-average molecular weight, n_1 R the gas constant (R = 8.3143 J/degree/mole/m^3), the number of monomer subunits per polymer molecule, V_2 the volume of 1 mole of solvent, d_1 the specific density of the polymer, and χ_{12} a dimensionless parameter (Flory-Huggins) which can be shown to be the product of the free energy of adhesion ΔG_{121} between two polymer molecules, immersed in the solvent and the minimum contactable surface area between two polymer strands (van Oss et al., 1989a).

The first term of eq. [18] represents van't Hoff's equation. It expresses the pressure exerted by the thermal activity of $\frac{c_1 N}{M_{n_1}}$ dissolved polymer molecules of energy k (= 1.38×10^{-21} J), at temperature T (in degrees K), where N = 6.022×10^{23} (R = kN). This part of Π is not included in ΔG_{121}, and may have to be taken into account in the exceptional contingency where M_{n_1} is small and c_1 is very high. To obtain the osmotic pressure sensu stricto (ΔG^{OS}_{ss}) contribution (in energy per unit surface area) from the first term of eq. [18], Π should be multiplied by the radius of gyration of the polymer molecules. The ΔG^{OS}_{ss} contribution to ΔG usually is exceedingly small.

The second term of eq. [18] ($B_2 \frac{c_1^2}{d_1^2} RT$) expresses the attraction, or the repulsion, between single monomeric units situated on different polymer molecules. That attraction, or repulsion, is already incorporated into ΔG^{LW} + ΔG^{AB}. The χ_{12} parameter in B_2 (eq. [19]) is proportional to the work of adhesion between subunits of opposing polymer molecules, and thus automatically comprises $\Delta G^{LW} + \Delta G^{AB}$. At high concentrations, and with monomeric polymer units of low molecular weight, which are endowed with strongly monopolar surface properties (which causes them to repel each other when immersed in water) the χ_{12} parameter is negative, and B_2 then is markedly positive. In such cases the second term of eq. [18] is by far the predonderant one. Thus, concentrated solutions of polyethylene glycol (PEG) have very high osmotic pressures, which are largely independent of the molecular weight of PEG observed (Evans and Needham, 1988; Arnold et al., 1988a; van Oss et al., 1989a). In the case of charged particles, cells, or polyelectrolytes, the electrostatic repulsion also is already incorporated in the total interaction, because then $\Delta G^{TOT} = \Delta G^{LW} + \Delta G^{AB} + \Delta G^{EL}$. In addition, with charged particles, cells, or polyelectrolytes, the excess counterions cause an increase in the first term of eq. [18], as the Donnan contribution to ΔG^{OS}_{ss}. The third term of eq. [18] may almost always be neglected, except possibly for very concentrated PEG solutions (van Oss et al., 1989a; see also Evans and Needham, 1988).

The free energy ΔG^{OS} engendered by the osmotic pressure Π can be obtained by multiplying Π by the radius of gyration of the polymer molecules, Rg:

$$\Delta G^{OS}_{\ell_o} = \Pi Rg \quad [20]$$

The decay vs. distance of ΔG^{OS}, as a first approximation, may be expressed as:

$$\Delta G^{OS}_{\ell} = \Delta G^{OS}_{\ell_o} \exp(-\ell/Rg) \quad [21]$$

Thus, for high molecular weight, asymmetrical polymers, i.e., polymers with a large Rg, the rate of decay of ΔG^{OS} with distance can be exceedingly shallow.

Steric Stabilization

The influence of attached or adsorbed nonionic polymers or surfactants on the stabilization of otherwise unstable hydrophobic particles in (usually aqueous) suspensions was originally explained by the exertion of a "steric repulsion" by the stabilizing nonionic polymers or surfactants (Mackor and van der Waals, 1952). The theoretical constructs which have evolved from this concept (Napper, 1983, has provided a very thorough treatment of these) predate the observation that, as a rule, these stabilizing nonionic monopolar polymers or surfactants strongly repel each other, especially when dissolved in a markedly bipolar solvent such as water (see the Section on Polar Interactions, above). There is one crucial property which all of these stabilizing molecules have in common: to be effective, they must be SOLUBLE in the liquid medium in which the particles are suspended (see, e.g., Napper, 1983). It thus is instructive to determine the conditions which convey solubility on these stabilizing nonionic polymers or surfactants in a given solvent; see van Oss and Good (1989).

If the polymer molecules are attached to the particles (or cells) by simple physical adsorption from a bulk polymer solution, two other things can happen (van Oss et al., 1989d):

I. Depletion Flocculation, when the concentrations of free polymer in solution is higher than the polymer concentration in the adsorbed layer, or;

II. Depletion Stabilization, when the concentration of adsorbed polymer is higher than the concentration of free dissolved polymer.

At low polymer concentration, flocculation due to polymer cross-binding may also occur, if the polymers are sufficiently asymmetrical, and sufficiently large. This is the principal mechanism of rouleau formation in erythrocyte suspension, e.g., in the presence of dextran, or fibrinogen (van Oss and Absolom, 1985; van Oss and Coakley, 1988; van Oss et al., 1989b). The designation "depletion" alludes to the existence of a polymer-free layer between the polymer layer adsorbed onto the particles and cells, and the polymer molecules in free solution (Bäumler and Donath, 1987; Arnold et al., 1988b; van Oss et al., 1989b); various aspects of depletion flocculation and stabilization of inert particles have been treated extensively by Napper (1983). The polymer-free (depletion) layer is maintained by the strong (polar) repulsion between the polymer molecules in aqueous solution; its thickness is of the order of the radius of gyration of the polymer molecules (Bäumler and Donath, 1987; van Oss et al., 1989b).

Except for preponderantly apolar systems, the role of steric stabilization sensu stricto thus is very minor. Nevertheless, the stability of cell suspensions in aqueous media (e.g., in blood) is largely due to the presence of the glycoprotein strands of the cells' glycocalices, which mutually repel each other, through a strong monopolar (often mainly non-electrostatic) repulsion (van Oss, 1989b). Although the polar faces of many varieties of phospholipid bilayers also repel each other, by the same mechanism, it is clearly too much to expect for naked phospholipid bilayer membranes, to provide a stabilizing protection for, e.g., erythrocytes, allowing them to remain in stable circulation for several months, without undergoing encounters at places where there are slight fluctuating impairments in the phospholipid membrane, which would then give rise to hydrophobic attraction, and thus to flocculation. The repulsive action between the glycoprotein strands of the glycocalices of opposing cells, totally obviates this risk, and one may well be tempted to designate this stabilizing action of the glycocalix as "steric stabilization". However, this is not steric stabilization sensu stricto. On the other hand, the glycocalix does represent the outer layer of the cells, and it is this very layer which is first involved in all intercellular interactions (or in their avoidance), and its surface and interfacial properties thus warrant much closer study.

Electrostatic (EL) Interactions

For low to medium low electrokinetic (ζ) potentials (i.e., $\zeta < 25$ mV), as a first approximation:

$$\Delta G^{EL} = \Delta G^{EL} \exp(-\kappa\ ) \quad [22]$$

where the Debye length $1/\kappa$, which is a measure of the effective thickness of the diffuse ionic double layer, expressed as:

$$1/\kappa = \sqrt{\epsilon kT/4\pi e^2 \Sigma v_i^2 n_i} \quad [23]$$

where ϵ is the dielectric constant of the liquid medium ($\epsilon = 80$ for water), k is Boltzmann's constant ($k = 1.38 \times 10^{23}$ J/°K), T the absolute temperature in degrees K, e is the charge of the electron ($e = 4.8 \times 10^{-10}$ e.s.u., or 1.6×10^{-19} C). The decay with distance of ΔG^{EL} is very much dependent on the ionic strength, μ, which is equal to $\frac{1}{2} \Sigma\, v_i^2 n_i$, where v_i is the valency of each ionic species and n_i the number of ions of each species per cm³ of liquid. The value of $1/\kappa$ is 1 nm at $\mu = 0.1$, and 10 nm at $\mu = 0.001$; at the physiological $\mu = 0.15$, $1/\kappa = 0.8$ nm.

For two flat parallel surfaces separated by a distance

$$\Delta G^{EL} = 1/k.64nkT\gamma_o^2 \exp(-\kappa\ell) \quad [24]$$

where n is the number of counterions per cm³ and:

$$\gamma_o = \frac{\exp\ [(v_c e\psi_o/2kT)-1]}{\exp\ [(v_c e\psi_o/2kT)+1]} \qquad [25]$$

where v_c is the valency of the counterions and ψ_o the electrostatic potential of the particle or cell surface, in electrostatic Volts. The value of ψ_o cannot be directly determined, but the surface potential at the slipping plane, or ζ-potential can be determined, by electrokinetic methods (see Methods of Determination - EL Interactions, below). ψ_o can be obtained from ζ according to:

$$\psi_o = \zeta\ (1 + z/a)e^{kz} \qquad [26]$$

where z is the distance from the charged particles' surface to the slipping plane (usually $z \approx 0.5$ nm) and a is the Stokes radius of the particles or cells. For 2 equal spheres, with radius R:

$$\Delta G^{EL} = \tfrac{1}{2}\epsilon R\psi_o^2\ n[1+\exp(-\kappa\ell)] \qquad [27]$$

and for a sphere, with radius R and a flat plate (as well as for two croned cylinders with Radius R):

$$\Delta G^{EL} = \epsilon R\psi_o^2\ n[1+\exp(-\kappa\ell)] \qquad [28]$$

In all these cases (eqs. [24,27,28]), it is not necessary first to calculate ψ_o, from ζ (eq. [26]). If ℓ is taken to start at the slipping plane, instead of ψ_o, the value of ζ (in electrostatic volts), may be used.

METHODS OF DETERMINATION

Interfacial (IF), or LW + AB Interactions

Interfacial interactions energies and their components are measured by (advancing) contact angle (θ) determinations with drops of a liquid L, deposited on a solid S (Good, 1979). Expressing Young's equation (Young, 1805) as:

$$(1 + \cos\theta)\gamma_L = -\Delta G_{SL}^{IF} \qquad [29]$$

which in the light of the additivity of apolar and polar forces (van Oss et al., 1987a, 1988a,b) may be written as:

$$(1 + \cos\theta)\gamma_L = -\Delta G_{SL}^{LW} - \Delta G_{SL}^{AB} \qquad [30]$$

inserting the expressions for ΔG_{SL}^{LW} (eqs. [2,5]) and for ΔG_{SL}^{AB} (eq. [12]) into eq. [30], Young's equation may be expressed in terms of surface tension components and parameters as:

$$(1 + \cos\theta)\gamma_L = 2\ (\sqrt{\gamma_S^{LW}\gamma_L^{LW}} + \sqrt{\gamma_S^{+}\gamma_L^{-}} + \sqrt{\gamma_S^{-}\gamma_L^{+}}) \qquad [31]$$

By contact angle measurement with three different lqiuids L, each with known values of γ_L^{LW}, γ_L^{+} and γ_L^{-}, the three equations thus obtained can be solved for the three unknown entities γ_S^{LW}, γ_S^{+} and γ_S^{-} (van Oss et al., 1987a; 1988a,b).

An extremely useful set of liquids, with known values for γ_L^{LW}, γ_L^{+}, and γ_L^{-} are given in Table I.

Table I. Apolar and Polar Surface Tension Components and Parameters of Liquids Used in Contact Angle Determination, in mJ/m², at 20°C

Liquid	γ	γ^{LW}	γ^{AB}	γ^{+}	γ^{-}
Diiodomethane	50.8	50.8	0	0[a]	0
α-Bromonaphthalene	44.4	44.4	0[a]	0[a]	0[a]
Water	72.8	21.8[b]	51[c]	25.5[d]	25.5[d]
Glycerol	64	34[c]	30[c]	3.92[e]	57.4[e]
Formamide	58	39[c]	19[c]	2.28[e]	39.6[e]
Ethylene Glycol	48	29[c]	19[c]	3.0[e]	30.1[e]

[a] These values may not be complete zero, but they are so small that they may be neglected in most cases, so that these two liquids may be taken as exclusively apolar, for all practical purposes. [b] From Fowkes (1963) calculations, based on the data compiled by Girifalco and Good (1957). [c] Chaudhury (1984). [d] These values cannot be measured with any known method; they are taken to be equal by convention (van Oss et al. (1987a), and the values of γ^{+} and γ^{-} for all other materials are expressed, relative to these standard values of water at 20°C. [e] van Oss, et al. (1989c).

Once γ_S^{LW}, γ_S^{+} and γ_S^{-} have been determined for a given substance S, the various interfacial tension and free energy value, can be obtained via eqs. [12,14,16, 3,4]. The apolar free energies, components and parameters can be obtained separately via eqs. [3,4,5-10], as can the polar free energies, components and parameters, using eqs. [12,13,15,17,18,3,4].

OS Interactions

The osmotic pressure sensu stricto (Π_{SS}) can be obtained directly from the concentrations and the molecular weight of the polymer, using the first (van't Hoff) term of eq. [18]. Osmotic pressures can be measured by membrane methods (Hermans, 1948), or by various approaches to measuring differences in vapor pressure (Hermans, 1948; Gawrisch, 1986). To obtain M_{n_1}, Π/RTc_1 should be plotted vs. c_1. The intercept on the ordinate then corresponds to $1/M_{n_1}$. If the curve Π/RTc_1, or Π/c_1 vs. c_1 is a straight line, this indicates that only the van't Hoff part of eq. [18] is operative, i.e., $\Pi = \Pi_{SS}$. If that curve is not linear, but curves upward with c_1, this indicates that B_2 is positive, i.e., that the polymer is dissolved in a "good" solvent (Napper, 1983).

EL Interactions

Cell electrophoretic mobilities can be measured by a variety of microelectrophoresis devices, see, e.g., Seaman and Brooks (1979) (microelectrophoresis at the stationary layer); van Oss et al. (1974); van Oss and Fike (1979) (microelectrophoresis with uniform electroosmotic backflow). For cells, with a fairly low ζ-potential and a rather large size, compared to the Debye thickness, $1/\kappa$, the ζ-potential is obtained from the electrophoretic mobility u, as:

$$u = \frac{\zeta \epsilon}{6\pi\eta} \qquad [32]$$

where η is the viscosity of the liquid medium, expressed in poises (Overbeek and Wiersema, 1967; van Oss, 1975, Overbeek and Bijsterbosch, 1979).

CELL INTERACTIONS

Influence and Thickness of the Glycocalix

Many cells, and especially most peripheral blood cells, have a substantial glycocalix. On blood cells the glycocalix consists of probably fairly evenly spaced glycoprotein and glycolipid strands, in which the carbohydrate moieties are mainly distal. The most important human blood group epitopes (A, B, H, Le, etc.) are well-characterized oligosaccharides (Cunningham, 1984); these are interspersed with sialic acid strands, which are the main bearers of the negative charge of blood cells.

Relatively few data exist concerning the length of the glycocalix strands of blood cells, in aqueous suspension. Winzler (1972) gives data on the isolated glycocalix glycoprotein: the sedimentation constant is 1.5S and the diffusion constant $D = 10.65 \times 10^{-7}$ cm^2/sec. From this a molecular weight of about 9,500 can be calculated. From these values for S and D one can also obtain the friction factor ratio $f/f_o = 1.507$, which yields (see Svedberg and Pedersen, 1940) an oblong asymmetry ratio ($1/\rho \approx 9.3$). From the specific volume $\upsilon = 0.632$ (Winzler, 1972) of the material, and the molecular weight, one would arrive at a cylinder with a diameter of 1.1 nm and a length, h, of 10.3 nm.

In confirmation of that value for h, recent work on the influence of IgG-class antibodies on the electrophoretic mobility of various mammalian blood cells (Cohly et al., 1985a; 1985b, Cohly, 1986) would indicate that the length h is slightly less than 13 nm. This can be deduced from the fact that IgG-class antibodies, which are about 13 nm long (Valentine and Green, 1967), have only a small influence on the cells' surface potential when attached to the cell membrane of blood cells, i.e., when largely hidden inside the glycocalix (Cohly, 1986). However, when a second IgG-class antibody is attached to the Fc moiety of the original antibody a reduction in cell electrophoretic mobility, u, of about 18% results (Cohly et al., 1985a; 1985b; Cohly, 1986). From the very slight (but not quite zero) influence on the cells' electrophoretic mobility, u, of IgG-class antibodies attached to cell membrane epitopes one arrives at the conclusion that the length, h, of the glycocalix strands is very close to the length of unopened IgG-class antibodies, i.e.; h is slightly larger than 10 nm. We, may thus accept the value for h obtained from hydrodynamic measurements, i.e., h ≈ 10.3 nm, as reasonably accurate.

Even with erythrocytes (ER), which, as mammalian cells go, have a fairly high surface or ζ-potential (for human ER, $\zeta = -18$mV), the influence of the electrostatic repulsion between cells accounts for only one or two percent of the total intercellular repulsion, which is responsible for cell stability (van Oss, 1989b). For leukocytes the relative influence of electrostatic repulsion is even smaller; the ζ-potential alone of leukocytes would not suffice to assure their stability in suspension: without the polar repulsion of their glycocalices, leukocytes would quickly aggregate irreversibly (van Oss, 1989b). Thus with cells it is indispensable, when devising an energy balance (i.e., a plot of ΔG^{TOT}_{121} _versus_ distance, ℓ), to include the polar interaction energy, ΔG^{AB}:

$$\Delta G^{TOT}_{121} = \Delta G^{EL}_{121} + \Delta G^{LW}_{121} + \Delta G^{AB}_{121} \qquad [32]$$

instead of the classical DLVO approach (Derjaguin and Landau, 1941; Verwey and Overbeek, 1948), which does not include the (in aqueous media all-important) ΔG^{AB}_{121} term. The ΔG^{OS}_{ss} term is almost invariably negligible; see above, under OS interactions.

Use of IgG Molecules for Estimating λ

The rates of decay with distance, ℓ, of ΔG^{EL} (eq.[22]) and ΔG^{LW} (eq.[9]) are well-known. The equation for the rate of decay with distance of ΔG^{AB} is also known (eq.[18]), but the parameter λ, the decay length of water at 20°C, which is somewhere between 0.2 and 1.0 nm, has not yet been more precisely established. However, the distance to which two erythrocytes can approach each other is, owing to the importance of ΔG^{AB} in that interaction, strongly linked to the value of λ. And that distance can be estimated, by means of the yardsticks provided by IgG and IgM molecules. Antibodies of the IgG-class, directed to the Rh_o (D) epitope (antigenic determinant) of Rh_o (D)-positive human ER have a distance between their too paratopes (antibody-active sites) that is just two small to achieve crosslinking two Rh_o (D)-positive ER. Given the known maximum distance between the two paratopes of IgG, i.e., ≈ 15 nm (Valentine and Green, 1967). The distance of closest approach between the cell <u>membranes</u> of two ER must be slightly more than 15 nm. However, the energy balance between two ER is determined primarily between the EL, LW and AB interactions between the distal edges of their glycocalices. It thus is essential to take the thickness of the glycocalix into account, when determining the closest distance of approach between cells. It is the distance between glycocalix rims, which is essentially governed by the three components of ΔG_{121}^{TOT} (see eq. [32]), and their rates of decay with distance (eqs. [8, 9, 17, 18, 24, 27]). The closest distance of approach is taken to be the distance at distance at which the repulsive energy between two cells is equal to their energy of Brownian motion, i.e., $\Delta G_{121}^{TOT} \approx + 1.5kT$. At that distance, the repulsive energy between opposing glycoprotein strands, in terms of energy per glycoprotein pair, is still relatively slight, but of sufficient importance to cause a significant reduction in the protruding length, h, of the glycocalix strands. Given that h is approximately 10.3 nm and that the minimum length of ℓ, even at the lowest value for λ (see Figure 1), is 3.7 nm, and given that only a slight decrease in ℓ suffices to allow IgG (with a "reach" of 15 nm, see Figures 2 and 3) to crosslink two red cells, the

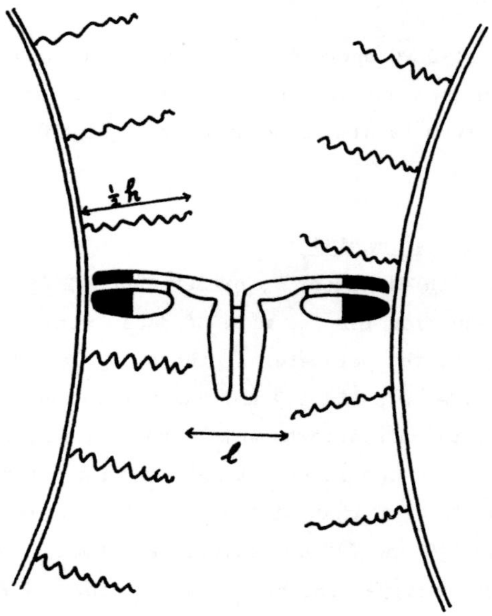

Figure 1. Schematic presentation of parts of two human red cells (A, Rh_o-positive), approaching each other, under the influence of their Brownian motion, to a minimum distance (between cell membranes) of slightly over 15 nm. Under these conditions the glycoproteins strands of the glycocalix (originally about h ≈ 10.3 nm long) are compressed to about half that length (= ½ h). This leaves an effective minimum distance between the outer tips of the glycoprotein strands of about ℓ = 5 nm. The dimensions of a completely opened IgG-class antibody molecule, with a maximum "reach" of 15 nm are just insufficient allow crosslinking.

distance $\ell + 2h$ must be at least reduced by about 8 or 9 nm (at the smallest value for λ), at closest normal encounter between two erythrocytes, which is about a 40% reduction of h, as a minimum. As a maximum, the strands are most unlikely to be shortened to less than their radius of gyration (Israelachvili, 1985), or a reduction to $1/\sqrt{12}$ times their original length (Hiemenz, 1984) i.e., a reduction of 71%. A reduction of 71% (which is only attainable under conditions of very high outside pressure, see Israelachvili, 1985), leading to h = 3 nm, would give rise to a total length of only slightly more than $\ell = 9$ nm, which indicates that for water, λ = 1.0 nm would be the highest possible value.

IgM, with a "reach" of about 27.4 nm (Dorrington and Mihaesco, 1970) should of course be able to cross-bind two erythrocytes under any circumstances, as is indeed the case. Thus only IgG, and not IgM, can be used as a feasible yardstick for erythrocyte interactions; it allows to estimate the upper limit λ (water) can have (i.e., 1.0 nm). The lower limit of λ is harder to measure. Nevertheless, a reasonable estimate can be made of the most likely value for λ (water). As the minimum reduction of the length, h, of the glycocalix strands (under the conditions of mean closest encounter of two erythrocytes, at $\Delta G_{121} \approx + 1.5$ kT) is 40%, and the maximum reduction of h is 71%, an actual reduction of h of ≈ 50% seems to be plausible; see Figure 2.

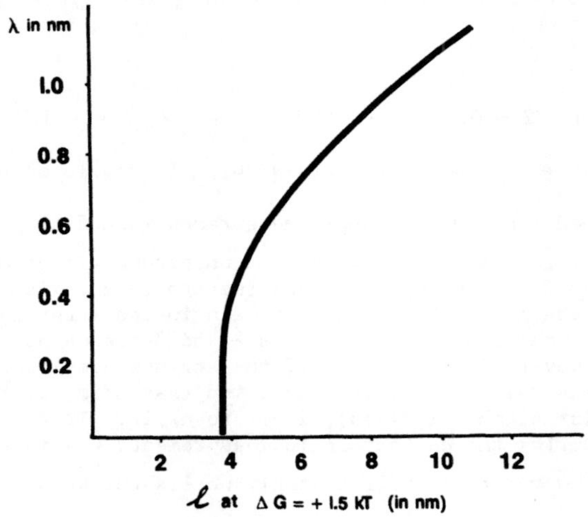

Figure 2. Plot of the putative decay length (λ) of water at 20°C vs. the corresponding distance ℓ between the outer edges of the glycocalices of human erythrocytes, where ℓ is computed taking ΔG^{EL}, ΔG^{LW} and ΔG^{AB} into account, at $\Delta G = + 1.5$ kT; see Figure 3. From the hemagglutination data with IgG-class antibodies ℓ may be estimated as having to be just over 5 nm, which would give a most likely value for λ of ≈ 0.6 nm (see text).

That would correspond to $\ell \geqslant 15 - 1/2 \times 2h \geqslant 15 - 10.3 \geqslant 4.7$ nm. Thus a value for λ (water) = 0.6 nm, which corresponds to $\ell \approx 5.1$ nm (see Figures 2 and 3), would be the most plausible one. A value of λ (water) = 0.6 nm is also plausible in view of the structure of liquid water. From the molecular dimensions of solitary water molecules, a value of λ (water) $\approx$ 0.2 nm would seem appropriate (Chan et al., 1979; Parsegian et al., 1979). However, considering that about 10% of the water molecules are linked by hydrogen bonds to one or several other water molecules, which causes the energy of cohesion of cohesion of water at 20°C to be for $\gamma_w^{AB} / \gamma_w = \frac{51}{72.8} = 70\%$ due to hydrogen-bonding, a considerable part of the water molecules in liquid water forms, at any one time, an array of several water molecules. Thus λ (water) = 0.6 nm is a value which also appears reasonable from the viewpoint of cohering water molecules. The much higher values for λ obtained by force balance measurements, which have been quoted recently (see e.g., Christenson, 1988), most likely are due to the influence of micro or macrosolutes that achieved a less than complete attachment to the mica or quartz surfaces used in these devices.

Energy Balance

Using the value of 0.6 nm for λ (water), estimated in the preceding section, and the values and conditions for normal human erythrocytes described earlier*, energy balances were constructed a) for erythrocytes approaching

---

*Ionic strength $\Gamma/2 = 0.15$, T = 293°K, pH = 7.2, ζ = -18mV, $\Delta G_{121}^{LW} = -0.56$ mJ/m^2 and $\Delta G_{121}^{AB} = +25$ mJ/m^2 (van Oss, 1989b). It should be noted that ΔG_{121}^{AB} was obtained from contact angle measurements on flat layers of human erythrocyte stromata, deposited by suction on porous silver membranes. It is not easily verified, during such measurements (e.g., with drops of water), whether the glycocalix strands remain in the stretched-out position, or are somewhat flattened. Figures 2 and 3 are based on the first presumption. However, if flattening of the strands does occur during contact angle measurement, it must be assumed that h is, at the most, reduced to Rg (Israelachvili, 1985), i.e., to $h/\sqrt{12}$. If the thickness of the glycocalix during the contact angle measurements was indeed as small as $h/\sqrt{12}$, then the value of ΔG_{121}^{AB} at a glycocalix thickness = h would be reduced by a factor $(1/\sqrt{12})^2 = 1/12$, because the value of ΔG_{121}^{AB} is governed by the second term of the osmotic pressure function, and thus is proportional to c_1^2 (see eq. [18]). Similarly, the value of ΔG_{121}^{AB} at 1/2 h then is $(\frac{2}{\sqrt{12}})^2 = 1/3 \times \Delta G_{121}^{AB}$ found by contact angle measurement on

each other via their convex edges of R = 1.5 μm and b) for erythrocytes approaching each other in the rouleau position, assuming that the area of the facing surfaces of closest approach is formed by a flat ring with an outer radius of 3.5 μm and an inner (doughnut) radius of 2 μm, leaving a contactable surface area S_c of 25.9 μm^2.

Figure 3 shows the energy balance diagrams for human erythrocytes in conformations a (left) and b (right). It is clear that at the secondary minimum of attraction (vertical arrows), red cells will seek to approach each other, at equilibrium, at a much higher level of attractive energy in the flat/flat conformation (b). Thus, at equilibrium red cells will "prefer" to approach each other in the flat/flat position, rather than with opposing convex edges, even though the distance, at equilibrium, between cells in the flat/flat position is greater than at the convex/convex confirmation. Nevertheless, various convective hydrodynamic disturbances will not, normally, allow the permanent formation of arrays, or stacks of red cells (or "rouleaux") in the flat/flat position. However, "rouleaux" do become readily observable in red cell suspensions to which long asymmetrical cross-binding polymers (e.g., fibrinogen, polymerized albumin, etc.) have been added (van Oss and Absolom, 1985). The polymers cross-bind the cells just sufficiently solidly in the flat/flat position, to stabilize them into rouleaux. Vigorous shaking, however, causes the rouleaux to disintegrate. Individual rouleaux formed out of red cell mixtures originating from different species (e.g., human and rabbit red cells mixed together), each turn out to consist of cells of only one species, while a neighboring rouleau then consist of only cells of the other species from the mixture (Sewchand and Canham, 1976). The explanation of this cell-sorting phenomenon lies in the fact that the energy levels at the wells of attraction at the secondary minimum differ significantly between the different species (van Oss and Absolom, 1985).

The differences between the two curves (Figure 3) may also serve to explain why, when outside pressure is exerted on red cell rouleaux, e.g., by

---

flattened glycocalix strands, in which case ΔG^{AB}_{121} at 1/2 h would have been 8.33 instead of 25 mJ/m^2. Using the value of 8.33 instead of 25 mJ/m^2 would only slightly displace the +1.5 kT point (on the left hand curve of Figure 3), from 4.3 to 4.0 nm. In Figure 2 this would displace the upper part of the curve very slightly to the left, and allow a maximum value of about 1.1 nm for λ (water), instead of 1.0 nm. The most plausible value for λ (water) then would still be about 0.66 nm. The smallness of the deviation, in the case of fairly different assumptions about the thickness of h during contact angle measurement, is due to the high value of ΔG^{AB}_{121}, compared to ΔG^{LW}_{121} and ΔG^{EL}_{121}.

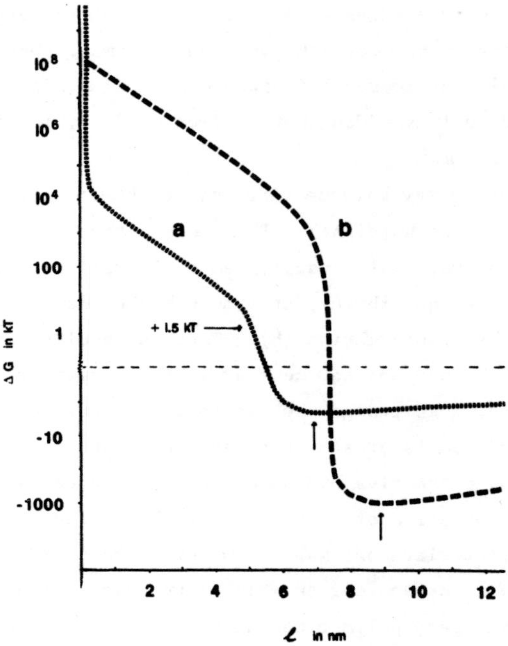

Figure 3. Energy balance of human erythrocytes, taking ΔG^{EL}, ΔG^{LW} <u>and</u> ΔG^{AB} into account. The decay length of water is taken to be λ = 0.6 nm (see Figure 2, and text). The left-hand curve (a) is computed by assuming a radius of curvature at the approach of two cells of R = 1.5 μm. The right-hand curve (b) is based on the interaction between two cells in the flat parallel slab mode, with a surface area of approach S = 25.9 μm^2. The vertical arrows indicate the secondary minima of attraction of the two modes a and b. A horizontal arrow indicates the place on curve (a) of ΔG = + 1.5 kT, corresponding to $\ell \approx$ 5.0 nm, which is the most likely minimum distance between glycocalix surfaces of two red cells approaching each other via their convex edges, under the influence of their Brownian motion. The relatively slight elastic repulsion engendered by the cell encounters at ΔG = + 1.5 kT is not taken into account here.

adding ≈ 2% (w/v) dextran (MW ≈ 500,000), the rouleaux break up, but a number of individual cells remain attached to each other in a hingeing fashion, usually in pairs, by their concave edges (Tilley <u>et al</u>., 1987; van Oss and Coakley, 1988). From Figure 3 it becomes evident that when cells which normally adhere in the flat/flat position, at the secondary minimum (right hand vertical arrow), instead are forced (by outside pressure) to get closer together, their most stable attractive configuration will shift to the edge/edge position at the convex/convex secondary minimum (left hand vertical arrow).

The lack of a primary minimum of attraction in either configuration (Figure 3) shows that however much of an exterior force is applied to human red cells, it will be impossible to make such cells adhere to each other. It could be demonstrated that it is indeed impossible to destabilize red cells by ultracentrifugation, exerting 296,000 times gravity (van Oss, 1985).*

The admixture of Ca^{++} to cell suspensions can bring the cells close enough together to cause cell fusion, by a combination of two mechanisms: 1) Ca^{++} causes a marked decrease in the cells' negative ζ-potential, and thus in the positive value of ΔG^{EL}; this effect does not, however, quantitatively account for the decrease in hydration pressure, engendered by the Ca^{++} ions (Prévost and Gallez, 1984), but: 2) Ca^{++} also makes the cells more hydrophobic (Ohki 1982), as Ca^{++} ions are electron-acceptors, which neutralize the electron-donor preponderance of the glycocalix, and thus also cause a strong decrease in the positive value of ΔG^{AB} (van Oss et al., 1988c). These two effects, combined, cause the cells to attract each other, thus favoring fusion.

REFERENCES

Arnold K, Herrmann A, Gawrisch K, Pratsch L (1988a) Water-mediated effects of PEG on membrane properties and fusion. In: Ohki S, Doyle D, Flanagan, TD, Hui SW, Mayhew E: Molecular Mechanisms of Membrane Fusion. Plenum Press, New York, pp 255-272.

Arnold K, Zschörnig O, Pratsch L, Donath E, Barthel D, Herrmann A (1988b) Exclusion of PEG for our membrane surfaces and aggregation of liposomes and lipoproteins. Studia Biophys. 127: 113-120.

Bäumler H, Donath E (1987) Does dextran indeed significantly increase the surface potential of human red blood cells? Studia Biophys 120: 113-122.

Chan DYC, Mitchell DJ, Ninham BW, Pailthorpe BA (1979) Solvent structure and hydrophobic solutions. In: Franks F (ed.): Water, Vol. 6, Plenum Press, New York, pp. 239-278.

Chaudhury MK (1984) Short-range and long-range forces in colloidal and macroscopic systems. Thesis, SUNY, Buffalo.

Christenson HK (1988) Non-DLVO forces between surfaces - solvation, hydration and capillary effects. J. Dispersion Sci. Technol. 9: 171-206.

Cohly H (1986) Cell separation by preparative electrophoresis and double antibody tagging. Thesis, SUNY, Buffalo.

---

*Due to the still relatively primitive techniques and methods of interpretation of interfacial interaction energies of biological materials in aqueous media prevailing in 1983 (van Oss, et al., 1983), the quoted LW interaction energy between cells (van Oss, 1985) still was significantly underestimated, which led to the hypothesis (van Oss, 1985) that the electrostatic repulsion alone was responsible for the lack of cell adhesion at high g-forces. Figure 3 comprises the more accurate values for these energies, showing that the prevention of cell adhesion really is mainly due to the close-range AB repulsion.

Cohly H, van Oss CJ, Weiser M, Albini B (1985a) Purification of intestinal intraepithelial lymphocytes by preparative electrophoresis. In: Schütt W, Klinkmann H (eds.): Cell Electrophoresis, DeGruyter, Berlin, pp. 603-609.
Cohly H, Albini B, Weiser M, Green K, van Oss CJ (1985b) Microelectrophoresis of epithelial cells and lymphocytes. In: Schütt W, Klinkmann H (eds.): Cell Electrophoresis, DeGruyter, Berlin, pp. 611-616.
Cunningham RK (1984) In: Atassi MN, van Oss CJ, Absolom DR (eds.): Molecular Immunology, Marcel Dekker, New York, pp. 53-70.
Derjaguin BV, Churaev NV, Muller VM (1987) Surface Forces, Pergamon Press, New York, p. 27.
Derjaguin BV, Landau L (1941) Acta Physicochim. URSS, 14: 633.
Dorrington KJ, Mihaesco C (1970) Subunit structure of human γM-globulins. Immunochemistry 7: 651-660.
Dupré A (1869) Théorie mécanique de la chaleur. Gauthier-Villars, Paris, pp. 367-370.
Edwards RB, Szamier RB (1977) Detective phagocytosis of isolated rod outer segments by RCS rat retinal pigment epithelium in culture. Science 197: 1001-1003.
Evans E, Needham D (1988) Intrinsic colloidal attraction/repulsion between lipid bilayers and strong attraction induced by non-adsorbing polymers. In: Ohki S, Doyle D, Flnaagan TD, Hui SW, Mayhew E: Molecular Mechanisms of Membrane Fusion. Plenum Press, New York, pp. 83-99.
Fowkes FM (1963) Additivity of intermolecular forces at interfaces. I. Determination of the contribution to surface and interfacial tensions of dispersion forces in various liquids. J. Phys. Chem. 67: 2538-2541.
Fowkes FM (1967) Intermolecular and interatomic forces at interfaces. In: Burke J.J., Reed NL, Weiss V (eds.): Surfaces and interfaces: Chemical and physical characteristics. Syracuse University Press, Syracuse, New York, pp. 197-224.
Fu SM, Winchester RJ, Rai KR, Kunkel HG (1974) Hairy cell leukemia: Proliferation of a cell with phagocytic and B-lymphocyte properties. Scand J. Immunol. 3: 847-851.
Gawrisch K (1986) Molekularmechanismen und Membranveränderungen bei der durch PEG induzierten Zellfusion. Thesis, Karl Marx University, Leipzig.
Girifalco LA, Good RJ (1957) A theory for the estimation of surface and interfacial energy. I. Derivation and application to interfacial tension. J. Phys. Chem. 64: 904-909.
Good RJ (1979) Contact angles and the surface free energy of solids. Surf. Colloid Sci. 11: 1-29.
Good RJ, Elbing E (1970) Generalization of theory of estimation of interfacial energies. Ind. Eng. Chem. 62: 54-78.
Goodall RJ, Thompson JE (1971) A scanning electron microscope study of phagocytosis. Exptl. Cell Res. 64: 1-8.
Gruen DWR, Marcelja S, Parsegian VA (1984) Water structure near the membrane surface. In: Perelson AS, DeLisi C, Wiegel FW (eds.): Cell Surface Dynamics: Concepts and Models. Marcel Dekker, New York, pp. 59-91.
Hermans JJ (1948) Thermodynamics of long-chain molecules. In: Kruyt HR (ed.): Colloid Science, Vol. II, Elsevier, Amsterdam, pp. 47-92.
Hiemenz P (1984) Polymers Chemistry, Marcel Dekker, New York, p. 707.
Israelachvili JN (1985) Intermolecular and Surface Forces. Academic Press, London.
Mackor EL, van der Waals, JH (1952) The statistics of the adsorption of rod-shaped molecules in connection with the stability of certain colloidal dispersions. J. Colloid Sci. 7: 535-550.
Napper DH (1983) Polymeric Stabilization of Colloidal Dispersions. Academic Press, New York.

Ohki S (1982) A mechanism of divalent ion-induced phosphatidylserine membrane fusion. Biochim. Biophys. Acta 689: 1-5.
Overbeek JThG, Bijsterbosch BH (1979) The electrical double layer and the theory of electrophoresis. In: Righetti PG, van Oss, Vanderhoff JW (eds.): Electrokinetic Separation Methods, Elsevier, Amsterdam, pp. 1-32.
Overbeek JThG, Wiersema PH (1967) The interpretation of electrophoretic mobilities. In: Bier M (ed.): Electrophoresis, Vol. II, Academic Press, New York, pp. 1-52.
Prévost M, Gallez D (1984) The role of repulsive hydration forces on the stability of aqueous black films: Application to vesicle fusion. J. Chem. Soc. Faraday Trans. 2, 80: 517-530.
Parsegian VA, Fuller N, Rand RP (1979) Measured work of deformation and repulsion of lecithin bilayers. Proc. Natl. Acad. Sci. USA 76: 2750-2754.
Parsegian VA, Rand RP, Rau DC (1985) Hydration forces: What next? Chem. Scripta 25: 28-31.
Pollack W, Hager HJ, Reckel R, Toren DA, Singher HO (1965) A study of the forces involved in the second stage of hemagglutination. Transfusion 5: 138-183.
Rabinovich YA, Derjaguin BV (1988) Interaction of hydrophobized filaments in aqueous electrolyte solutions. Colloids Surfaces 30: 243-251.
Seaman GVF, Brooks DE (1979) Analytical cell electrophoresis. In: Righetti PG, van Oss CJ, Vanderhoff JN (eds.): Electrokinetic Separation Methods. Elsevier, Amsterdam, pp. 95-110.
Sewchand LS, Canham PB (1976) Induced roulean formation in interspecies populations of red cells. Can. J. Physiol. Pharmacol. 54: 437-442.
Svedberg T, Pedersen KO (1940) The Ultracentrifuge. Clarendon Press, Oxford, pp. 38-44.
Tilley D, Coakley WT, Gould RK, Payne SE, Hewison LA (1987) Real time observations of polylysine, dextran and polyethylene glycol induced mutual adhesion of erythrocytes held in suspension in an ultrasonic standing wave field. Eur. Biophys. J. 14: 499-507.
Valentine RG, Green NM (1967) Electron microscopy of an antibody-hapten complex. J. Molec. Biol. 27: 615-617.
van Oss CJ (1975) The influence of the size and shape of molecules and particles on their electrophoretic mobility. Separ. Purif. Meth. 4: 167-188.
van Oss CJ (1978) Phagocytosis as a surface phenomenon. Ann. Rev. Microbiol. 32: 19-39.
van Oss CJ (1985) Stability of human red cell suspensions at 300,000 x G. J. Dispersion Sci. Technol. 6: 139-146.
van Oss CJ (1986) Phagocytosis: An overview. Meth. Enzymology 132: 3-15.
van Oss CJ (1989a) Coacervation, complex-coacervation and flocculation. J. Dispersion Sci. Technol. in press.
van Oss CJ (1989b) Energetics of cell-cell and cell-biopolymer interactions. Cell Biophys. 14: 1-16.
van Oss CJ, Absolom DR (1983) Zeta potentials, van der Waals forces and hemagglutination. Vox Sang. 44: 183-190.
van Oss CJ, Absolom DR (1985) Influence of cell configuration and potential energy equilibria in rouleau phenomena. J. Dispersion Sci. Technol. 6: 131-137.
van Oss CJ, Coakley WT (1988) Mechanisms of successive modes of erythrocyte stability and instability in the presence of various polymers. Cell Biophys. 13: 141-190.
van Oss CJ, Fike RM (1979) In: Righetti PG, van Oss CJ, Vanderhoff JW (eds.): Electrokinetic Separation Methods. Elsevier, Amsterdam, pp. 111-120.
van Oss CJ, Good RJ (1988a) Orientation of the water molecules of hydration of serum albumin. J. Protein Chem. 7: 179-183.

van Oss CJ, Good RJ (1988b) On the mechanism of "hydrophobic" interactions. J. Dispersion Sci. Technol. 9: 355-362.
van Oss CJ, Good RJ (1989) Surface tension and the solubility of polymers and biopolymers. The role of polar and apolar interfacial free energies. J. Macromolec. Sci. Chem., in press.
van Oss CJ, Fike RM, Good RJ, Reinig JM (1974) Cell microelectrophoresis simplified by the reduction and uniformization of the electroosmotic backflow. Anal. Biochem. 60: 212-251.
van Oss CJ, Absolom DR, Neumann AW (1983) Interaction of phagocytes with other blood cells and with pathogenic and nonpathogenic microbes. Ann. N.Y. Acad. Sci. 416: 332-350.
van Oss CJ, Good RJ, Chaudhury MK (1986) Solubility of proteins. J. Protein Chem. 6: 385-405.
van Oss CJ, Chaudhury MK, Good RJ (1987a) Monopoles. Advan. Colloid Interface Sci. 28: 35-65.
van Oss CJ, Chaudhury MK, Good RJ (1987b) The mechanism of partition in aqueous media. Separ. Sci. Technol. 22: 1515-1526.
van Oss CJ, Chaudhury MK, Good RJ (1988a) Interfacial Lifshitz-van der Waals and polar interactions in macroscopic systems. Chem. Revs. 88: 927-941.
van Oss CJ, Good RJ, Chaudhury MK (1988b) Additive and nonadditive surface tension component and the interpretation of contact angles. Langmuir 4: 884-891.
van Oss CJ, Chaudhury MK, Good RJ (1988c) Polar interfacial interactions, hydration pressure and membrane fusion. In: Ohki S, Doyle D, Flanagan TD, Hui SN, Mayhew E (eds.): Molecular Mechanisms of Membrane Fusion, Plenum Press, New York, pp. 113-122.
van Oss CJ, Arnold A, Good RJ, Gawrisch K, Ohki S (1989a) Interfacial tension and the osmotic pressure of polar polymers. J. Dispersion Sci. Technol. (in press).
van Oss CJ, Arnold K, Coakley WT (1989b) Depletion flocculation and depletion stabilization of erythrocytes. Submitted
van Oss CJ, Good RJ, Busscher HJ (1989c) Estimation of the polar surface tension parameters of glycerol and formamide, for use in contact angle measurements on polar solids. Submitted
Verwey EJN, Overbeek JThG (1948) Theory of the stability of lyophobic colloids. Elsevier, Amsterdam.
Winzler RJ (1972) Glycoproteins of plasma membranes: Chemistry and function. In: Gottschalk A (ed): Glycoproteins, Part B. Elsevier, Amsterdam, pp. 1268-1293.
Young T (1805) An essay on the cohesion of fluids. Phil. Trans. Roy. Soc. 95: 65-87.

General thermodynamic description of elastic properties of membranes

V. S. Markin, M. M. Kozlov
A. N. Frumkin Institute of
Electrochemistry, USSR
Academy of Sciences,
Leninsky prosp.31, Moscow
117071, USSR

CONTENTS

Introduction

The mechanical properties of membranes have diverse manifestations in biology. Suffice it to mention such essential processes as cell spreading on a support, ejection by the cell of protrusions transformed into lamellas, tank-treading cell motion which is a morphogenetic stage [Trinkaus 1986]. Of special significance the mechanical properties of membranes are for blood cells, specifically for erythrocytes. The activities of these cells are associated with their travel in the blood flow and strong deformation at the movement through narrow capillaries [Canham and Burton 1968; Jay 1973].

The complexity and significance of the cell membrane mechanical properties were the reason for their experimental and theoretical studies by biophysicists in the last twenty years [Bessis 1973; Evans and Skalak 1979; Chien 1985; Hochmuth and Waugh 1987]. Simultaneously, the interest in the problems of the general thermodynamic description of the mechanical properties of interfaces of arbitrary structure, a particular case of which are biological membranes, was revived in the researchers dealing with surface physical chemistry [Buff 1956; Melrose 1968; Boruvka and Neumann 1977; Rowlinson 1983; Markin, Kozlov and Leikin 1988; Kozlov, Leikin and Markin 1989; Kozlov and Markin 1989]. The biophysical and physico-chemical trends of research have developed virtually not intersecting. At the same time, the results obtained within the framework of these trends complemented each other. The bulk of the studies on membrane mechanics dealt with the red blood cell. Original methods have been developed to study erythrocyte deformation in response to external effects [Rand and Burton 1964; Chien 1977; Waugh and Evans 1979; Evans and Skalak 1979; Waugh 1983; Kucera, Lerche and Meier 1984]. The most fruitful method became suction of part of the red cell membrane in a microcapillary and measurement of the length of the sucked part which varied depending on the value of suction pressure [Rand and Burton 1964; Evans and Skalak 1979]. Experiments have shown that the red cell membrane resists three basic deformations: extention-compression, bending and shear (the last two deformations mean the change in the form of the membrane at a constant area). The elasticity moduli corresponding to all three deformations have been determined.

The theory of membrane mechanical properties was developed mainly by two biophysical teams: Evans' group and Helfrich's group. The ideas developed by Evans and Skalak and collected in a well-known monograph [Evans and Skalak 1979] are based on the theory of thin shells. Main attention in those works is given to force factors occurring in the membrane during its axisymmetric deformation. Analyzed in detail were isotropic tensions due to the change in the membrane area, anisotropic tensions caused by shearing deformations, bending moment, cutting forces occurring at the nonuniform bending. The authors have found membrane equilibrium equations linking the force factors of axisymmetric deformation. The least attention in the works was given to the membrane elasticity moduli: actually analyzed was only the membrane elasticity with respect to extension-compression.

The theoretical approach to the membrane mechanics by Helfrich and coworkers [Helfrich 1973, 1975, 1978; Harbich, Servuss and Helfrich 1978; Helfrich and Mutz 1988] is based on the ideas developed earlier to describe liquid crystals. Within the framework of this approach intended to describe membrane elasticity with respect to bending deformations, the membrane is characterized by two material constants: bending elasticity modulus and saddle-splay bending elasticity modulus. Besides that, Helfrich introduced the structural-geometrical characteristics of the membrane - the spontaneous curvature. This value corresponds to the membrane form at which the bending elastic energy (without the saddle-splay bending) is minimal. The ideas of spontaneous curvature and bending elasticity moduli have been developed in the works of the Petrov team [Mitov 1978; Petrov and Mitov 1982; Petrov and Bivas 1984].

Two above-mentioned theoretical approaches to the description of membrane mechanics led to the emergence of a new important trend in biophysics. Both approaches, however, lacked the general thermodynamic description of the membrane which would make it possible to strictly define all notions used, generalize and expand the description of the membrane mechanical properties. The thermodynamic approach to the description of arbitrary form interfaces was first formulated by Gibbs [1928]. In the Gibbsian approach the interface is considered as a surface characterized by excessive thermodynamic values. In the 50s, the interest in this problem was revived due to the appearance of new experimental systems, which contained surfactants and were characterized, therefore, by low surface tension. A number of researchers [Buff 1956; Melrose 1968; Boruvka and Neumann 1977; Rowlinson 1983; Markin, Kozlov and Leikin 1988; Kozlov, Leikin and Markin 1989; Kozlov and Markin 1989] developed the Gibbsian theory of capillarity with respect to the description of the system in which tension and bending effects play an equal role. The local thermodynamics approach has been formulated within the framework of which the interface (unlike the Gibbs' theory) is considered as a transitive layer characterized by the pressure tensor in each of its points. Local thermodynamics made it possible to interpret the thermodynamic values, introduced by Gibbs, in terms of the microscopic properties of the interface. This is especially important for such interfaces as surfactant layers and biological membranes the microscopic structure of which is largely known.

The aim of this review is a generalized thermodynamic description of the membrane mechanical properties using the Gibbsian approach enriched with the local thermodynamics.

II. Membrane Structure

Material and geometrical characteristics of the membrane are determined by its structure, much about which is known at present [Ivkov and Berestovsky 1981]. The basis of a cell membrane is a membrane bilayer consisting of lipids and proteins. Lipid molecules, each of which has a hydrophilic head and a hydrophobic hydrocarbon tail, form a lipid bilayer: two adjacent monolayers of lipid molecules the hydrophobic tails of which make up the interior of the bilayer and the polar heads cover its surfaces. The proteins included in the membrane bilayer are referred to as integral proteins; they span the lipid bilayer due to its hydrophobic part which plays the role of an 'anchor' retaining the protein in the lipid 'sea'. The other constituent of the cell membrane characteristic, specifically, of red cells is the membrane skeleton connected with the cytoplasmic side of the membrane bilayer.

The membrane structure is complex in both directions: normal to its surface and lateral (tangential). The lateral properties of the membrane are summarized in the Singer and Nicolson liquid-mosaic model [1969]. Lipid and protein molecules of the membrane bilayer can move changing places. This process is referred to as lateral diffusion. As a result, the membrane bilayer has the properties of a two-dimensional liquid: the bilayer components can be redistributed within its plane and upon deformations no lateral shearing stresses occur.

The structural complexity of the membrane bilayer in the direction of its normal is primarily determined by the polar heads of the lipid molecules and the hydrophilic parts of integral proteins. Firstly, the hydrophilic parts of membrane proteins forming the so-called glycocalyx are, as a rule, electrically charged. It means that at the membrane surface there is a layer with a space charge which can be thicker than the lipid bilayer [Lerche 1984]. The glycocalyx charge is supplemented with that of the polar heads of the lipid molecules, focussed in an area on the bilayer surface. The membrane charges interact, the interaction being distributed throughout the region of the glycocalyx and the polar heads.

Secondly, the polar heads of the lipid molecules interact with water, structuring it within a thin layer near the surface of the membrane [Gruen, Marcelja and Parsegian 1984]. As a result, there occurs the interaction between the heads of the lipid molecules which is mediated by the structured water layer. It is distributed in the layer whose effective thickeness is several angstroms.

Thirdly, besides the charge, the polar heads of the lipid molecules also possess the dipole moments able to interact with one another. This interaction is effectively distributed in the layer, whose thickness is of the same order of magnitude as that of the polar heads, i.e., approximately 5-7 A.

Fourthly, the hydrophobic interior of the membrane is a layer approximately 20-30 A thick. The components of the links inside each hydrocarbon chain interact. Besides, there are interactions between the hydrocarbon chains. As a result, a complex distribution of mechanical tensions can occur inside the hydrophobic part of the membrane.

Thus, a detailed consideration of the cell membrane shows it to be a system of adjacent parallel layers characterized by space interactions. Therefore, the membrane is a complex structure interface; an efficient method of its description is local thermodynamics.

III. Fundamental Equations

The small thickness space between two uniform bulk phases having specific physical properties has been the subject of studies for more than a century. Pioneer work in this field was made by J W Gibbs [1928]. Because of its peculiar structure the transient zone is described as a pure interface, thin film or thick film [Rusanov 1978]. The more general term 'interface', covers any of the above-mentioned types of transient zone.

1. Gibbsian Approach. To describe the interface Gibbs introduced the concept of a dividing surface the normal to which in each point coincides with the direction of the physical value gradients (densities of the components etc) (Fig.1). The geometry of the surface element is characterized by its area A and two principal curvatures c_x and c_y. Therefore, the termodynamic variables of the surface are, besides entropy S^s, temperature T, number of components n_i^s and chemical potentials μ_i, also the values related to the geometrical charateristics of the surface: surface tension γ_G and moments C_{xG} and C_{yG} [Gibbs 1928]. The change in the internal energy of the interface can be described as

$$dU^s = TdS^s + \sum \mu_i dn^s + \gamma_G dA + C_{xG} dc_x + C_{yG} dc_y \qquad (1)$$

Surface tension γ_G is a pressure analog (of the opposite sign) for the bulk phase. Moments C_{xG} and C_{yG} have no analogs in the thermodynamics of liquid bulk phases. Expression (1) describing the change in the internal surface energy was given in the fundamental work by Gibbs [1928]. The terms in the right side

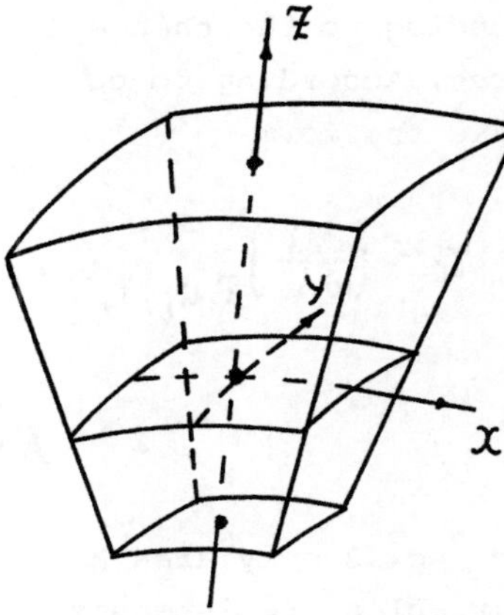

Fig.1. Piece of interface

of eq.1 proved to be in a sense inequivalent. The definitions of entropy S^s and mass of components n_i^s relevant to the surface element were introduced by Gibbs on the basis of the bulk phases thermodynamics using the concepts of dividing surface and surface excesses. As for the last three terms in eq.1, in the Gibbsian approach they are introduced by postulating. Henceforth, we shall refer to surface tension γ_G and momemts C_{xG} and C_{yG} as well as their combinations as force factors of an interface.

Describing the geometry of a surface element it is convenient to use instead of the principal curvatures c_x and c_y their combinations: mean curvature $J = c_x + c_y$ and the Gaussian curvature $K = c_x c_y$. Besides that, in the Gibbsian approach the bending moments C_{xG} and C_{yG} are extensive variables, whereas the approach in which the force factors are intensive values is more customary. In many cases the most convenient thermodynamic potential of an interface is the grand canonical potential which is a natural characteristic of a uniform surface. The expression for the change in the grand canonical potential of a surface in the Gibbsian reference system [Buff 1965; Melrose 1968; Rowlinson 1983] has the following form

$$d\Omega^s = -S^s dT - \Sigma n_i^s d\mu_i + \gamma_G dA + \mathbb{C}_1 A dJ + \mathbb{C}_2 A dK \qquad (2)$$

Values C_1 and C_2 referred to, respectively, as the first moment and the second moment are intensive force factors; they determine the changes in the grand canonical potential corresponding to the change in the mean (J) and Gaussian (K) curvatures. According to eq. 2, the force factors of the interface take the form

$$\gamma_G = \left(\frac{\partial\Omega^s}{\partial A}\right)_{T,\mu_i,J,K}; \qquad C_1 = \frac{1}{A}\left(\frac{\partial\Omega^s}{\partial J}\right)_{T,\mu_i,A,K}$$

$$C_2 = \frac{1}{A}\left(\frac{\partial\Omega^s}{\partial K}\right)_{T,\mu_i,A,J} \tag{3}$$

In eq. 2 only area A is an extensive variable; all the other variables are intensive. Therefore, the use of the Euler theorem in this case leads to equation

$$\Omega^s = \gamma_G A \tag{4}$$

The relationship between the first and second moments of the surface and Gibbs' moments is given by the expressions

$$C_{xG} = A\,(C_1 + C_2\, c_y); \qquad C_{yG} = A\,(C_1 + C_2\, c_x) \tag{5}$$

Along with the Gibbs' surface tension γ_G yielding the change in free energy of a surface region during the change in its area at constant curvatures, it may be convenient to use zero moment C_0 referred to also as the Boruvka-Neumann surface tension [Markin et al.1988]. This value is determined by the formula

$$C_0 = \left(\frac{\partial\Omega^s}{\partial A}\right)_{T,\mu_i,\tilde{J},\tilde{K}} \tag{6}$$

where $\tilde{J} = \iint J\,dA$, $\tilde{K} = \iint K\,dA$; the domain of integration is the membrane region area. The Gibbs' surface tension is presented as

$$\gamma_G = C_0 + C_1 J + C_2 K \tag{7}$$

It follows from eq.7 that for plane surfaces (J = 0, K = 0) the Boruvka-Neumann surface tension coincides with the Gibbsian value.

2. Local Thermodynamics Approach. The Gibbsian method provides no possibility to establish the relationship between force factors γ_G, C_1, C_2 since it makes no use of the microscopic properties of the transient zone. For membranes we shall use the concepts of local thermodynamics: for each point of the transient zone one can set all three-dimensional thermodynamic variables linked by the usual relationships. The magnitude of thermodynamic variables depends on the situation inside the film. The three-dimensional force factors of the interface are described with the aid of the pressure tensor, the values of which depend on their coordinates. Knowledge of the pressure tensor in each point of a film makes it possible to derive the expressions for surface tension and moments of the dividing surface which characterizes the interface [Buff 1956; Melrose 1968; Rowlinson 1983]. From these expressions naturally follows the relationship between the force factors.

The pressure tensors can be conveniently defined in Cartesian coordinates with the z-axis directed along the normal to the dividing surface of the interface (Fig.1).

$$P = \begin{pmatrix} P_{xx} & P_{xy} & P_{xz} \\ P_{yx} & P_{yy} & P_{yz} \\ P_{zx} & P_{zy} & P_{zz} \end{pmatrix} \tag{8}$$

Each component P has the sense of a force (in terms of an area unit) applied from the side of a selected space element to the environment. We shall consider the liquid interfaces which means that shear stresses in the lateral directions do not arise if the interface is in equilibrium: $P_{xy} = P_{yx} = 0$. The other components of the pressure tensor, both diagonal and nondiagonal, are a priori unequal to zero in the films consisting of amphiphilic surfactants. Examples of such films are biological membranes and lipid bilayers. Inside the lipid bilayer, pressures normal to its surface and tangential pressures can arise which are the diagonal components of the pressure tensor. If the bilayer undergoes cutting deformations along the z-axis, the nondiagonal cutting components of the pressure tensor, P_{xz}, P_{yz}, can arise due to interactions between the hydrocarbon tails of the neighbouring molecules and the hydrophobic effect on the surfaces of the monolayers. The relative shift of the bilayer elements lying one under the other, which we shall subsequently refer to as the layer-by-layer shift, should also result in the emergence of a corresponding nondiagonal component of the pressure tensor, P_{zx}, P_{zy}, due to the deformation of the hydrocarbon chain. The pressure tensor elements are related by the internal equilibrium equations $P_{xx} = P_{yy} = P_T$; $P_{xz} = P_{zx}$; $P_{yz} = P_{zy}$. The force factors of the interface are expressed via the components of the microscopic pressure tensor:

$$\mathbb{C}_0 = \int_{z^\alpha}^{z^\beta} (P^{\alpha\beta} - P_T)\, dz + \int_{z^\alpha}^{z^\beta} z\, \mathrm{div}_2 L\, dz \tag{9}$$

$$\mathbb{C}_1 = \int_{z^\alpha}^{z^\beta} z\, (P^{\alpha\beta} - P_T)\, dz + \frac{1}{2} \int_{z^\alpha}^{z^\beta} z^2\, \mathrm{div}_2 L\, dz \tag{10}$$

$$\mathbb{C}_2 = \int_{z^\alpha}^{z^\beta} z^2\, (P^{\alpha\beta} - P_T)\, dz + \frac{1}{3} \int_{z^\alpha}^{z^\beta} z^3\, \mathrm{div}_2 L\, dz \tag{11}$$

In formulae (9)-(11) integration is carried out with respect to the z-axis: the coordinates corresponding to the integration limits of z^α and z^β are in the homogenous phases below (z^α) and above (z^β) the interface. Components of auxiliary vector L are 'cutting' constituents of the pressure tensor, $L_x = p_{xz}$; $L_y = p_{yz}$. Function $p^{\alpha\beta}$ is the step function taking the pressure values in the bulk phases above (p^β) and below (p^α) dividing surface where the origin of the coordinate is placed (Fig.1)

$$P^{\alpha\beta} = \begin{cases} P^\alpha, & z < 0 \\ P^\beta, & z > 0 \end{cases} \tag{12}$$

Operator div_2 stands for two-dimensional divergence in plane (x, y) perpendicular to the z-axis. The difference $p^{\alpha\beta} - -p_T$ is the excessive tangential pressure (as compared with the bulk pressure) taken with the opposite sign. When the cutting forces in the system are absent, L = 0, the force factors C_0, C_1, C_2 are, respectively, the zero, first and second moments of the excessive tangential pressure component (taken with the opposite sign). The expression for the Gibbs' surface tension can be presented as eq.(7) with eqs.(9)-(11) taken into account.

3. Equations for Mechanical Equilibrium. In the equilibrium state the force factors of an interface - surface tension γ_G, the first and second momemts C_1 and C_2 - are interconnected [Kozlov and Markin 1989]. The equation of mechanical equilibrium of an interface region in a direction normal to the dividing surface can be presented as

$$J\,\mathbb{C}_0 + 2K\,\mathbb{C}_1 + \int_{z^\alpha}^{z^\beta} (1 + Jz + Kz^2)\, div_2 L\, dz = P^\alpha - P^\beta \tag{13}$$

The cutting components of the pressure tensor are absent

in the system, L = 0, if the curvatures are homogenous. In this case the equilibrium equation (13) takes the simple form [Buff 1956; Melrose 1968; Rowlinson 1983]

$$J\mathbb{C}_0 + 2K\mathbb{C}_1 = P^\alpha - P^\beta$$

or

$$J\gamma_G - \mathbb{C}_1(J^2 - 2K) - \mathbb{C}_2 JK = P^\alpha - P^\beta$$

referred to as the generalized Laplace equation.

The equilibrium equation of the interface in the lateral directions has the form

$$\frac{\partial}{\partial x}(\mathbb{C}_0 + \mathbb{C}_1 J + \mathbb{C}_2 K) + c_x \int_{z^\alpha}^{z^\beta} (1 + Jz + Kz^2) L_x \, dz + \qquad (14)$$

$$+ \frac{1}{2}\left(\int_{z^\alpha}^{z^\beta} z^2 \, div_2 L \, dz\right) \frac{\partial}{\partial x} J + \frac{1}{3}\left(\int_{z^\alpha}^{z^\beta} z^3 \, div_2 L \, dz\right) \frac{\partial}{\partial x} K = 0$$

$$\frac{\partial}{\partial y}(\mathbb{C}_0 + \mathbb{C}_1 J + \mathbb{C}_2 K) + c_y \int_{z^\alpha}^{z^\beta} (1 + Jz + Kz^2) L_y \, dz$$

$$+ \frac{1}{2}\left(\int_{z^\alpha}^{z^\beta} z^2 \, div_2 L \, dz\right) \frac{\partial}{\partial y} J + \frac{1}{3}\left(\int_{z^\alpha}^{z^\beta} z^3 \, div_2 L \, dz\right) \frac{\partial}{\partial y} K = 0$$

where x and y are the local Cartesian coordinates along the principal directions of the surface. For a surface with the homogenous curvature (L = 0) eq.14 takes the form

$$\nabla \gamma_G = 0 \tag{15}$$

The equilibrium equations of a surface region with respect to the rotation appears as

$$\frac{\partial C_1}{\partial x} + c_y \frac{\partial C_2}{\partial x} = - \int dz\, L_x (1 + J z + K z^2) \tag{16}$$

$$\frac{\partial C_1}{\partial y} + c_x \frac{\partial C_2}{\partial y} = - \int dz\, L_y (1 + J z + K z^2) \tag{17}$$

From eqs.13 and 14 it follows that on a surface of the variable curvature all the force factors (surface tension and both moments) change from point to point.

IV. Elasticity of Interfaces

The interface force factors - surface tension γ_G, the first and second moments C_1 and C_2 - determine the work done by the system deformation up to the first order of the deformation value. Recently the scope of physico-chemical research has been supplemented by a class of the systems a distinctive feature of which is a very small or practically zero value of surface tension [De Gennes and Taupin 1982]. These systems include first of all microemulsions - suspensions of water drops in oil or oil drops in water the size of which is in the order of 10 nm. The small value of the surface tension at the boundary of oil and water, equal approximately $\gamma = 10$ dyn/cm, is provided by the presence of surfactants and cosurfactants in the system. Besides microemulsions, the widely studied systems are mixtures of lipids and water at various concentrations of both components [Kirk, Gruner and Stein 1984]. The amphiphilic nature of the lipid molecule (the presence of the hydrophilic polar head and the hydrophobic hydrocarbon tail) determine the type of the structures formed by lipid in water. One of the most widely distributed structures is the lipid bilayer. The bilayers form the closed bubbles-liposomes or plane multilamellar structures. Certain types of lipids form in a minor amount of water various nonbilayer structures: systems packed into hexagonal lattice cylinders, (normal and inverted) micelles etc. In the absence of the transmembrane osmotic pressure the closed membrane bilayer also represents a system with zero surface tension. Description of such systems requires additional physical concepts not traditional for the physical chemistry of the capillary systems. In this case it is useful to consider the elastic energy of deformation.

1. Spontaneous Geometric Characteristics. The work of deformation is related in the first order of magnitude with the value of this deformation through eq.1. The force factors of the surface - Gibbs' surface tension γ_G, the first and second momemts C_1 and C_2 - determine the work for each type of deformation: for extension dA, change in the mean curvature dJ, change in the Gaussian curvature dK. If any force factor equals zero, the corresponding work of deformation equals zero in the first order of magnitude.

The force factors of any interface change in the process of deformation. The character of these changes depends on two circumstances: firstly, on the material properties of the interface; secondly, the conditions under which the deformation occurs. We shall dwell on the second circumstance. What is understood as conditions is the behaviour of the thermodynamic quantities - external pressure, temperature, distribution and number of all components making up the interface - during the deformation. We shall consider the process as occurring at constant temperature and constant values p^α and p^β of external pressure conforming to the homogenous phases located below and above the interface (Fig. 1). As for the distribution and number of components of the interface, their behaviour depends on the character of deformation. Depending on the velocity of deformation the interface region considered may behave as an open, partially open or closed system [Rusanov 1977]. In the first case the thermodynamic equilibrium between the region considered and the environment is maintained with respect to each component of the system. The number of components inside the interface region, naturally, changes during the deformation. In the second case the equilibrium is established not for all but only some components of the system. And, finally, in the third case the amount of each component of the interface region remains constant during the deformation. Examples of open systems are the interfaces of immiscible liquids without surfactants, while partially or completely closed systems are represented by various films and, specifically, biological membranes.

The state in which the surface tension equals zero $\gamma_G = 0$ is referred to as spontaneous with respect to extension-compression. Similarly, the spontaneous states are determined relative to the change in the mean curvature J and Gaussian curvature K. In these spontaneous states the first moment $C_1 = 0$ and the second moment $C_2 = 0$ are equal to zero. The equation determining the geometric characteristics of spontaneous states - spontaneous area A_s, spontaneous mean curvature J_s and spontaneous Gauss curvature K_s - may be presented as

$$\gamma_G(A, J, K) = 0 \tag{18}$$

$$C_1(A, J, K) = 0 \tag{19}$$

$$C_2(A, J, K) = 0 \tag{20}$$

Each of eqs.(18)-(20) determines the spontaneous geometric characteristics with respect to a corresponding type of deformation (extension-compression, the first and second bending moments). The values of surface tension and moments are determined through the microscopic characteristics of the system according to eqs.(9)-(11).

The state in which all three equations (18)-(20) are satisfied is referred to as an absolutely spontaneous one. This state has spontaneous area A_s, spontaneous mean curvature J_s and spontaneous Gaussian curvature K_s. In an absolutely spontaneous state the work of arbitrary deformation of a surface element equals zero in the first order of magnitude.

2. Elasticity Moduli of an Interface. Knowing the dependence of the force factors of an interface on its geometric characteristics, one can determine the elasticity moduli of the system. Gibbs [1928] introduced the elasticity of a film conforming to the deformation of extension-compression $E = A(\partial^2 F/\partial A^2)$. The elasticity moduli of extension-compression for membranes was considered in the papers by Evans and coworkers [Evans and Skalak 1979]. In the works by Helfrich [1978] the elasticity modulus was introduced with respect to the bending due to the change in the mean curvature. Thus, until recently the elasticity moduli conforming to only two types of deformation - extension-compression and the first bending - have been considered. At the same time, of interest are also the elasticity moduli conforming to the second bending and mixed deformations. Consistent determination of all elasticity moduli of the interface has been carried out in our recent paper [Kozlov, Leikin and Markin 1989]. The elasticity moduli can be conveniently presented as a matrix

$$E = \begin{pmatrix} E_{AA} & E_{AJ} & E_{AK} \\ E_{JA} & E_{JJ} & E_{JK} \\ E_{KA} & E_{KJ} & E_{KK} \end{pmatrix} \qquad (21)$$

The diagonal elements of matrix E conform to the elasticity moduli of the following 'pure' deformations modulus of extension $$E_{AA} = A\left(\frac{\partial^2 F}{\partial A^2}\right)_{J,K} \qquad (22)$$

first bending modulus $$E_{JJ} = \frac{1}{A}\left(\frac{\partial^2 F}{\partial J^2}\right)_{A,K} \qquad (23)$$

second bending modulus $$E_{KK} = \frac{1}{A}\left[\frac{\partial^2 F}{\partial K^2}\right]_{A,J} \qquad (24)$$

The nondiagonal elements of matrix E conform to the elasticity moduli of mixed deformations

$$E_{AK} = E_{KA} = \left[\frac{\partial^2 F}{\partial A\, \partial K}\right]_J \qquad (25)$$

$$E_{AJ} = E_{JA} = \left[\frac{\partial^2 F}{\partial A\, \partial J}\right]_K \qquad (26)$$

$$E_{JK} = E_{KJ} = \left[\frac{\partial^2 F}{\partial J\, \partial K}\right]_A \qquad (27)$$

The elasticity moduli can be expressed through the interface force factors as

$$E_{AA} = \left(\frac{\partial \gamma_0}{\partial\, \ln A}\right)_{J,K} \qquad E_{JJ} = \left[\frac{\partial C_1}{\partial J}\right]_{A,K} \qquad E_{KK} = \left[\frac{\partial C_2}{\partial K}\right]_{A,J} \qquad (28)$$

$$E_{AJ} = E_{JA} = \frac{\partial \gamma_G}{\partial J} = \mathbb{C}_1 + \frac{\partial \mathbb{C}_1}{\partial \ln A} \tag{29}$$

$$E_{AK} = E_{KA} = \frac{\partial \gamma_G}{\partial K} = \mathbb{C}_2 + \frac{\partial \mathbb{C}_2}{\partial \ln A} \tag{30}$$

$$E_{JK} = E_{KJ} = \left(\frac{\partial \mathbb{C}_1}{\partial K}\right) = \left(\frac{\partial \mathbb{C}_2}{\partial J}\right) \tag{31}$$

Using expressions (28)-(31) and formulae (9)-(11), the interface elasticity moduli can be related to the derivatives of the microscopic pressure tensor components.

Determination of the elasticity moduli makes it possible to find the work of deformation of the interface up to the terms of the second order of magnitude using the value of deformation with respect to the initial values of the geometric characteristics (A^o, J^o, K^o):

$$\Delta F = \gamma_G \Delta A + A\,\mathbb{C}_1 \Delta J + A\,\mathbb{C}_2 \Delta K + \frac{1}{2} E_{AA} \left(\frac{\Delta A}{A}\right)^2 \tag{32}$$

$$\frac{1}{2} A E_{JJ} (\Delta J)^2 + \frac{1}{2} A E_{KK} (\Delta K)^2 + E_{AJ} \Delta A \Delta J +$$

$$+ E_{AK} \Delta A \Delta K + A E_{JK} \Delta J \Delta K$$

The expression for the change in elastic energy during the bending deformation was first introduced by Helfrich [1973]. The assumption was that the curvature radius of an interface significantly exceeds its thickness. Therefore, the expression deduced by Helfrich contained only the terms of the first order in the ratio of the membrane thickness to the curvature radius. In this work, which generalizes the results of Helfrich, there are no suggestions on the magnitude of the ratio of the interface thickness and curvature radius.

If one considers the deformations relative to the spontaneous state, the corresponding first order term in eq.(32) is absent. At the deformation relative to the absolutely spontaneous state the change in free energy is contributed to by only the second order terms the coefficients in which are the elasticity moduli. From eq.32 it follows that in the case of 'pure' deformation, at which only one geometric characteristic changes and the other two remain constant, the second order work is determined by the corresponding diagonal elasticity modulus. At the pure extension-compression ($\Delta J = 0$, $\Delta K = 0$) the work is determined by the extension modulus E_{AA}; at the pure bending with the Gaussian curvature preserved ($\Delta A = 0$, $\Delta K = 0$) the first bending modulus E_{JJ} is essential, at the pure bending with the mean curvature preserved ($\Delta A = 0$, $\Delta J = 0$) the second bending moment E_{KK}. If the deformation is of a 'mixed' character and includes the change in two geometric parameters, the second order work is also determined, besides the elasticity moduli of pure deformations, by crossed moduli. From eqs.(9)--(11), (28)-(31) it follows that the values of elasticity moduli depend on the position of the dividing surface. Appropriate selection of the dividing surface can turn one of the crossed moduli to zero. Equality to zero of a respective expression from eqs.(29)-(31) serves as the equation for the coordinate of such dividing surface. The dividing surface, for which the elasticity modulus conforming to mixed deformation of extension and the first bending ($E_{AJ} = 0$) goes to zero, is analogous to the neutral surface [Landau and Lifshits 1987].

We have considered a general thermodynamics approach to the problem of membrane elastic properties which made possible the strict definitions of membrane elasticity moduli and spontaneous geometric characteristics. The equilibrium equations for a membrane region relative to various displacements have been obtained. The results given in the review make it possible to deal with numerous problems in experimental manifestatins of the mechanical properties of membranes.

REFERENCES

Bessis M (1973) Living Blood Cells and their Ultrastructure. Springer Berlin
Boruvka L, Neumann AW (1977) J.Chem.Phys. 66: 5464
Buff FP (1956) J.Chem.Phys. 25: 146
Canham PB, Burton AC (1968) Circ.Res. 22: 405
Chien S (1985) Microvasc.Res. 29: 129
De Gennes PG, Taupin C (1982) J.Chem.Phys. 86: 2295
Evans EA, Skalak R (1979) Mechanics and Thermodynamics of Biomembranes. CRC Crit.Rev.Bioenerg. 3: 181
Gibbs JW (1928) The Collected Works. Longmans-Green, New York, 2
Gruen DWR, Marcelja S, Parsegian VA (1984) In: Cell Surface Dynamics (eds.Perelson AS, Delisi C and Wiegel F) Marcel Dekker New York
Harbich W, Servuss RM, Helfrich W (1978) Z.Naturforsch. 33a: 1013
Helfrich W (1973) Z.Naturforsch. 28C: 693
Helfrich W (1975) Z.Naturforsch. 30c: 841
Helfrich W (1973 Z.Naturforsch. 33a: 305
Helfrich W, Mutz M (1988) In: Fluctuations and Pattern Growth: Experiments and Theory (eds.Stanley HE, Ostrovsky N) Kluwer, Dordrecht
Hochmuth RM, Waugh RE (1987) Ann.Rev.Physiol. 49: 209
Ivkov VB, Berestovky GN (1981) The dynamic structure of lipid bilayer. Nauka Publishers, Moscow
Yay AWL (1973) Bioph.J. 13: 1166
Kirk GL, Gruner SM, Stein DL (1984) Biochemistry 23: 1093
Kozlov MM, Leikin SL, Markin VS (1989) J.Chem.Soc.Faraday Trans.2 (in press)
Kozlov MM, Markin VS (1989) J.Chem.Soc.Faraday Trans.2 (in press)
Kucera VW, Lerche D, Meier WZ (1984) Med.Labor.Diagn. 25: 334
Landau LD, Lifshits EM (1987) Theory of Elasticity. Nauka Publishers, Moscow (in Russian)
Lerche D (1984) Biorheol. 21: 477
Markin VS, Kozlov MM, Leikin SL (1988) J.Chem.Soc.Faraday Trans.2, 84: 1140
Melrose JC (1968) Ind.ng.Chem. 60: 53
Mitov MD (1978) Comptes rendus de l'Academie Bulgare des sciences, 31: 513
Petrov AG, Mitov MD (1982) Stud.Biophys. 90: 223
Petrov AG, Bivas IM (1984) Progr.Surf.Sci. 16: 389
Rowlinson JC (1983) J.Chem.Soc.Faraday Trans.2, 79: 77
Rusanov AI (1978) Phasengleigewichte und Grenzflachemer Scheinungen. Academie Verlag, Berlin
Singer SJ, Nicolson GL (1972) Science 175: 720
Trinkaus JP 1 985) In: Experimental Biology and Medicine (ed. Wolsky A) Karger, Basel 10: 130
Waugh RE, Evans EA (1979) Biophys. J. 26: 115

The influence of membrane electric field on cellular functions

R. Glaser

Bereich Biophysik
Sektion Biologie
Humboldt-Universität
Invalidenstr. 42
1040 Berlin
DDR

1. Introduction

Bioelectric phenomena have been known since the classical experiments of Aloisio Galvani in Bologna in 1786. These investigations led to a dramatic controversy between Galvani and Volta. Really this reflects two directions of bioelectricity existing even today: on one hand the generation of membrane electric fields in the living cell, and on the other hand the excitation of cells and their manipulation by external electric fields for biotechnological (electrofusion, electroperforation) or therapeutical reasons (various electrostimulation techniques e.g.).

In this paper we will touch neither the distribution and registration of bioelectric fields in the body, nor the influence of biological mechanisms by outside applied fields, but we will rather concentrate our attention on the membrane electric field itself, its generation and modification and finally we will ask the question: Is the membrane electric field a parameter controlling membraneous functions in vivo, or is this just a side-effect of cellular metabolism?

As we will see later, the electric field, which is determined by the electric potential $\psi(x,y,z)$ is a true biophysical structure which is partly in an equilibrium state (as for example electrical double layers), and partly in nonequilibrium state (e.g. diffusional potential). These non equilibrium compounds of the field in particular enable triggering to take place, as in nerve excitation. This property qualifies the membrane electric field to become an important part in a number of cellular feedback mechanisms.

TAB.1
Examples for cellular functions which correlate with transmembrane potential modifications.

CARRIER MEDIATED ION TRANSPORT:	Bassilana et al. (1985), Glaser et al. (1983), Lauger (1987), Lauger and Apell (1988), Serpersu and Tsong (1984), Zade-Oppen et al. (1979)
METABOLITE TRANSPORT:	Chay (1986), Kessler (1983), Villeral and Cook (1977)
ENZYMATIC REACTIONS:	Honig et al. (1986), Engasser and Horvath (1975), Riddell and Arumugam (1988)
LIPID COMPOSITION:	Clementz and Christiansson (1986)
CELL SHAPE:	R. Glaser(1979, 1982), R. Glaser et al (1988), McCann et al. (1983)
ENDOCYTOSIS, EXOCYTOSIS, MEMBRANE FUSION:	DosReis et al. (1977, 1979), A. Glaser (1988), Nagura et al. (1977), Okada et al. (1981), Tsuchiya et al. (1981), Williams (1981)
PLATELET AGGREGATION:	Friedhoff et al. (1981), Pales (1988)
MITOSIS:	Bingelli u. Weinstein (1986), Cone (1970), Kiefer et al. (1980), Sachs et al. (1974)
FERTILISATION, CAPACITATION, POLYSPERMY BLOCK:	Eppel (1980), Gould-Somero and Jaffe (1984), Jaffe (1979), Jaffe et al 1982), Taglietti (1979)
CELL GROWTH:	Jaffe (1981), Jaffe and Stern (1979), McDonald et al. (1972), Larter et al. (1979)
CELL MOVEMENT:	Dolowy (1987), Sugimoto and Hagiwara (1979), Van Houten (1979)
ROTATION OF BACTERIAL FLAGELLAE:	Jou et al. (1986)

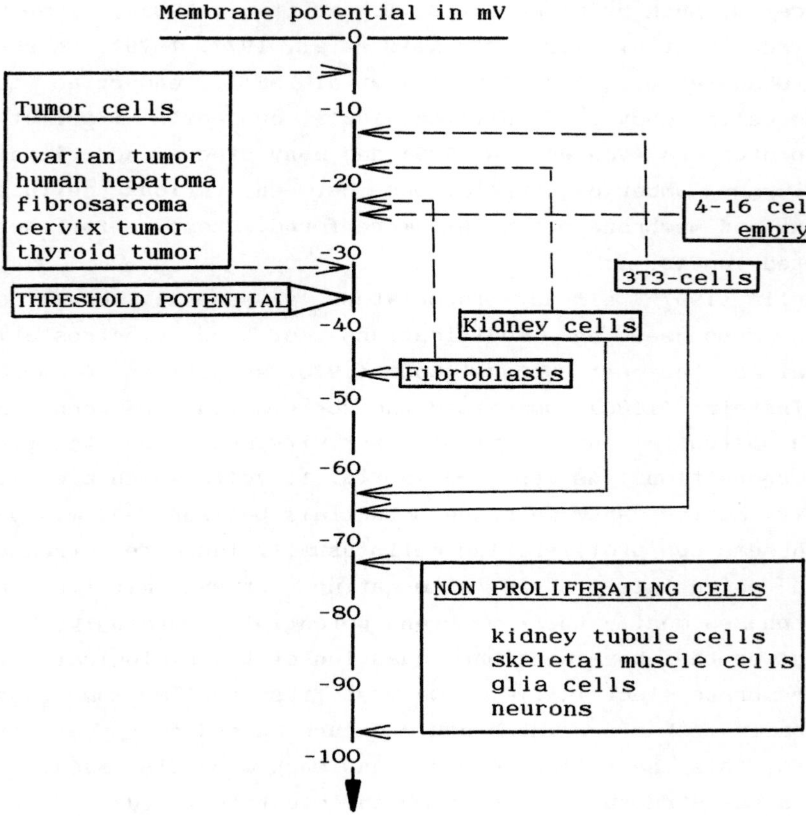

FIGURE 1:
The transmembrane potential of mitotic (— — —) and non-mitotic (———) cells (Data from Bingelli and Weinstein 1986).

Continuing the above mentioned story, the followers of Galvani and Volta, namly Giovanni Aldini and Carlo Matteucci from Pisa continued their work and transfered these ideas to physiologists and physicists in Berlin, namely to Johannes Müller, Emil Du Bois-Reymond and Herrmann von Helmholtz. This eventually lead to the investigation of the nerve impulse by Julius Bernstein, and therefore at first time to the indication of an electrochemically controlled membrane process.

Apparently the electric field in and near a biological membrane plays a significant role in a number of other cellular processes. In our days action potentials are observed in many other cells, such as plant cells (Takeda et al. 1983), lymphocytes (McCann et al. 1983, Dos Reis et al. 1977, 1979), cancer cells (Okada et al. 1979, Tsuchiya et al 1981), endocrine and exocrine cells (Chay 1986, Williams 1981), egg cells (Taglietti 1979), protozoons (Van Houten 1979) and many others. Additionally a large number of correlations between various cellular functions and membrane potentials were found. Examples for these are listed in TAB. 1.

Gingell (1967) already formulated a general scheme of cellular responses on modifications of the electrostatic potential of the cell surface. Cone (1970) as well as Bingelli and Weinstein (1986) summarised the correlations between the membrane potential and the mitotic activity and tried to find logical connections. As depicted in FIG. 1, cells which are mitotically active show membrane potentials between -37 mV and 0 mV, whereas non proliferating cells usually indicate stronger negative potentials. Cells alternating between mitotic and resting phases modify their membrane potential accordingly.

To answer the above mentioned question of the biological role of the membrane electric field, we must first analyse some possible basic mechanisms which could occure in and near the membrane. For this the following questions should be discussed:

- What is the structure of the membrane electric field?
- How can it be modified by in vivo processes?
- How does this field affect membrane structure and membrane processes?

2. The general structure of the membrane electric field

The electric field $\vec{E}$ is a vector which is determined by the gradψ(x,y,z). The Poisson-equation allows to calculate the potential (ψ) as a function of the space charge density σ(x,y,z) and the dielectric constant which in heterogeneous media also can be a function of geometrical coordinates. In the case of biological membranes the analysis of the electric field is split off into two problems:

- What is the profile of the electrical potential perpendicular to the plane of the membrane (ψ(x))?

- What is the electric field in the plane parallel to the surface of the membrane ($\psi(z,y)$)?

Let us start with the first question, which is investigated much more in detail (see for example: Blank 1985, Heinrich et al. 1982, McLaughlin 1989, Tsong and Astumian 1987, Ohki 1981).

In FIG. 2 a cross section through the cell membrane is depicted schematically. On the outer surface the membrane is covered by a glycoprotein layer, the so called glycocalyx, carrying negatively charged sialic acid residues. The membrane itself consists of a lipid bilayer with proteins embedded in it. A few lipid head groups are net charged as for example phosphatidylserine; mostly they are dipoles. The dielectric constant of the hydrophobic part of the lipid domain is very low in contrast to the region of the polar head groups.

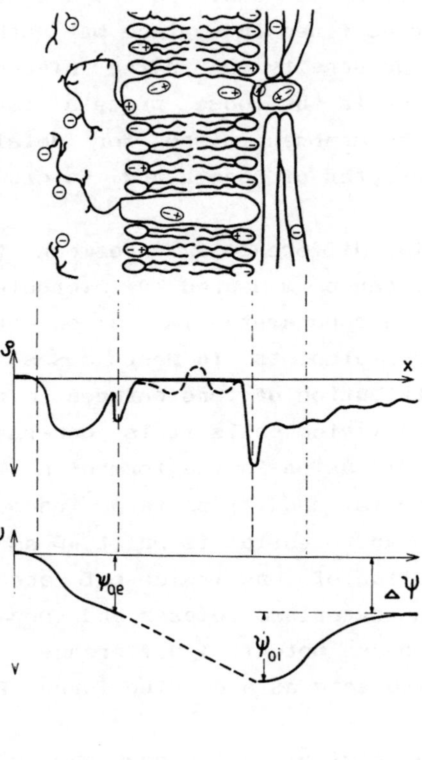

FIGURE 2:

The schematic representation of the molecular membrane structure with the glycoprotein layer (glycocalyx) outside (left) carrying negative charges, and the cytoskeleton on the inner surface (right). Below is the charge density $\rho(x)$. (The dashed line indicates possible charge distribution in case where the profile passes through a protein). The lower graph shows the resulting potential function. (Inside the membrane the potentials are just connected by a dashed line because the real function in this region is unknown).

The dielectric heterogenity of the membrane is even larger at points where proteins are located. In general the potential profile in this area is strongly influenced by various fixed charges and dipoles. This touches a special problem, namely the potential profile inside a so called ionic channel, which by now has been more or less hypothetical (Läuger and Apell 1988).

There have been various attempts to determine the potential profile in the membrane and in the membrane near space in x-direction. In general one has to differentiate between the surface potentials ψ_{oi} and ψ_{oe} on both sides of the membrane on one hand, and the phase potential difference $\Delta\psi$ on the other hand. The surface potentials of the membrane reflect the potential differences between the potential directly at the surface and the potential far away in the corresponding phase. Basically these are electrical double layers which are equilibrium structures predicted by the Poisson-Boltzmann-equation. In contrast to the Gouy-Chapman-model, however, these double layers are not simply the result of surface charges fixed in a plane but rather they are built up by a space-charge density (ρ) in the glycocalyx near the external surface, or in the loose protein layer near the internal surface of the membrane (see for detaile Donath and Voigt 1983 and the chapter of Donath and Voigt in this book!).

The electrical phase potential difference $\Delta\psi$ between the cytoplasm and the outside medium can be measured by microelectrodes or fluorescence probes. This parameter is called "the membrane potential" by electrophysiologists. In most cases it reflects the nonequilibrium distribution of ions between cytoplasm and the external medium. In living cells it is generated by electrogenic pumps or by ion diffusion in the form of a liquid junction potential. Under special conditions (e.g. ionophores, electroporation etc.) a Donnan potential is built up as a result of an equilibrium distribution of ions inside-out according to the charge density of the cytoplasm (Glaser and Donath 1984). In general, however, the phase potential difference reflects a non-equilibrium state and acts as a driving force for ions and charged molecules.

The field $\psi(y,z)$ in the plane parallel to the membrane surface depends on the mosaic of proteins and lipids in the membrane (McLaughlin and Poo 1981). A simple electrostatic membrane mosaic can be demonstrated by electron microscopy using positively charged particles of heavy metal (Nicolson and Painter 1973). This reflects the surface distribution of charged membrane compounds and its functional pattern. On the other hand, the membrane mosaic of electrogenic pumps and leaks additionally generates nonequilibrium fields and ionic currents (Fromherz 1988 a,b,c, 1989). These currents can be measured directly using vibrating electrodes (Betz et al. 1980, Jaffe 1979). Little is known about these fields, but certainly they do exist and obviously they are important for various membraneous processes.

3. Alterations of the membraneous field in vivo

Before discussing possible mechanisms which could be responsible for membrane field effects, we will check whether this field will be modified under various in vivo conditions. A number of such possibilities are listed in TAB. 2. In some cases local potentials simply depend on local conditions. Additionally, however, a number of trigger processes can occur in molecular structures which are in metastable states, as well as in nonequilibrium situations, especially in those which are far from equilibrium.

The best known triggering-process, which has been studied for almost a century, is the process of nerve excitation. A field sensitive channel switches permeabilities and alters the potential and therefore the electric field. As we already mentioned before, action potentials meanwhile have been detected not only in neurons but in many other cells too. They can be triggered by receptors reacting specifically on external signals, or by a large number of ionophore-antibiotics, toxins, viruses etc. This probably answers the question, why cells use metabolic energy to charge up an electrochemical battery, accumulating potassium and extruding sodium. The electrochemical gradient can easily be transformed into electrical energy by such trigger processes. Thus the electrochemical gradient really keeps ready the energy for inducing and turning on various membrane processes, even mechanical, such as vesiculations or membrane extrusions.

In this connection it should be mentioned, that the diffusional potential doesn't only depend on concentration differences of ions and their permeabilities but additionally on the particular functions $c(x)$ and $\psi(x)$. The integration of the Nernst-Planck equation, leading to the calculation of fluxes and subsequently to the diffusional potential, requires a knowledge of these functions. Change of surface charge near a channel therefore may modify the resulting diffusional potential (Ohki 1981, Glaser et al. 1983).

TAB. 2
Conditions causing alterations of the membrane electric field in vivo

ALTERATION OF ELECTROCHEMICAL CONDITIONS:
pH-modifications
Variation of local ion concentrations
ALTERATIONS OF MOLECULAR MEMBRANE STRUCTURE:
Lipid exchange, lipid flip-flop, phospholipase activity
Protein translocation
Adsorption or binding of charged compounds
Sialic acid removed by neuraminidase
ALTERATION OF FUNCTIONAL MEMBRANE PROPERTIES:
Modifications of membrane permeability
Modifications of electrogenic pumps
membrane bending
ALTERATIONS BY INTERACTION WITH OTHER SURFACES
Cell-cell interaction
Particle-cell interaction
Interaction with inorganic surfaces

An important way of modifying the electric field is the approach between the membrane and another surface (another cell, a particle, a virus etc.). These changes were already calculated and discussed by Gingell (1967). In some cases special mechanisms diminish the number of surface charges in order do make the contact easier (neuraminidase activity occasionally virus--cell contact or during sperm capacitation etc.). In any case contacts between a cell and other surfaces lead to changes of the potential profile and possibly even to a lateral charge displacement.

We will not reiterate the information presented in TAB. 2 but simply underline the fact that alterations of the membrane potential are not only an interesting experimental trick but rather a process which accomplishes many in-vivo processes and which plays a significant role in various cellular mechanisms. The influence of cells by outside electric fields is not the subject of this paper.

4. Mechanisms whereas the electrical membrane field influences membrane properties.

All interactions of physical factors with living structures have to be considered under two aspects: On one hand unspecific influences occure, usually needing large energies, while on the other hand in many cases, special sensory adaptations have been developed, reacting on these influences very sensibel and specific.

The same holds in the case of the interaction of membrane electric field with the extremly dynamic molecular structure of the membrane. On one hand the electric field modifies the membrane structure by a number of unspecific effects, while on the other hand special field sensitive molecules (especially potential gated channels) can switch molecular functions. A survey of these effects is listed in TAB. 3.

From the electrochemical point of view, two main field effects have to be considered in general: On one hand the field as driving force, controling the distribution of ions and other mobile charges, and on the other the shift of the positions of chemical or structural equilibria under the influence of the electric field in the sense of Le Chatelier's principle. The latter is responsible in the most cases of lipid and protein reactions, which will be discussed later. (We will not touch here the concept of the role of the electric field as a general chemical energy, according to Mitchell's theory).

TAB.3
Influences of the membrane electric field on membrane and cell properties

INFLUENCE ON ION- AND pH-CONTROL:

- Shift of the inside-out equilibrium of ions
- Production of local ion concentrations and concentration gradients
- Alteration of the driving force in processes of electrodiffusion

INFLUENCE ON MOLECULAR MEMBRANE COMPOUNDS:

- Membrane compression
- Orientation and translocation of lipid head groups
- Modification of lipid microviscosity
- Modification of the glycocalyx structure
- Translocation of mobile charged molecules
- Orientation of polarized or charged protein compounds
- change in lateral orientation of membrane compounds

FUNCTIONAL CONTROL OF FIELD SENSIBLE MOLECULES

- field sensitive gated transporters

As mentioned before, the membrane electric field does not act in all cases as a driving force. It partly reflects equilibrium situations, as for example in the case of electrical double layers. In some situations, however, it is the result of an electrochemical non-equilibrium and is therefore a driving force for all ions and charged molecules.

In the last years, an increasing number of co- and antiport systems with highly specific stochiometric relation of transported species were found. This demonstrates the existence of a sofisticated system of charge transfer through the membrane and therefore limits free electrodiffusion. Subsequently the transmembrane potential in several cells was found to be independent of the external potassium concentration (Bashford and Pasternak 1986). Furthermore, sodium-transporters in erythrocytes and many other cells are independent of the transmembrane potential (Kracke and Dunham 1987). The ions in these cases do not follow simple electrodiffusional processes but cross the membrane by electrically silent transporters. These highly stochiometric transport mechanisms obviously are special adaptations. They help the cells in keeping fluxes and concentrations constant even if theire transmembrane potential changes by functional reason. Later we will see that even electrosilent transporters may be affected by the membrane electric field.

For erythrocytes, the simplest animal cell, the interplay of membrane potential and ion distribution has been calculated and described in the form of electrochemical models (Brumen et al. 1979, Freedman and Hoffman 1979, Glaser 1979, Glaser and Donath 1984). In this case equations for water, chloride- and pH-equilibration, titration properties of hemoglobine etc. have been solved numerically. In the future, such models have to be built up, considering not only electroneutrality, but additionally a strong stochiometry of all fluxes.

The electrical double layers and ionic clouds in the vicinity of fixed charges can be considered as equilibrium structures even if the whole cell is far from equilibrium. In this case local concentrations ($c_i(x)$) are governed by the Boltzmann relationship modified by the Nernst equation:

$$c_i(x) = c_i(o) \exp \frac{z_i F \psi(x)}{RT}$$

($c_i(o)$ - concentration of the species i in the bulk phase, z_i - charge number, F - Faraday, R - gas constant, T - Temperature) For a surface potential $\psi(x)$=60 mV, this implies a local concentration of univalent ions 10 times higher (or lower depending on the charge) than in the bulk phase. For bivalent ions this difference will be reached at local potentials of 30 mV. McLaughlin (1989) points out the role of local potentials on the concentration of charged substrates (for example polypeptides with z=+6.).

These local concentrations can simply act as surface compartments (Blank and Kavanaugh 1982, Blank et al. 1982, Gingell 1967, Glaser et al. 1983) or they can create specific ionic conditions, including pH and a specific combination of mono- and polyvalent cations (Berith and Nir 1980, Engasser and Horvath 1975, Riddell and Arumugam 1988, Ravindran and Moczydlowski 1989, Wojtczak and Nalecz 1979). This specificity is given by the charge number z_i according to this equation as well as by specific adsorption constants which are especially important for polyvalent ions (McLaughlin 1982, 1989).

It is not surprising that lipid bilayers with charged or dipolar head groups will be influenced by membrane electric field, since the electric field influences the orientation of head groups, and compresses films with charged surfaces (like biological membranes in general). This strongly influences the mobility and position of the hydrocarbon parts of the lipids. Träuble and Eibl (1974), Jähnig (1976) and Forsyth et al (1977) indicated that an increase in electric field strength leads to phase transitions in artificial lipid membranes, in the same manner as is known for a rise in temperature. Corda et al. (1982) and O'Shea et al. (1984) found potential- dependent changes in the microviscosity of artificial phospholipid vesicles. Stulen (1981) showed for multiple bilayers of phosphatidylcholine and potassium oleate, that up to 10^6 V/m the field greatly affects the conformation of the headgroups, while no influence on the structure and dynamics of the hydrocarbon chains could be detected.

Biological membranes usually comprise a mixture of lipids without a distinct phase transition point and without a well defined viscosity. Nevertheless, even in cells, field dependent modifications of lipid mobility have been detected (Muller and Herrmann 1987). These results lead to the conclusion that a specific lipid composition in the membrane may affect protein properties, either creating local ionic conditions near the proteins, or changing the fluidity of the protein surroundings. (Woolley and Eibl 1977).

Proteins of course may interact directly with the electrical field in a similar way as lipids. The most flexible parts of membrane proteins are the glycoprotein chains, forming the glycocalyx (see FIG. 2). This structure is of course strongly influenced by electrostatics. This is important in processes of cell-cell and cell-surface interactions (Donath and Voigt 1983, Lerche 1987).

Kempf et al. (1982) discussed whether the transmembrane potential may influence the disposition of proteins in the membrane. After being included in the membrane, the proteins, however, are in a quite stable position. Even dramatic alterations of the electric field are unable to flip the whole molecule across the membrane. For this the hydrophilic protein regions on both sides of the membrane has to be transferred through the hydrophobic phase of the lipids. Smaller membrane compounds, however, including various drugs can be shifted by the field and therefore translocated by field alterations. All these field induced modifications of the membrane may lead to membrane bending and other mechanical alterations. This may be the basic mechanism which underlies the transmembrane dependent transformations of the red cell shape (Glaser 1979, 1982, Glaser et al. 1987, 1988).

Investigations of field effects on proteins inserted in artificial membranes indicate topological and structural alterations (Blumenthal et al. 1983, Brumfeld and Miller 1988). In general, the electric field generates displacements of fixed charges and permanent dipols of these molecules. If the field interacts with induced dipols the resulting effects are independent of the polarity of the field. This explains the observation that the Na-K-transport in human erythrocytes has a lowered activity at zero transmembrane potential (Glaser et al. 1983, Krake and Dunham 1987). In the same sense Brumfeld and Miller (1988) showed field dependent conformations in rhodopsin-molecules which depend strongly on field strength, but are independent of the field orientation.

In many cases permanent and induced dipols show a resonance behaviour in alternating fields. The application of AC fields in kHz frequencies, for example, indicate an electroconformational coupling especially to transport ATPases (Tsong and Astumian 1987). In the same context one has to understand that even electrically silent transport systems may become field dependent. Apart from direct structural modifications of the transport protein even binding equilibria can be shifted according to the Le Chatelier's principle (Läuger 1987, Läuger and Apell 1988).

It is hard to draw a borderline between membrane proteins which are occasionally influenced by the electric field and channels which are specially adopted for voltage gating. Classical examples of field sensitive proteins are the Na- and K-channels in neurons. In recent years a large number of voltage gated channels have been found in various cells. A similarly large number of models exist to explain these mechanisms (see for example: Catterall 1988, Edmonds 1988, Jan and Jan 1989, Levitt 1989).

In contrast to the multitude papers on the functional role of the electric field in x-direction, few investigations have been reported on electric field effects in the plane of the membrane. Mostly observations of lateral translocations of proteins under the influence of an externally applied electric field have been undertaken in order to study the mobility of proteins. Nevertheless it is quite possible that lateral electrophoresis of functional proteins can occur in vivo. This could be fundamental to understanding the membrane mosaic as a true dynamical system (for detaile see: Edidin and Wei 1977, Gross 1988, McLaughlin and Poo 1981, Poo et al. 1979, Ryan et al. 1988).

Recently Fromherz (1988 a, b, c, 1989) described the distribution of channels and the profile of the potential in a linear cable like cell with a fluid mosaic membrane. He demonstrated the formation of periodic patterns as dissipative structures, and pointed out that in a biological membrane the coherent patterns of channels and potential may be of direct functional relevance.

5. The membrane electric field as a cellular control system

The question raised at the beginning of this chapter now has to be answered: are the correlations, listed in TAB. 1 merely side-effects or does a change in the transmembrane potential trigger a functionally important sequence of cellular processes? The same question is to be asked concerning the correlation of $\Delta\psi$ and the mitotic activity of cells as depicted in FIG. 1.

Despite the fact that the list of these correlations includes quite heterogeneous functions, the following considerations speak for the functional role of the membrane electric field:

- Alteration of the electric field leads to a series of definite membrane and compartmental consequences. In some cases it is shown that membrane alterations, correlating with particular electrical potentials, occur independently of the way and under which ionic conditions these potentials have been induced (see for example R. Glaser et al. 1988)
- The electrochemical gradient of ions across all biological membranes, especially the Na/K-nonequilibrium between the internal and external media, which is built up by metabolic energy, can only be understood as a storage battery. There is no other reason explaining its function.
- At least in the case of excitable cells, the mechanism of permeability controlled and triggered transmembrane potential is evident and has been well known for a long time. It seems reasonable to assume that other cells, building up the same electrochemical grandient use it in a similar way.

The variety of the phenomena (see TAB. 1), the diversity of the mechanisms (see TAB. 3) and finally the complexity of the field (see FIG. 2), are reasons why we are still far from being able to formulate a general conclusion.

The increasing insight into the ionic metabolism of animal cells, especially the investigations of the variety of groups of extremely specific transporters with definite electrical properties, have supported strongly our understanding of the electrophysiological phenomena of the cell. The classical view, considering an active transport of ^{+}Na and ^{+}K and subsequently an induced diffusional potential, has to be changed dramatically. It seems that we are now just beginning to understand electrical membrane phenomena, regardless to the fact that these investigations trace back to Galvani and Volta two centuries ago, as we mentioned in the beginning of this paper.

REFERENCES

Bashford C.L., Pasternak C.A. (1986) Plasma membrane potential of some animal cells is generated by ion pumping, not by ion gradients. Trends in Biochem. Sciences 11: 113-116

Bassilana M., Damiano-Forano E., Leblanc G. (1985) Effect of membrane potential on the kinetic parameters of the Na or H melibiose symport in Escherichia coli membrane vesicles. Biochem. Biophys. Res. Commun. 129: 626-631

Bentz J., Nir S. (1980) Cation binding to membranes: Competition between mono-, di-, and trivalent cations. Bull. Mathem. Biol. 42: 191-220

Betz W.J., Caldwell J.H., Ribchester R.R., Robinson K.R., Stump R.F. (1980) Endogenous electric field around muscle fibres depends on the K -Na -pump. Nature 287: 235-237

Bingelli R., Weinstein C. (1986) Membrane potentials and sodium channels: Hypotheses for growth regulation and cancer formation based on changes in sodium channels and gap junctions. J. Theoret. Biol. 123: 377-401

Blank M. (Ed.) (1985) Electrical double layers in biology, Plenum Press, New York, London

Blank M., Kavanaugh W.P. (1982) the surface compartment model (SCM) during transients. Bioelectrochem. Bioenerget. 9: 427-437

Blumenthal R., Kempf Ch., Van Renswoude J., Weinstein J., Klausner R.D. (1983) Voltage-dependent orientation of membrane proteins. J.Cell Biochem. 22:55-67

Brumen M., Glaser R., Svetina S. (1979) Osmotic states of the red blood cell. Bioelectrochem. Bioenerget. 104: 227-241

Brumfeld V., Miller I.R. (1988) Effect of membrane potential on the conformation of bacteriorhodopsin resonstituted in lipid vesicles. Biophys. J. 54: 747-750

Catteral W.A. (1988) Structure and function of voltage-sensitive ion channels. Science 242: 50-61

Chay T.R. (1986) On the effect of the intracellular calcium-sensitive K channel in the bursting pancreatic β-cell. Biophys. J. 50: 765-777

Clementz T., Christiansson A. (1986) Transmembrane electrical potential affects the lipid composition of Acholeplasma laidlawii. Biochemistry 25: 823-830

Cone C.D.Jr. (1970) Variation of the transmembrane potential level as a basic mechanism of mitosis control. Oncology 24: 438-470

Corda D., Pasternak C., Shinitzky M. (1982) Increase in lipid microviscosity of unilamellar vesicles upon the creation of transmembrane potential. J. Membrane Biol. 65: 235-242

Dolowy K. (1987) Membrane-potential-controled lipid vesicle recycling and amoeboid locomotion. J. Bioelectricity 6: 109-128

Donath E., Voigt A. (1983) Charge distribution within cell surface coats of single and interacting surfaces - a minimum free electrostatic energy approach. Conclusions for eletrophoretic mobility measurements. J. Theoret. Biol. 101: 569-584

DosReis G.A., Oliveira-Castro G. M. (1977) Electrophysiology of phagocytic membranes. I. Potassium-dependent slow membrane hyperpolarization in mice macrophages. Biochim. Biophys. Acta 469: 257-263

DosReis G.A., Persechini P.M., Marros J., Ribeiro C., Oliveira--Castro G.M. (1979) Electrophysiology of phagocytotic membranes II. Membrane potential and induction of slow hyperpolarization

in activated macrophages. Biochim. Biophys. Acta 552: 331-340
Edidin M., Wei.T. (1977) Diffusion rates of cell surface antigens of mouse human heterokaryons. II. Effect of membrane potetial on lateral diffusion. J. Cell. Biol. 75: 483-489
Edmonds D.T. (1988) An alternative electrostatic model for membrane ion channels. In: Markov M., Blank M.: Electromagnetic Fields and Biomembranes. Plenum Press, New York 27-35
Engasser J.M., Horvath C. (1975) Electrostatic effects on the kinetics of bound enzymes. Biochem. J. 145: 431-435
Epel D. (1980) Ionic triggers in the fertilization of sea urichin eggs. Ann. N.Y. Acad. Scinces 339: 74-85
Forsyth P.A., Marcelja S., Mitchell D.J. (1977) Phase transition in charged lipid membranes. Biochim. Biophys. Acta 469: 335-344
Freedman J.C., Hoffman J.F., (1979) Ionic and osmotic equilibria of human red blood cells treated with nystatin. J. Gen. Physiol. 74: 157-185
Friedhoff L.T., Kim E., Priddle M., Sonenberg M. (1982) The effect of altered transmembrane ion gradients on membrane potential and aggregation of human platelets in blood plasma. Biochem. Biophys. Res. Commun. 102: 832-837
Fromherz P. (1988a) Self-organization of the fluid mosaic of charged channel proteins in membranes. Proc. Natl. Acad. Sci. USA 85: 6353-6357
Fromherz P. (1988b) Spatio-temporal patterns in the fluid-mosaic model of membranes. Biochim. Biophys. Acta 944: 108-111
Fromherz P. (1988c) Dissipative structures of ion channels in the fluid mosaic model of a membrane cable. Ber. Bunsenges. Phys. Chem. 92: 1010-1016
Fromherz P. (1989) Dissipative spinodal condensation of membrane proteins. (1989) Chemical Physics Letters 154: 146-150
Gingell D. (1967) Membrane surface potential in relation to possible mechanism for intercellular interactions and cellular responses: A physical basis. J. Theoret. Biol. 17: 451-482
Glaser A, Donath E (1988) Osmotically induced vesicle-membrane fusion in plant protoplasts. studia biophysica 127: 129-138
Glaser R. (1979) The shape of red blood cells as a function of membrane potential and temperature. J. Membrane Biol. 51: 217-228
Glaser R. (1982) Echinocyte formation induced by potential changes of human red blood cells. J. Membrane Biol. 66: 79-85
Glaser R., Bernhardt I., Donath E., Heinrich R., Gaestel M. (1983) Ion transport in human red blood cells as a function of electric potential profile. In: G. Spach (Editor): Physical chemistry of transmembrane ion motion, Elsevier Publ. Amsterdam 635-644
Glaser R., Donath J. (1984) Stationary ionic states in human red blood cells. Bioelectrochem. Bioenergetics 13: 71-83
Glaser R., Fujii T., Müller P., Tamura R., Herrmann A. (1987) Erythrocyte shape dynamics. Influence of electrolyte conditions and membrane potential. Biomed. Biochem. Acta 46: 327-333
Glaser R., Gengnagel C., Donath J. (1988) Membrane electric field and erythrocyte shape. studia biophysica 127: 201-206
Gould-Somero M., Jaffe L.A., Holland L.Z. (1979) Electrically mediated fast polystermy block in eggs of the marine worm, Urechis caupo. J. Cell Biol. 82: 426-440
Gross, D. (1988) Electromobile surface charge alters membrane potential changes induced by applied electric fields. Biophys. J. 54: 879-884
Heinrich R., Gaestel M., Glaser R. (1982) The electric potential

profile across the erythrocyte membrane. J. Theoret. Biol. 96: 211-231
Hoffmann E.K., Simonsen L.O., Sioholm C. (1979) Membrane potential, chloride exchange and chloride conductance in Ehrlich mouse ascites tumor cells. J. Physiol. (London) 296: 61-84
Honig B.H., Hubbell W.L., Flewelling R.F. (1986) Electrostatic interactions in membranes and proteins. Ann. Rev. Biophys. Chem. 15: 163-193
Jaffe L.F. (1979) Control of development by ionic currents. In Cone R.A., Dowling J.E. (Editors): Membrane transduction mechanisms, Rawen Press, New York, 199-231
Jaffe L.F. (1981) Control of development by steady ionic currents. Federation proceedings 40: 125
Jaffe L.A., Gould-Somero M., Holland L.Z. (1982) Studies on the mechanism of the electrical polyspermy block using voltage clamp during cross-species fertilization. J.Cell Biol. 92: 616-621
Jaffe L.F., Stern C.D. (1979) Strong electrical currents leave primitive streak of chick embryos. Science 206: 569-571
Jähnig F. (1976) Electrostatic free energy and shift of the phase transition for charged lipid membranes. Biophys. Chem. 4: 309-318
Jan L.Y., Jan Y.N. (1989) Voltage-sensitive ion channels. Cell 56: 13-25
Jou D., Perez-Garcia C., Clebot J.E. (1986) Bacterial flagellar rotation as a nonequiligrium phase transition. J. Theoret. Biol. 122: 453-458
Kempf Ch., Klausner R.D., Weinstein J.N., Van Renswoude J, Pincus M., Blumenthal R. (1982) Voltage-dependent trans-bilayer orientation of melittin. J. Biol. Chem. 257: 2469-2476
Kessler M., Semenza G. (1983) The small-intestinal Na^+,D-glucose cotransporter: An asymmetric gated channel (or pore) responsive to $\Delta\psi$. J. Membrane Biol. 76: 27-56
Kiefer H., Blume A.J., Kaback H.R. (1980) Membrane potential changes during mitogenic stimulation of mous spleen lymphocytes. Proc. Natl. Acad. Sci. U.S.A. 77: 2200-2204
Kracke G.R., Dunham P.B. (1987) Effect of membrane potential on furosemid-inhibitable sodium influx in human red blood cells. J. Membrane Biol. 98: 117-124
Larter R., Schmidt S., Ortoleva P. (1979) Electrical effects in nonlinear physico-chemical systems: field chemical wave interaction and bio-self-electrophoresis. In: Haken, H. (Editor): Pattern formation by dynamic systems and pattern recognition, Springer Berlin, Heidelberg, New York, 144-154
Läuger, P. (1987) Voltage dependence of sodium-calcium exchange: Predictions from kinetic models. J. Membrane Biol. 99: 1-11
Läuger P., Apell H-J. (1988) Voltage dependence of particle reactions of the Na /K pump: Prediction from microscopic models. Biochim. Biophys. Acta 945: 1-10
Lerche D., (1987) Electrostatic field and their influence on surface structure, shape and deformation of red blood cells. In: M. Blank and E. Findl (Editors): Mechanistic Approaches to Interactions of Electric and Electromagnetic Fields with Living Systems. Plenum Press, New York, 349-365
Levitt D. G. (1989) Continuum model of voltage-depending gating. Macroscopic conductance, gating current, and single-channel behaviour. Biophys. J. 55: 489-498
McCann F.V., Cole J.J., Guyre P.M., Russel J.A.G. (1983) Action po-

tentials in macrophages derived from human monocytes. Science 219: 991-993
McDonald T.F., Sachs H.G., Orr C.W., Ebert J.D. (1972) External potassium and baby hamster kidney cells: intracellular ions, ATP, growth, DNA synthesis and membrane potential. Develop. Biol. 28: 290-303
McLaughlin S. (1982) Divalent cations, electrostatic potentials, bilayer membranes. In: A. N. Martonosi (Editor): Membranes and Transport, Plenum Press, Vol. 1, 51-55
McLaughlin S. (1989) the electrostatic properties of membranes. Ann. Rev. Biophysics 18 (in press)
McLaughlin St., Poo M.-M. (1981) The role of electro-osmosis in the electric-field-induced movement of charged macromolecules on the surface of cells. Biophys. J. 34: 85-93
Nagura H., Asai J., Kojima K. (1977) Studies on the mechanism of phagocytosis. I Effect of electric surface charge on phagocytic activity of macrophages for fixed red cells. Cell Structure and Function 2: 21-28
Müller P., Herrmann A. (1987) Ionic strength dependent alterations of membrane structure of red blood cells. Bioscience Reports 6: 1007-1015
Nicolson G.L., Painter R.G. (1973) Anionic sites of human erythrocyte membranes. II. Antispectrin-induced transmembrane aggregation of the binding sites of positively charged colloidal particles. J. Cell Biol. 59: 395-406
Ohki, S. (1981) Membrane potential, surface potential, and ionic permeabilities. Physiol. Chem. Phys. 13: 195-210
Okada Y., Tsuchiya W., Inouye A. (1979) Oscillations of membrane potential in L cells. IV. Role of intracellular Ca in hyperpolarizing excitability. J. Membrane Biol. 47: 357-376
Okada Y., Tsuchiya W., Yada T., Yano J., Yawo H. (1981) Phagocytic activity and hyperpolarising responses in L-strain mouse fibroblasts. J. Physiol. (London) 313:101-119
O'Shea P.S., Feuerstein-Thelen S., Azzi, A. (1984) Membrane-potential dependent changes of the lipid microviscosity of mitochondria and phospholipid vesicles. Biochem. J. 220: 795-801
Pales J., Lopez H., Gual A. (1988) Platelet membrane potential as a modulator of aggregating mechanisms. Biochim. Biophys. Acta 944: 85-89
Poo M., Lans J.W., Orida N., Chao A.W. (1979) Electrophoresis and diffusion in the plane of the cell membrane. Biophys. J. 26: 1-21
Ravindran A., Moczydlowski E. (1989) Influence of negative surface charge on toxin binding to canine heart Na channels in planar bilayers. Biophys. J. 55: 359-365
Ridell F.G., Arumugam S. (1988) Surface charge effects upon membrane transport processes: the effect of surface charge on the monensin-mediated transport of lithium ions through phospolipid bilayer studied by Li-NMR-Spectroscopy. Biochim. Biophys. Acta 945: 65-72
Ryan T.A., Myers J., Holowka D., Bird B., Webb W.W. (1988) Molecular crowding on the cell surface. Science 238: 61-64
Sachs H.G., Stambrook P.J., Ebert J.D. (1974) Changes in membrane potential during the cell cycle. Exp. Cell Res. 83: 362-366
Serpersu E.H., Tsong T.Y. (1984) Activation of electrogenic Rb transport of (Na,K)-ATPase by an electric field. J. Biol. Chem. 259: 7155-7162

Stulen G. (1981) Electric field effects on lipid membrane structure. Biochim. Biophys. Acta 640: 621-627

Sugimoto Y., Hagiwara A. (1979) Cell locomotion on differently charged substrates. Effect of substrate charge on locomotive speed of fibroblastic cells. Exp. Cell Res. 120: 245-252

Taglietti V. (1979) Early electrical responses to fertilization in sea urchin eggs. Exp. Cell Res. 120: 448-451

Takeda J., Abe S., Moeikawa H., Senda M. (1983) Membrane potentials of heterotrophically cultured tobacco cells. Plant and Cell Physiol. 24: 667-676

Träuble H., Eibl H. (1974) Electrostatic effects on lipid phase transitions: membrane structure and ionic environment. Proc. Natl. Acad. Sci. U.S.A. 71: 214-219

Träuble H., Teubner M., Wooley P., Eibl H. (1976) Electrostatic interaction at charged lipid membranes. Biophys. Chem. 4: 319-342

Tsong T.Y., Astumian R.D. (1987) Electroconformational coupling and membrane protein function. Progr. Biophys. molec. Biol. 50: 1-45

Tsuchiya W., Okada Y., Yano J., Murai A., Miyahara T., Tanaka T. (1981) Membrane potential changes associated with pinocytosis of serum lipoproteins in L cells. Exp. Cell Res. 136: 271-278

Van Houten J. (1979) Membrane potential changes during chemokinesis in Paramecium. Science 204: 1100-1103

Vilereal M.L., Cook J.S. (1977) Role of the membrane potential in serum-stimulated uptake of amino acid in a diploid human fibroblast. J. Supramol. Struct. 6: 179-189

Weinstein J.N., Blumenthal R. van Renswoude J., Kempf C., Klausner R.D. (1982) Charge Clusters and the orientation of membrane pro teins. J. Membrane Biol. 66: 203-211

Williams J.A. (1981) Electrical correlates of secretion in endocrine and exocrine cells. Federation Proceedings 40: 128-134

Wojtczak L., Nalecz M.J. (1979) Surface charge of biological membranes as a possible regulator of membrane-bound enzymes. Europ J. Biochem. 94: 99-107

Wooley P., Eibl H. (1977) Electrostatic interaction of charged lipid membranes. FEBS-Letters 74: 14-16

Zade-Oppen A.M.M., Schooler J.M.Jr., Cook P., Tosteson D.C. (1979) Effect of membrane potential and internal pH on active sodium--potassium transport and on ATP content in high potassium sheep erythrocytes. Biochim. Biophys. Acta 555: 285-298

Phospholipid vesicle fusion induced by cations and poly-(ethylene glycol)

S. Ohki and K. Arnold*

Department of Biophysical Sciences
State University of New York at Buffalo
224 Cary Hall
Buffalo, New York 14214
USA

INTRODUCTION

Membrane fusion is one of the essential events involved in many biological cellular processes such as exocytosis, endocytosis, membrane recycling, cell membrane assembly, fertilization, viral and bacterial infectious processes, etc. (Poste and Allison 1973, Rasmussen 1970, Blumenthal 1987, Ohnishi 1988). In spite of extensive research on membrane fusion, its molecular mechanism is not well understood. This is probably due to the fact that membrane fusions are attained through various molecular pathways depending on different biological fusion events (Duzgunes and Bronner 1988), and also biological membranes are composed of complex molecular structures (Houslay and Stanley 1982).

In order to gain insight into the molecular mechanism of such biological membrane fusion events, model lipid membranes with defined molecular components and simpler molecular structures, have been utilized for membrane fusion studies for the last one and a half decades (Poste and Nicolson 1978, Sowers 1987, Ohki et al 1988).

For lipid membrane fusion studies, a number of methods to induce fusion have been found (Sowers 1987). Some cations can induce fusion of acidic lipid membranes (Papahadjopoulos 1974, Breisblatt and Ohki 1976). Macromolecules such as poly-(ethylene glycol) (Knutton 1979) and certain polypeptides (Gad et al 1982, Morgan et al 1983, Blumenthal 1987) are found to induce fusion of not only negatively charged lipid membranes, but also neutrally charged lipid membranes. Short-chain fatty acids could be fusogenic agents (Ahkong et al 1973b, Kantor and Prestegard 1975, Lucy 1970, 1978). Temperature either increasing or decreasing (Ahkong 1973a, Schullery et al 1980, Hui et al 1981) also induces membrane fusion. Osmotic pressure enhances or inhibits membrane fusion

*Present address
Institute of Biophysics, Karl-Marx University
27 Liebig Str., Leipzig 7010 GDR

induced by fusogenic agents (Miller et al 1976, Cohen et al 1980, Lucy and Ahkong 1986, 1988). Certain lipids, themselves, could serve or make membranes susceptible to fusion (Cullis and Hope 1978, Verkleij et al 1980, and others), and recently an applied electric field of either D.C. or A.C. is utilized to induce membrane fusion (Zimmermann 1982, Sowers 1986).

In this chapter, we will discuss two of the above mentioned membrane fusion topics: ion-induced lipid membrane fusion and poly-(ethylene glycol) induced lipid membrane fusion. Among membrane systems, our discussion will be limited on fusion in lipid vesicle systems. It is known that Ca^{2+} is involved in most biological membrane fusion, especially in exocytotic processes (Eckert and Randall 1978). One of such biological membrane fusion events is the process occurring at the presynaptic terminal, where the synaptic vesicles in the nerve terminal exocytose their internal contents by Ca^{2+}-induced fusion with the presynaptic membrane (Miledi 1973, Heuser and Reese 1973). However, the fusion reactions seen in such exocytotic processes induced by Ca^{2+} may involve certain proteins which interact with Ca^{2+} (Creutz et al 1978, Blitz and Fine 1974, Gratzl et al 1980). In spite of these facts many electron microscope studies resolve the fusing area between two cells to be closely apposed lipid bilayers and membraneous particles to be cleared away from these regions (Zakai et al 1977, Lucy 1978). The effect of divalent cations on lipid membrane fusion has been extensively investigated in relation with other ions in the vesicle medium.

It has been found that acidic lipid vesicles can be fused by not only Ca^{2+}, but also other divalent and polyvalent metal cations. There is a critical or "threshold" concentration for each fusogenic cation which is necessary to induce membrane fusion. The extent of fusion is found to be well correlated with the increase in interfacial tension of the monolayer, as well as the decrease in surface dielectric constant of the membrane induced by these fusogenic cations. These changes in physico-chemical quantities are the consequence of increased hydrophobicity of the membrane surface due to the special ion binding to the membrane.

The critical or threshold concentration of cations necessary for exocytotic vesicle fusion to the plasma membrane in vivo can be much less than those determined in vitro under comparable ionic conditions, except for the absence of other macromolecules. In order to compare the critical concentration of cation necessary for vesicle fusion in vivo the modification of ion-induced lipid vesicle fusion by introducing certain proteins has been attempted and some have been successful in lowering the critical concentration of fusogenic ions (Hong et al 1981, 1982, Papahadjopoulos 1988).

Poly-(ethylene glycol), PEG, has surface properties very similar to most proteins and polysaccharides (van Oss et al 1987). Therefore, PEG may be used to mimic the behavior of other proteins in a solution, and is used here for lipid vesicle fusion. Unlike other non-ionic polymers, PEG itself is able to induce the fusion of lipid vesicles to a greater extent at concentrations larger than about 20 wt% (Blow et al 1978, Tilcock and Fisher 1979, Boni et al 1981, 1984, MacDonald 1985, Parente and Lentz 1986, Hoekstra 1982, Arnold et al 1987, 1988). It has also been used to fuse biological cells for practical purposes (e.g. hybridoma cells for production of monoclonal antibodies.

It was shown that surfaces of proteins, liposomes and erythrocytes are depleted of PEG (Arnold et al 1987, 1988, Pratsch and Donath 1988). This depletion is the origin of the high flocculating power of PEG for lipid vesicles and for the facilitation of lipid vesicle fusion. An enhancement of the cation-induced fusion of PS vesicles was described by Hoekstra (1982). Beside the high ability of PEG to bind water (Tilcock and Fisher 1982, Molynex 1975) and the strong aggregation of particles, the decrease in the polarity and the dielectric constant of the aqueous phase (Arnold et al 1983, 1985) and the high osmotic pressure (Arnold et al 1988, van Oss et al 1989), are important components of the action of PEG on fusion process.

In this chapter, we will describe some of our experimental results of the above mentioned ion-induced acidic lipid vesicle fusion and poly-(ethylene glycol)-induced lipid vesicle fusion and how the increase in membrane hydrophobicity induced by ion binding to the membrane and the depletion of interlayer water by PEG are important for membrane fusion. The similarities and differences of the molecular mechanisms of these two types of membrane fusion will be discussed as well.

LIPID VESICLE FUSION

A. Fusion of acidic lipid vesicles induced by cations.

A-1) Fusion

For most cases of ion-induced lipid vesicle membrane fusion, acidic lipid vesicles and cation systems are used (Papahadjopoulos 1978). Positively charged or neutrally charged lipid vesicles are rather difficult to be fused by either positively charged or negatively charged ions. Fig. 1 shows some of the experimental results of small unilamellar phosphatidylserine (PS) vesicle fusion induced by various cations, where the PS vesicles were suspened in 0.1 M NaCl buffer, pH 7.4 having various concentrations of fusogenic ions. Fusion was monitored by the fluorescence content mixing assay ($TbCl_3$ and Dipicolinic acid (DPA) (Wilschut and Papahadjopoulos 1979, Ohki 1982)

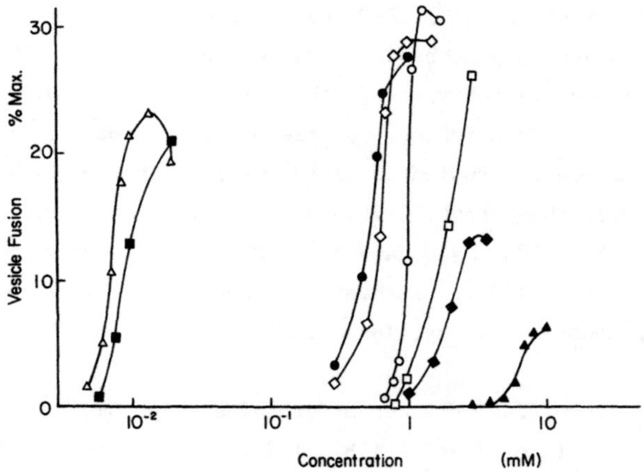

Figure 1. Fusion of phosphatidylserine vesicles in 0.1 M NaCl with various cation concentrations. The ordinate refers to the fluorescence intensity at 492 nm and the abscissa refers to concentration of various cations. The fluorescence intensity was corrected for the background scattered light by subtracting the scattered light intensity at 492 nm. The intensity is given in percentage of the maximum fusion. The wavelength of excitation light was 272 nm. PS vesicles were suspended in 0.05 mM for all divalent cation cases and 5 μM for the trivalent cases. (△) La^{3+}; (■) Tb^{3+}; (●) Mn^{2+}; (◇) Ba^{2+}; (○) Ca^{2+}; (◆) Sr^{2+}; (▲) Mg^{2+}; (□) H^+. (Ohki 1982, 1988a)

There is a specific concentration for each fusogenic cation, at which the rate of vesicle fusion is the greatest. Such a concentration is defined as the "fusion threshold" concentration of the ion. From Fig. 1, such fusion threshold concentrations are 4 ~ 6 μM for Tb^{3+} and La^{3+}, 0.6 mM for Mn^{2+}, 0.75 mM for Ba^{2+}, 1 mM for Ca^{2+}, 1.6 mM for Sr^{2+}, 6 mM for Mg^{2+} and about 1 ~ 3 mM for H^+. It appears that a special type of ion binding to negatively charged polar groups of the membrane is responsible for the induction of such membrane fusions. Further details about the special types of ion binding will be discussed in the latter section of this chapter. Not all cations induce fusion of the PS vesicles. Monovalent cations, except for H^+, do not induce vesicle fusion, or polyamines, which are multi-valent cations in the solution, do not induce fusion. For the phosphatidic acid (PA) vesicle-cation systems, vesicle fusion similar to those mentioned in the above is observed (Papahadjopoulos et al 1976, Sundler and Papahadjopoulos 1982, Ohki and Ohshima 1985) and the fusion threshold concentrations of fusogenic ions are a little smaller in value than those obtained for the PS-cation system. For other lipid membrane systems, ion-induced membrane fusion has been studied and the results are summarized in Table 1.

The fusion threshold concentrations of ions vary depending on the types

Table 1 Ion-Induced Fusion of Various Lipid Vesicles and Their Threshold Concentrations

Lipids	Types	Threshold Conc.*	References
PS	SUV	>1 mM Ca^{2+}	Papahadjopoulos et al (1974)
PS	LUV ($0.1\mu\phi$)	2.4 mM Ca^{2+}	Wilschut et al (1981)
PS	LUV ($0.1\mu\phi$)	no fusion for Mg^{2+}	Wilschut et al (1981)
PS/PC(1:1)	SUV	$Mg^{2+} > Ca^{2+}$	Ohnishi & Ito (1974)
PS/PC(1:1)	SUV	>4 mM Ca^{2+}	Papahadjopoulos et al (1974)
PS	SUV/LUV($0.1\mu\phi$)	1.2 mM Ca^{2+}	Duzgunes & Ohki (1981)
PS/PC(1:1), PS	SUV(PS/PC)/ LUV(PS)	3-5 mM Ca^{2+}	Duzgunes & Ohki (1981)
DPPG	SUV	5 mM Ca^{2+}	Papahadjopoulos et al (1974)
PA	SUV	0.2 mM Ca^{2+} Mg^{2+}	Papahadjopoulos et al (1976)
PA	LUV ($0.1\mu\phi$)	0.03-0.1mM Ca^{2+}	Sundler & Papahadjopoulos (1982)
PA	LUV ($0.1\mu\phi$)	0.07-0.15mM Mg^{2+}	Sundler & Papahadjopoulos (1982)
PI	LUV ($0.1\mu\phi$)	no fusion up to 50 mM Ca^{2+}	Sundler & Papahadjopoulos (1982)
CL	SUV	Ca^{2+} Mg^{2+}	Vail & Stollery (1979)
PC/PS(1:1)	SUV	5 mM Ca	Hoekstra (1979)
PS,PS/PC, Ps.PE	SUV	Ca(PS/PC)>Ca (PS/PE)>Ca(PS)	Vanderwerf & Ullman (1980)
PS,PS/PC, PS/PE	LUV ($0.1\mu\phi$)	Ca(PS/PC)>Mg (PS/PE)	Duzgunes et al (1981)
PS,PS/PC, PS/PE	LUV ($0.1\mu\phi$)	Mg(PS)>Mg(PS/PE)	Duzgunes et al (1981)
PS/PC, PS/PC/Chol	SUV	Ca(PS/PC)>Ca (PS/PC/Chol)	Uster & Deamer (1981)
PS/PC/PE	SUV	Ca(PS/PC)>Ca (PS/PC/PE)	Uster & Deamer (1981)
CL/PC	SUV	Ca^{2+}	Verkleij et al (1980)

PS: phosphatidylserine, PC: phosphatidylcholine, PE: phosphatidylethanolamine, PA: phosphatidic acid, PG: phosphatidylglycerol, PI: phosphatidylinositol, CL: cardiolipin, SUV: small unilamellar vesicle, LUV: large unilamellar vesicle. *In most cases, the vesicles were suspended in 0.1 M NaCl.

of lipids used for vesicles (shown in Table 1) and also are different depending on the size of vesicles. Fig. 2 shows some of the experimental results for the PS vesicles having different diameters. The large unilamellar lipid vesicles are prepared by a modified reverse phase evaporation method (Szoka and Papahadjopoulos 1978, Ohki 1984a). The average sizes of vesicles in different elutions from a sepharose CL-2B column, are determined by use of a photon-correlation spectrophotometer (Submicron particle analyzer, Coulter N-2).

For vesicles having a given diameter, the order of the strength of fusogenic cations to induce fusion is the same for all cations examined: $Mn^{2+} > Ca^{2+} > Mg^{2+}$. However, the magnitude of the fusion threshold concentration of cations increases monotonically with the increase in the diameter

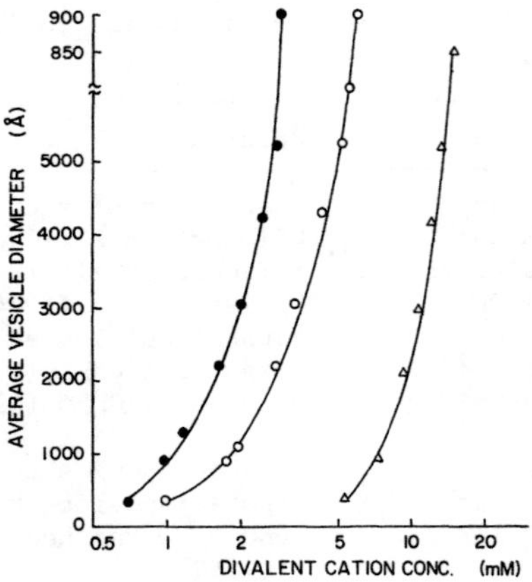

Figure 2. Fusion threshold concentrations of Mn^{2+} (●), Ca^{2+} (O), and Mg^{2+} (△) versus the sizes of phosphatidylserine vesicles in 0.1 M NaCl buffer, pH 7.4. The internal content mixing (Tb-DPA) fusion assay was used.

of vesicles. Earlier results showed (Wilshut et al 1981, Ohki 1984) that Mg^{2+} did not induce acidic lipid vesicles with a diameter greater than approximately 0.1 μm. However, our recent study (Ohki and Donath 1989) shows that Mg^{2+} also induced fusion of larger unilamellar PS vesicles, but in a limited extent at high threshold concentration (>8 mM Mg^{2+}) (see Fig. 2). Since the extent of fusion is small, the signal indicating fusion could easily be buried in the background signals. This may be the reason why the previous study could not determine the fusion event.

A-2) Interfacial tension

Among various physico-chemical properties of the membrane, the following three quantities have a good correlation with the extent of negatively charged vesicle fusion; i) Ion binding of certain types of cations, ii) Increase in interfacial tension, and iii) Decrease in surface dielectric constant of the membranes.

So far, the best correlation between the above physico-chemical properties and the extent of vesicle fusion is seen with the increase in interfacial tension of the membrane (Fig. 3). As the concentration of a fusogenic ion in the solution is increased, the interfacial tension of the monolayer increases. When the increased magnitude in the interfacial tension reaches a certain value, the rate of membrane fusion becomes the greatest. In other words, at the fusion threshold concentration of each fusogenic cation, the increased quantities in the interfacial tension of the acidic lipid membranes are approximately the same for all fusogenic cations. It should be noted, however, that the interfacial tension measured is for a lipid monolayer formed either at the air/water or oil/water interface whose surface area is constant and the fusogenic cation concentration of the subphase of the monolayer is altered (Ohki 1982, Ohki and Duax 1986). There-

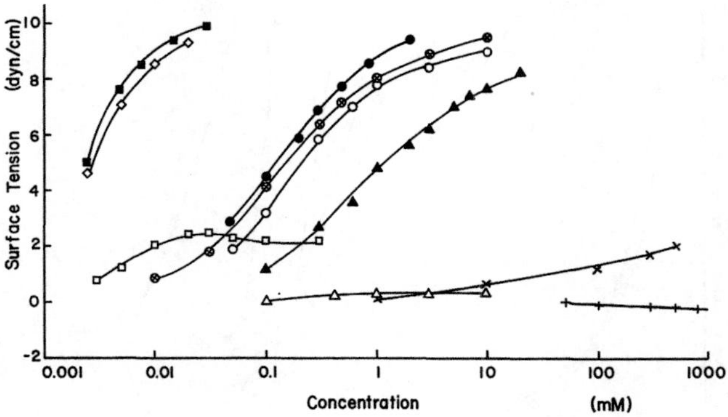

Figure 3. The increase in interfacial tension of a phosphatidylserine monolayer (70 Å^2 per molecule) formed at the air-water (0.1 M NaCl/1 mM TES/0.01 mM EDTA (0 mM EDTA for La^{3+} and Tb^{3+})/pH 7.4) interface as a function of various metal cation and polyamine concentrations. (■) Tb^{3+}; (◇) La^{3+}; (●) Mn^{2+}; (O) Ca^{2+}; (▲) Mg^{2+}; (⊗) H^+; (x) Li^+; (+) Na^+; (□) Spermine^{4+}; (△) Spermidine^{3+}. (Ohki 1988a)

fore, the actual interfacial tension of the vesicle membrane should be different from that obtained in this experiment. Non-fusogenic cations such as monovalent metal ions, and multivalent polyamines do not alter the interfacial tension (if it does, the change is not appreciable) even if the equivalent amount of the charged groups of the acidic lipid membranes are neutralized by these ions. The mere ion binding or adsorption to reduce or screen the surface charge of the membrane is not sufficient to alter the interfacial tension, and consequently to induce membrane fusion (Ohki and Ohshima 1984).

A-3) Surface dielectric constants

Another physico-chemical property of the membrane, which correlates well with the extent of fusion, is the surface dielectric constant (Kimura and Ikegami 1985). The surface dielectric constants measured by use of Dansyl-phosphatidyl-ethanolamine (DPE), incorporated in the PS vesicle membrane with respect to various fusogenic cation concentrations in the vesicle suspension are shown in Fig. 4. As the fusogenic ion concentration increases, the surface dielectric constant decreases. At the fusion threshold concentration of each ion, the surface dielectric constant decreases to about 8. Similar experimental results were obtained for other acidic lipid (e.g. phosphatidic acid) vesicles. According to earlier work (Waggoner and Stryer 1970), the fluorescence probe of DPE is located about the glycerol backbone in the lipid membrane and its fluorescent signal detects the local dielectric constant of that region. When the local dielectric

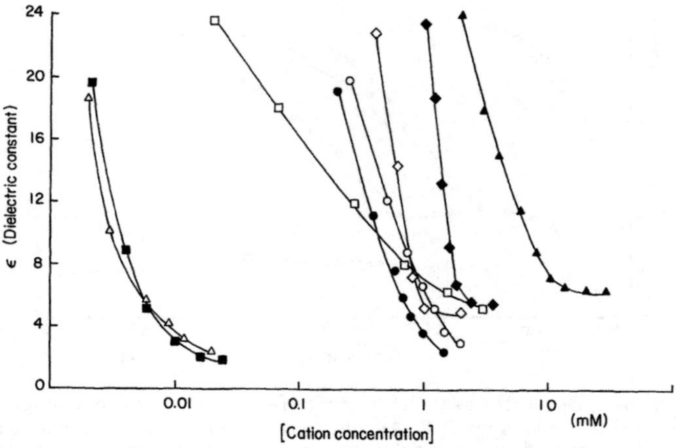

Figure 4. Surface dielectric constants of small unilamellar PS vesicles as a function of various cation concentrations suspended in 0.1 M NaCl buffer, pH 7.4. The dielectric constant was determined from the position of the maximum of the DPE fluorescence excited at 340 nm. The value of dielectric constant was calibrated from the signals of DPE in the known dielectric media (Ohki 1988b, Ohki and Arnold 1989). The molar ratio of PS/DPE was 200-300. The concentration of the lipids suspended was 0.1 mM. △ : La^{3+}, ■ : Tb^{3+}, ●: Mn^{2+}, ◇: Ba^{2+}, O: Ca^{2+}, ◆: Sr^{2+}, ▲: Mg^{2+}, □ : H^+.

medium of the fluorescence probe becomes more hydrophobic, the fluorescent spectrum will be shifted to the shorter wave length. The magnitude of such shifts can be calibrated from the signals of the fluorescence placed in the known dielectric media. It is noted that in the cases of strong fusogenic cations such as La^{3+}, Tb^{3+}, Mn^{2+} and Ca^{2+}, at or above the threshold concentrations of these fusogenic cations, the fluorescence intensity of DPE increased significantly and the maximum position of the emission spectrum shifted to further lower wave length. Consequently, the surface dielectric constant reduced to lower values, as low as 2. These results suggest that when these strong fusogenic cations interact with acidic lipid membranes, they may form nearly anhydrous complexes of cation and lipid polar groups (Hauser et al 1977, Portis et al 1979, Feigenson 1986).

The surface dielectric constant of the PS membrane in 0.1 M NaCl buffer is about 32, and it is not altered appreciably by the change in concentration of Na^+, K^+ or polyamines (e.g. spermine) up to 1 M, 1.5 M and 1 mM, respectively, although in their ionic concentration ranges, the surface potentials of the PS membrane are altered greatly (i.e. -80 mV to -20 mV) and a great deal of vesicle aggregation occurs (Ohki and Duax 1986).

A-4) Ion binding/membrane surface charge

The degree of ion binding has some correlation with the extent of membrane fusion. For this reason, a number of studies have been done to relate ion binding to membrane adhesion or fusion. However, a mere reduction in surface negative charge of the membrane by either binding or screening charges by the counter ions does not seem to be sufficient to induce membrane fusion. The apparent surface charge density of acidic lipid membranes can be obtained from the study of the vesicle electrophoretic mobility. The PS vesicle electrophoretic mobility in 0.1 M NaCl/pH 7.4 is about -4.6 μm sec^{-1} cmV^{-1} which corresponds to the ζ-potential of -64 mV and the surface potential is approximately -83 mV (Eisenberg et al 1979, Ohki 1984b). As the cation concentration increases, the vesicle electrophoretic mobility decreases in magnitude for all cations examined. For fusogenic cations, such as Tb^{3+}, La^{3+}, Ca^{2+} and H^{+}, when the surface potential is reduced to -40mV, the vesicle fusion occurs. However, for non-fusogenic cations (e.g. Na^{+}, K^{+}) and polyamines (spermine or spermidine), even when vesicle surface potential is reduced far below (in magnitude) -40 mV by these ions, the vesicle fusion is not observed. The details are published elsewhere (Ohki et al 1982, Ohki and Duax 1986).

On the other hand, the membrane surface charge is rather well correlated with the extent of vesicle aggregation. Cation or polyamine dependent phosphatidylserine vesicle aggregation is followed by measuring the

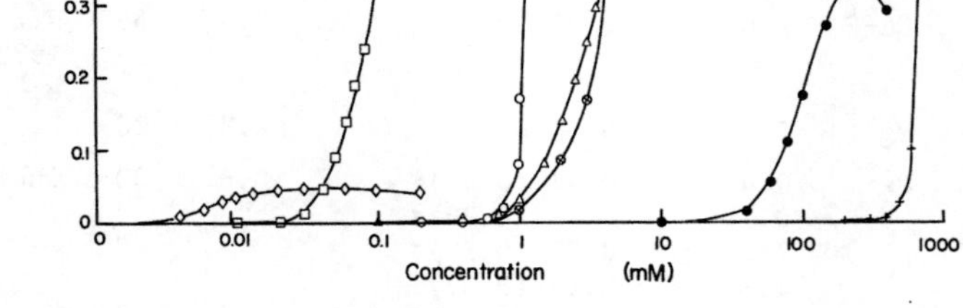

Figure 5. Turbidity of small unilamellar phosphatidylserine vesicles (0.1 μmoles lipid/ml of the NaCl buffer) for all cases except the La^{3+} case (0.01 μmoles lipid/ml of the NaCl buffer) suspended in 0.1 M NaCl buffer solution as a function of various metal cation and polyamine concentrations (◇) La^{3+}; (0) Ca^{2+} (⊗) H^{+}; (+) Na^{+}; (□) $spermine^{4+}$; (Δ) $spermidine^{3+}$; (●) $putrescine^{2+}$. (Ohki and Duax 1986)

turbidity of lipid vesicle suspension (Fig. 5). The sharp increase in turbidity of phosphatidylserine vesicles suspended in 0.1 M NaCl occurred at about 6 μM La^{3+}, 90 μM spermine, 1 mM Ca^{2+}, 3 mM spermidine, 3 mM H^{+}, 0.7 M Na^{+} and 0.8 M Li^{+}. The concentration corresponding to these points are somewhat similar to those at which the vesicle mobility is reduced to -2.5 - -3.0 μm $sec^{-1}cmV^{-1}$, which corresponds to the range of approximately -35 - -45 mV in surface potential.

From the above experimental results, it is likely that the membrane surface charge density has a rather good correlation with the extent of vesicle aggregation but not with vesicle fusion in the cation-acidic membrane systems.

However, the amounts of divalent cations bound to the lipid membrane surface seem to have some correlation with the extent of membrane fusion. In earlier studies, the fraction of the bound divalent cations for lipid molecules at the fusion threshold concentration of each divalent cation was the same for all divalent cations in the same concentration of monovalent salt solution. The value of the fraction depended upon the ionic strength. However, according to our recent study taking into account the ionic activity, the fraction of divalent cation bound per lipid molecule is independent of the ionic strength (Sukumaran and Ohki 1988). See Table 2.

Table 2 Divalent Cation Binding to Phosphatidylserine Membranes

M^{+}	M^{2+}	C^{2+}_{Th}	Bound divalent cation/lipid	K_1	K_2	Ref
		(mM)		(M^{-1})	(M^{-1})	
0.1 M NaCl	Ba^{2+}	0.7	0.32	0.6	37	Ohki 1982
" "	Ca^{2+}	1.0	0.32	0.6	30	"
" "	Sr^{2+}	1.6	0.33	0.6	25	"
" "	Mg^{2+}	6.0	0.35	0.6	10	"
" "	Ca^{2+}	2.0	0.34	0.8	35	Nir et al 1983 & Bentz et al 1983
" "	Mg^{2+}	5.0	0.35	0.8	20	"
0.01 M NaCl	Ca^{2+}	0.4	0.42 (0.41)*	0.6	30	Ohki 1982
0.05 M "	"	0.7	(0.39)*	"	"	"
0.1 M "	"	1.0	0.32 (0.38)*	"	"	"
0.3 M "	"	1.8	0.25 (0.37)*	"	"	"
0.55 M "	"	2.7	(0.38)*	"	"	"
0.8 M "	"	3.1	(0.40)*	"	"	"

M^{+} and M^{2+} are mono-and di-valent cations, respectively. C^{2+}_{Th} is the fusion threshold concentration of divalent cations. K_1 and K_2 are the association constants of mono- and di-valent cations to the PS membrane in M^{2+}: PS = 1:2 binding mode. ( )*: Sukumaran and Ohki 1988

B. Fusion of lipid vesicles induced by poly-(ethylene glycol).

B-1) Vesicle fusion

Macromolecules such as poly-(ethylene glycol), polypeptides and other macromolecules also induce vesicle fusion. Positively charged polypeptides such as polylysine were used to induce acidic lipid membranes. In such cases, both electrostatic interaction and hydrophobic interaction between the polymers and the membrane surface or its core may be involved in fusion. Uncharged macromolecules, such as poly-(ethylene glycol) and other macromolecules can induce both fusion of neutrally charged and charged lipid vesicles. Parents and Lentz suggested that vesicle aggregation and close interbilayer contact induced by PEG is sufficient to induce fusion of the PC vesicles. The initial rate of fusion was found to be dependent on the PEG concentration, while the extent of fusion was not. A substantial enhancement of the fusion rate was observed around 30 wt% PEG (Boni et al 1981, MacDonald 1985, Parente and Lentz 1986). Dextran can induce limited large-vesicle fusion, but at a much lower rate than PEG (Boni et al 1981, Parente and Lentz 1986).

Figure 6 shows the phosphatidylcholine vesicle fusion induced by poly-(ethylene glycol), where the PC vesicles are suspended in 0.1 M NaCl containing various amounts of PEG. 15～20 wt% of PEG-6000 induces the fusion of PC vesicles without having any fusogenic cations. The extent of fusion is time-dependent but definitely concentration dependent; the higher wt% of PEG induces a greater extent of the vesicle fusion (see Fig. 6). PEG is considered to remove the intermembraneous water including hydrated water depending on the PEG concentration and induce very strong aggregation of lipid vesicles. Fusogenic metal cations and PEG act on acidic lipid vesic-

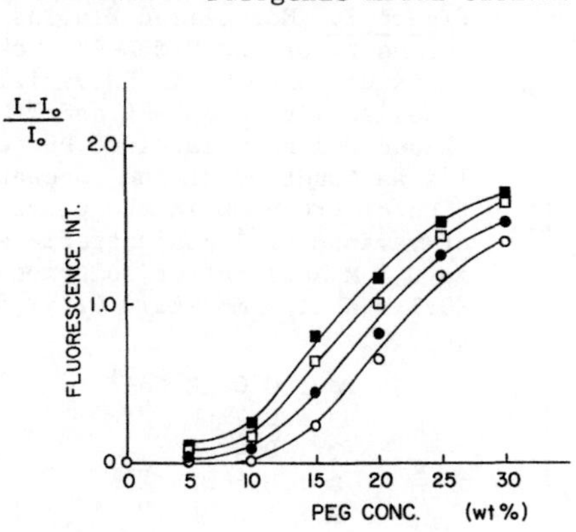

Fig. 6. Time dependent fusion of PC vesicles induced by PEG. Normalized fluorescence intensity (at 525 nm) of N-NBD-PE probe in a mixture of 0.05 mM PC vesicles labelled with N-NBD-PE and N-Rh-PE and 0.1 mM unlabelled PC vesicles as a function of PEG-6000 concentration in 0.1 M NaCl/4 mM TES/0.02 mM EDTA, pH 7.0.
I and I_o are the fluorescence intensities of the NBD probe in the presence and absence of fusogens in 0.1 M NaCl buffer.

○ : 1 min, ● : 5 min
□ : 10 min, ■ : 25 min

le fusion by enhancing each other with their effects. Hoekstra (1982) has studied the fusion of PS and PS/PE vesicles under the influence of Ca^{2+}, Mg^{2+} and PEG. As shown in the preceding sections, Ca^{2+} alone induced fusion of the PS vesicle for concentrations higher than or equal to 1 mM Ca^{2+} in the presence of 0.1 M NaCl. We have studied the fusion of the PS vesicles over a wide range of Ca^{2+} and PEG concentrations. The result of the fusion assay (Struck et al 1981) which makes use of the probe dilution method with PS/N-NBD-PE/N-Rh-PE vesicles and PS vesicles are given in Fig. 7. The measurements show that fusion occurs for subthreshold concentrations of Ca^{2+} in the presence of PEG. For 0.6 mM Ca^{2+}, a strong increase is observed between 5 and 10 wt% of PEG, and in the presence of 15 wt% of PEG, a strong fusion occurs at such a low concentration of Ca^{2+} as 0.2 mM. We have found that in the absence of fusogenic cations, fusion of the PS vesicles starts at about 25% of PEG-6000. These results are also confirmed by the fusion assay of the internal content mixing with $TbCl_3$ and DPA (dipicolinic acid). For both cases, the PEG effects are very similar. A facilitation of the Ca^{2+}-induced fusion by PEG was also found for didodecylphosphate vesicles (Rupert et al 1988). It can be concluded that the concentration of Ca^{2+} necessary for fusion is shifted to lower values when the PEG concentration is increased. This behavior is very similar to those observed in the vesicle aggregation process (see Fig. 9).

B-2) Surface dielectric constants

A good correlation between the extent of cation-induced vesicle fusion and the degree of decrease in dielectric constant of the DPE environment is found and is described in the preceding section. In Fig. 8, the change in

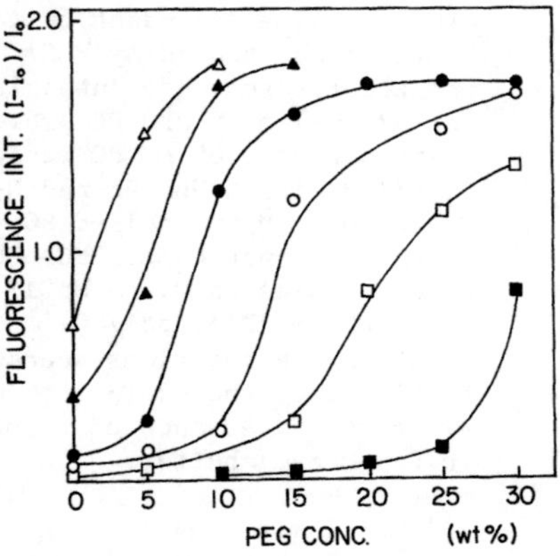

Figure 7. Normalized fluorescence intensity of the N-NBD-PE probe in a mixture of 0.05 mM PS vesicles labelled with N-NBD-PE and N-Rh-PE and 0.1 mM unlabelled PS vesicles as function of the concentration of PEG-6000 in the presence of various Ca^{2+} concentrations in 0.1 M NaCl buffer solution (0.1 M NaCl/4 mM TES/0.02 mM EDTA pH 7.0).

■ : 0.0 mM Ca^{2+}
□ : 0.2 mM "
○ : 0.4 mM "
● : 0.6 mM "
▲ : 1.0 mM "
△ : 1.2 mM "

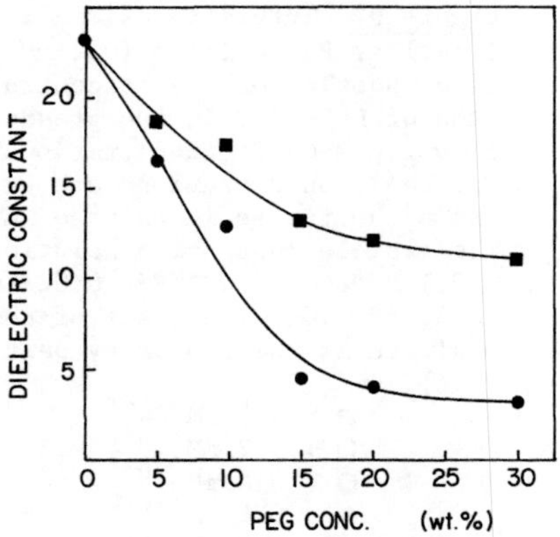

Figure 8. Dielectric constants of PS vesicles in the presence and absence of 0.6 mM Ca^{2+} as a function of the concentration of PEG-6000. The dielectric constant was determined from the position of the maximum of the fluorescence signal of Dansylphosphatidylethanolamine (DPE) in corporated in the PS vesicles at a molar ratio PS/DPE = 300, suspended in 0.1 M NaCl/4 mM TES/0.02 mM EDTA, pH 7.0

■ : 0.0 mM Ca^{2+}
● : 0.6 mM "

dielectric constant is given in the absence or presence of 0.6 mM Ca^{2+} as a function of the PEG concentration. The good correlation of the decrease in the dielectric constant with the fusion properties given in Fig. 8 is remarkable. However, it is evident that fusion appears at higher dielectric constants (12 ~ 14) compared to the dielectric constants of about 8 for the cation-induced fusion. This indicates that additional mechanisms occur in the PEG-lipid vesicle system where PEG destabilizes the region of contacting membranes in addition to the Ca^{2+} effect on the hydrophobicity of the surface. The withdrawal of free water from the vesicles by PEG could cause them to flatten. Thus, two lipid vesicles might fuse due to their stress on the highly curved boundaries as proposed by MacDonald 1985 and Parente and Lentz 1986. The detergent-like properties of PEG are another possibility of the bilayer destabilization (Arnold et al 1985).

B-3) Aggregation of lipid vesicles

Significant progress in the understanding of the mode of action of PEG on the aggregation process is the finding that PEG is excluded from lipid vesicle surfaces. The first experimental evidence for an exclusion of PEG was found by ^{2}H-NMR studies of heavy water in multilamellar vesicles (Arnold et al 1982, 1988). Electrophoresis measurements have shown that this phenomenon also occurs for individual unilamellar vesicles (Arnold et al 1987, 1988). Evans and Needham (1988) developed a depletion-based theory to explain the PEG-induced aggregation of giant phospholipid vesicles. PEG exerts an additional attractive force on the vesicles which is proportional to the difference between the osmotic pressures of the polymer in the bulk phase and in the gap between the vesicles. For the relatively high concentration of PEG used for the fusion of lipid vesicles, the membranes of appos-

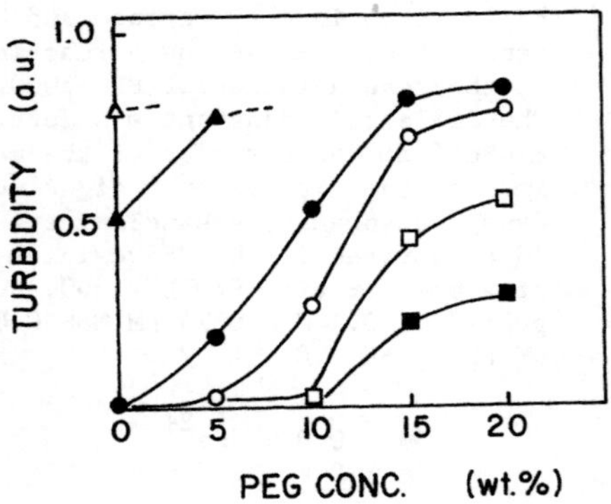

Figure 9. Turbidity of small unilamellar PS vesicles (0.2 mM PS) as a function of the concentration of PEG-6000 in the presence of various Ca^{2+} concentrations. The Ca^{2+} concentrations are given in the figure in units of mM. The vesicle suspension solution = 0.1 M NaCl, 4 mM TES, 0.02 mM EDTA, pH 7.0. The scale of the ordinate is the arbitrary unit (a,u).

■ : 0.0 mM Ca^{2+}
□ : 0.2 mM "
○ : 0.4 mM "
● : 0.6 mM "
▲ : 1.0 mM "
△ : 1.2 mM "

ed vesicles are intensely compressed due to the high osmotic pressure of the PEG molecules in the suspension and the PEG depletion in the gap. The close contact of the membranes is considered to be a necessary condition for two membranes to fuse. The aggregation of small unilamellar PS vesicles in the 0.1 M NaCl buffer, pH 7.0 occurs for Ca^{2+} concentration larger than or equal 1 mM (Ohki et al 1982). Addition of PEG reduces the threshold concentration of Ca^{2+} for aggregation. The change in turbidity as a function of PEG-6000 is shown for different Ca^{2+} concentations in Fig. 9. The addition of PEG increases the aggregation. Without Ca^{2+}, vesicle aggregation starts at about 15 wt%.

Compared to the PEG-induced aggregation of the PC vesicles, the PEG concentrations for the aggregation of PS vesicles are higher due to the electrostatic repulsion of PS vesicles. The combined effect of Ca^{2+} and PEG results from the reduction of the electrostatic repulsion by Ca^{2+} and the increase of the attractive component from the depletion flocculation of PEG. Similar influences of surface charges and ions are found for many other negatively charged vesicles and serum lipoproteins (Arnold and Zschornig 1988).

B-4) Surface potential of lipid membranes

It was shown by several authors that the fluorescence probe 4-Heptadecyl-7-Hydroxy coumarin can be used to measure surface potential (Pal et al 1983). Measurements are performed with the dependence on pH and the ratio of the ionized and the unionized form is determined from the spectrum. The surface potential ψ is calculated from these measurements by use of the following equation:

$$\psi = -(pK - pK_o) \times 2.3 \times \frac{RT}{F} ,$$

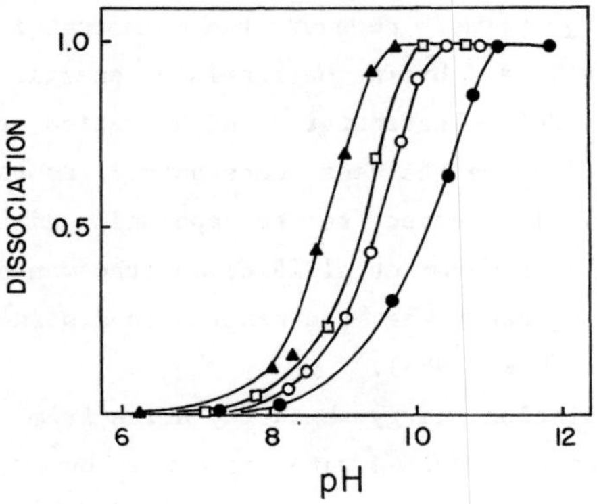

Figure 10. Dissociation of the fluorescence probe 4-Heptadenyl-7-Hydroxycoumarin in PC and PS vesicles at a molar ratio of 1: 300 as a function of pH. The vesicles were suspended in 0.1 M NaCl/4 mM TES/0.02 mM EDTA, pH 7.0.

▲ : PC
□ : PS + 0.6 mM Ca^{2+} + 5 wt %PEG
○ : PS + 0.6 mM Ca^{2+}
● : PS

where pK_o is the pK value of the fluorescence probe incorporated in an uncharged lipid vesicle. Results of measurements for PC and PS vesicles in dependence on pH for different PEG concentrations and the addition of 0.6 mM Ca^{2+} are given in Fig. 10. When PC vesicles are taken as uncharged vesicles, a surface potential of -80 mV is determined for the PS vesicle. This value agrees with surface potential calculated from electrophoresis measurements (Eisenberg et al 1979, Ohki 1984b). Addition of 5 and 10 wt% PEG-6000 does not influence the surface potential of either PC or PS vesicles. When 0.6 mM Ca^{2+} is added to the PS vesicles, the surface potential is reduced to -40.3 mV. Addition of 5 wt% PEG leads to a further decrease in the surface potential to -28.7 mV. This shows a stronger binding of Ca^{2+} to the PS surface in the presence of PEG. This effect is very important for the enhancement of the Ca^{2+}-induced fusion by PEG. This influence of PEG in the Ca^{2+} binding was not observed for the vesicles of PC/PS (1:1) mixtures which do not fuse at 0.6 mM Ca^{2+} in the presence of 5 wt% PEG.

It can be concluded from these results that the PEG-induced aggregation of lipid vesicles creates improved conditions for a stronger binding of Ca^{2+} to the bilayer surface. This stronger binding is directly related to the increase of the hydrophobicity of the membrane surface as an important step in the fusion events (Ohki and Ohshima 1984). In this sense, the influence of Ca^{2+} and PEG are not simply additive actions, but they exhibit a synergistic effect.

THEORETICAL CONSIDERATION

In this section, we will consider the interaction forces involved in two interacting membranes which are responsible for the membrane close apposition (aggregation) and fusion.

The total interaction energy of the two interacting membranes can be expressed by several interaction energy terms. However, the main contributing terms would be both electrostatic and hydration repulsive energies and van der Waals attractive energy. Both electrostatic and hydration repulsive energies decay exponentially (having the decay constants, 1/Debye constant, and 2 - 2.5 Å, respectively) with respect to the separation distance, R, for the two membranes (Hiemenz 1977, LeNeveu et al 1976) and the van der Waals attractive energy has a $\frac{1}{R^2}$ dependency when the separation distance is small compared to the vesicle size (Ohki 1985).

Among these three terms, the hydration energy which may arise from the dipole interaction forces of the oriented adsorbed water molecules on two interacting membrane surfaces, is usually much greater in magnitude compared with the energies due to other forces when two membrane surfaces are closely apposed (<20 Å) and are also well hydrophilic or polar in nature.

These interaction energies can be calculated as follows (Ohki 1985, 1988 a): Here, we consider that the two interacting membranes are represented by the two hydrocarbon plates having infinite thickness and also having hydrophilic surface layer of thickness h, which are separated by an aqueous phase at a distance R. If we assume proper interaction coefficients, λ, (= ρ (molecular density) x α(molecular polarizability)), which is the quantity similar to the so-called "Hamaker constant", for the aqueous and hydrocarbon phases, the van der Waals interaction energy can be calculated as a function of the interaction coefficient λ_p of the surface polar layer for a given separation distance and a given thickness of the surface polar layer.

The electrostatic interaction energy is difficult to evaluate. However, we use the following approximation formula (Hiemenz 1977):

$$V^{el} = \frac{64\ n\ kT}{\kappa} B^2 \exp\ (-2\kappa R),$$

$$B \doteq \frac{\exp(Ze\psi_o/2kT) - 1}{\exp(Ze\psi_o/2kT) + 1}\ ,$$

by assuming that the surface potential of the membrane is relatively small (less than |50| mV). Here, n is the number of density of ions, κ the Debye constant and ψ_o the surface potential. From the vesicle electrophoretic mobility measurements and vesicle aggregation studies, a strong vesicle aggregation occurs at the surface potential of about -40 mV (Ohki et al 1982, Ohki and Duax 1986) in the ionic strength of 0.1 M NaCl. By using this surface potential value and with a given separation distance R, and a given ionic strength of the solution, we can calculate the electrostatic interaction energy of the two interacting membranes. These calculated energies are shown in Fig. 11, as a function of the interaction coefficient λ_p for the

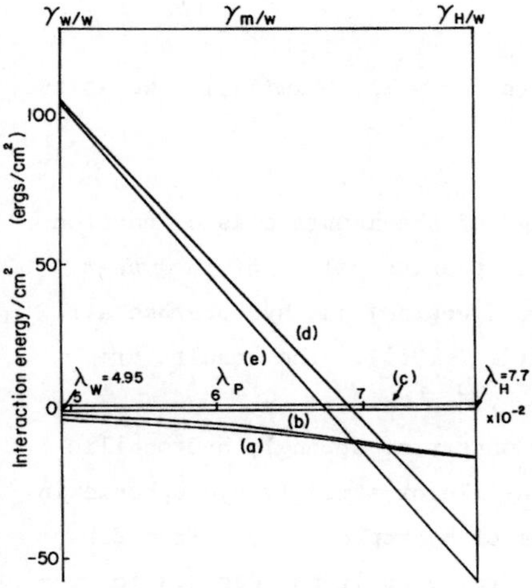

Figure 11. Van der Waals and electrostatic interaction energies between the two membranes and the dehydration energy per unit area. The van der Waals energy for the cases (R= 2.5 Å and (a) h = 6 Å and (b) h = 10 Å) are shown as curves (a) & (b) respectively, with respect to the interaction coefficient λ_p of the surface layer. λ_w (4.96×10^{-2}) and λ_m (7.7×10^{-2}) correspond to those of the aqueous (water) and hydrocarbon phases, respectively. Curve (c) is the electrostatic energy for the case of the membrane surface ($^-e/75$ Å^2) and the bulk aqueous solution (0.1 M NaCl + 1 mM Ca^{2+}). Curve (d) is the effective dehydration energy V^{dh}, with respect to hydrophobicity parameter p. Curve (e) is the total energy of the interaction energy = $U + V^{dh} + V^{el}$.

surface polar layer.

The hydration energy would be the association energy of water onto the membrane. In other words, the hydration energy is the same magnitude but a different sign of the energy required to dehydrate water from the membrane surface. We estimate such an energy in the following way: the dehydration energy is the sum of the energies of first separation of the membrane and water phases, and secondly removal of the water phase away from the intermembraneous space (Ohki 1985, 1988a). The energy to separate the membrane/aqueous interface, ΔG^{sep} is,

$$G^{sep} = \gamma_w + \gamma_m - \gamma_{m/w}$$

where γ is the surface tension and $\gamma_{m/w}$ the interfacial tension of the membrane/water interface. Thus, the dehydration energy for one membrane/water interface is,

$$\Delta G^{dh} = \gamma_m - \gamma_{m/w} ,$$

Since these are two such interfaces for the interacting membranes, and the intermembraneous space has a finite separation distance R, the dehydration energy of water from the intermembraneous space is,

$$\Delta G^{dh}_{int} (R) = 2(\gamma_m - \gamma_{m/w}) + \Delta(R),$$

where Δ is the correction term for the water removal energy of the intermembraneous space with a finite separation distance. Although $\gamma_{m/w}$ should be an experimentally obtainable quantity, and as the knowledge of both $\gamma_{m/a(air)}$ and $\gamma_{m/w}$ are not well compiled, we assume that the surface tension of the membrane surface may be expressed as a linear sum of water (w) and hydrocarbon (H) portions.

$$\gamma_m = p\,\gamma_H + q\,\gamma_w \quad ,$$

where p or q is the fraction of each phase (p + q = 1). Similarly, we assume,

$$\lambda_p \equiv \lambda_m = p\,\lambda_m + q\,\lambda_w \quad .$$

Then, we can calculate the dehydration energy of the membrane as a function of the fraction p or q, using the interfacial tension values of the water/air (~72 dynes/cm), water/hydrocarbon (~50 dynes/cm) and hydrocarbon/air (~20 dynes/cm), respectively (Davies and Rideal 1961). The results are shown as curve d in Fig. 11, where $G^{dh}_{int} \equiv V^{dh}$.

When p = 0, the membrane is completely watery or strongly hydrophilic and when p = 1, it is completely hydrocarbon-like or strongly hydrophobic in nature. The value p may indicate the degree of hydrophobicity. From Fig. 11 it is seen that when the value of p changes from 0 to 1, the dehydration energy gradually reduces from a large positive value to a negative value.

When the sum of the attractive van der Waals energy and the repulsive dehydration and electrostatic energies ($U + V^{dh} + V^{el}$), which is curve e shown in Fig. 11, becomes negative (at a point of the interfacial tension of 23 dynes/cm), the two interfaces could come close to the molecular adhesion. In such a case, if the membrane surface does not have structural constraint, the two membranes can fuse.

From the above argument, the dehydration energy is related to the membrane hydrophobicity which relates to the interfacial tension of the membranes.

DISCUSSION

Experimental results on the cation-induced fusion of phospholipid membranes and the related physico-chemical properties of the membrane indicate that the degree of membrane fusion correlates well with the degree of increased surface tension or increased hydrophobicity of the membrane caused by cations (Ohki 1982, 1984, 1988a). A massive fusion of small unilamellar phosphatidylserine (PS) vesicles occurs at 1 mM Ca^{2+} in 100 mM NaCl, where the increase in interfacial tension of the membrane is about 8 dynes/cm from the reference state (0 mM Ca^{2+} and 0.1 M NaCl) in the monolayer system. This is consistent with other cation cases. We may consider that this increase in interfacial tension is solely due to the change in nature of the surface hydrophilic layer. Our recent studies suggest that this energy increase arises mainly from a conformational change in the interfacial polar groups of the membrane molecules upon the chelation like-binding of divalent cation, rendering the hydrocarbon portion of the membrane molecule open to the water phase, but not due to the change in the electrostatic energy of the surface by the reduction of surface charges (Ohki and Ohshima 1984, Ohshima and Ohki 1985).

When the interfacial tension of the membrane increases and reaches to the point where the total interaction energy becomes negative, close adhesion of membranes occurs. This point corresponds to the state of about 23 dynes/cm in the interfacial tension of the monolayer membrane. The interfacial tension of a small unilamellar vesicle membrane in 0.1 M NaCl would be about 12 dynes/cm because the area per molecule is about 85 $Å^2$, and an additional increase of about 8 dynes/cm would be provided by the binding of fusogenic ions to the lipid polar groups of the membrane surface. These could make the total interfacial tension of the membrane to be about 20 dynes/cm. This change in interfacial tension of the membrane may satisfy the energy condition for two membranes to come close to adhesion (or molecular contact of the membranes). It should be pointed out, however, that the interfacial tension measured in the monolayer is not the same magnitude of the mechanical tension measured for the bilayer membrane because the mechanical elastic tension of the bilayer usually equilibrates the interfacial tension of the membrane. Therefore, we should understand that the fusogenic ion binding to the lipid vesicle surface makes the membrane surface more hydrophobic, the quantity of which corresponds to the value of the interfacial tension measured in the monolayer system. The membrane surface does not become hydrophobic enough by non-fusogenic ion binding.

The curvature of the membrane surface is another factor for the surface hydrophobicity. It is seen from Figure 2, that the larger the radii of curvature of phosphatidylserine vesicles are, the greater are the threshold concentrations of divalent cations. As the radius of curvature of the membrane becomes larger, the area per lipid molecule will be small and therefore the surface tension will be smaller than that of the vesicle having a smaller radius of curvature. A vesicle having a larger size has less ability to make the membrane surface more hydrophobic by the same amounts of fuosgenic ion binding compared to that having a smaller size. It is also found that there is a systematic relation between the dielectric constants of various substances and their interfacial tensions against water (Ohki 1988b): the lower the dielectric constant is, the larger is its interfacial tension. This also indicates that our results with the lipid membranes with respect to various ions have a similar correlation (see Figs. 3 and 4). The earlier study of the time-resolved fluorescence measurements of DPE (Kimura and Ikegami 1985) supports the estimate of the local dielectric constant around the DPE environment in lipid membrane from the Stokes shift of DPE fluorescence since the diffusion constant of the DPE environment is not affected much by the presence of Ca^{2+} in the vesicle suspension solution (1.8 mM Ca^{2+} in the PS vesicle suspension).

Although it is not clear which region of dielectric environment in the lipid membrane is measured by the DPE probe method, the earlier work suggests that the signal of the DPE probe indicates the dielectric medium around the glycerol backbone in lipid membranes (Waggoner and Stryer 1970). We may say that in order to have a sufficient degree of surface hydrophobicity to induce the membrane close adhesion (or molecular contact of membrane), it is necessary for the interfacial tension to become greater than 20 dynes/cm for the monolayer or the dielectric constant of the certain region of the membrane surface to become sufficiently low for the case of ion-induced membrane fusion. The above conditions for the surface properties are necessary for the two membranes to adhere closely (or membrane molecular contact) but not necessarily sufficient to induce membrane fusion as long as the surface molecules are not freely movable. This may be the case of divalent and trivalent cation-induced membrane fusion of acidic lipid membranes. Multivalent metal cations bind the lipid polar groups tightly together which make the membrane surface molecules to be physically rigid or constrained. In such cases, the membrane molecules at the boundary between the close contact and non-contact region may receive a greater physical stress due to bending pressure or molecular expansion, which leads to further increase in interfacial tension of such membrane areas. Another possibility is the intermembraneous molecular exchanges mediated by fusogenic ions since such fusogenic ions, while bound on one membrane, may interact with the other membrane. Therefore, these molecules in such regions which are affected by the mechanical or chemical stress, would become a site of membrane fusion in the case of polyvalent metal cation-induced fusion (see Fig. 12).

The process involved in PEG-induced membrane fusion is quite different from that in ion-induced membrane fusion discussed above. However, the removal of water from the intermembraneous space and consequently making the membrane surface more hydrophobic in nature and the possible fusion sites being at the highly curved boundaries of the membrane surface are common physical facts for both cases.

The action of PEG in facilitating vesicle fusion results from the ability to force vesicles together. Relatively low concentrations of about 10 wt% of PEG are able to aggregate PC vesicles. A rough estimate of the critical concentration of PEG for vesicle aggregation was given by application of the concept of excluded volume (Tilcock and Fisher 1982). Concentrations of PEG-6000 of about 4 to 8 wt% would leave no free volume for other particles

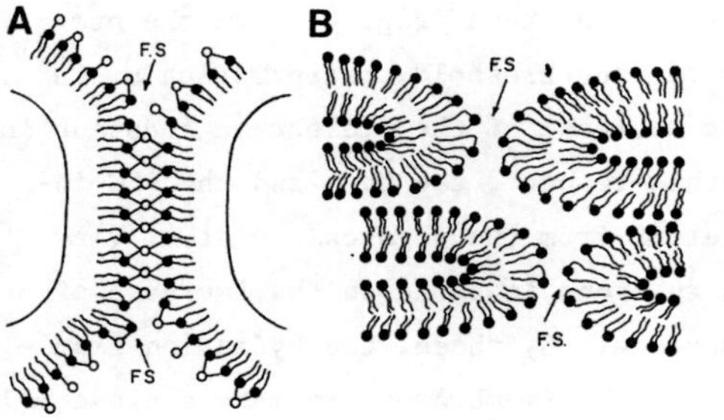

Figure 12. Schematical diagrams of the close contacting lipid vesicles induced by either fusogenic cations (A) or PEG (B). FS stands for the fusion site where the membrane molecules are energetically unstable and may induce inter-membraneous molecular mixing.

unless they are sufficiently small to penetrate between the polymer coils. These values are in the order of concentrations which lead to the aggregation of PC vesicles. In the same range of concentration, non-interacting proteins should influence the stability of the vesicle suspensions under in vivo condition.

An elaborated analysis of the destabilization of a vesicle suspension desires the calculation of the repulsive forces of PEG-PEG, PEG-vesicle and vesicle-vesicle pairs in dependence on the PEG concentration (van Oss et al 1988). Due to the electrostatic repulsion, the critical concentration of PEG for aggregation is shifted to higher values for the charged lipid vesicles (see Fig. 10). Once the vesicles are aggregated, the approaching of membranes increases with the increase of the PEG concentration as a result of the depletion of PEG in the gap and the action of an additional attractive force due to the PEG depletion. For 30 wt% PEG-6000, the water layer between the PC vesicles is reduced to about 10 water molecules per phospholipid molecule (Arnold et al 1988). This process is accompanied by the change in the physico-chemical properties of water and membrane structure (Arnold et al 1982, 1985). This bilayer membrane contact seems to be sufficiently high so that the contacting PC as well as PS bilayers can fuse. Because the repulsive force is dominated by the hydration repulsion of distance smaller than 20 Å, the fusion of PS vesicles occurs in the same range of PEG concentration of about 30 wt%, despite the higher concentration of PEG for aggregation of PS vesicles compared to PC vesicles.

The mechanisms which lead to fusion from a strongly approaching membrane are not well understood. Non-bilayer structures are not formed. A disruption of bilayers at the highly curved boundaries is supported by many authors (MacDonald 1985, Parente and Lentz 1986). The PEG-induced aggregation of PS vesicles is facilitated by the addition of Ca^{2+} because of the decrease of the electrostatic repulsion. However, Ca^{2+} does some other things . It increases the hydrophobicity of the bilayer surface as the decrease of the dielectric constant has been shown (Fig. 8). In the presence of PEG this decrease occurs at the subthreshold concentration of Ca^{2+}. In the absence of Ca^{2+}, this dramatic decrease of the dielectric constant in addition to PEG does not occur. In this respect, the Ca^{2+} and the PEG-induced fusion of PS vesicles are different from the physical point of view. However, the combined effect of both fusogens is based on the increase of the hydrophobicity of the membrane surface. By these, the hydration repulsion is sufficiently reduced to a degree that membranes can make a close molecular contact and membrane fusion becomes possible as was discussed in the theoretical text.

In both ion-induced fusion and PEG-induced fusion, lipid vesicles are first brought into a close adhesion state by quite different physico-chemical processes, and then the contacting membrane areas having a low radium curvature created by the close apposition, become susceptible sites for membrane fusion which is a similar process for both cases. These low radium curved regions are more hydrophobic than the rest of the membrane surface. This situation may also be analogous to the phase defect or irregularity formed in the membrane which is considered to be the fusion site (Schullery et al 1980, Verkleij et al 1980, Hui et al 1981, Ellens et al 1986, Siegel 1984). These non uniform molecular structures can be caused by either the freeze-thaw action, mixing of non uniform molecular components in the membrane, or non-bilayer structure formation. The pinching off point for the endocytotic particles to internalize into the other cell interior also has a similar molecular structure of the low radium curvature (Haywood and Boyer 1981, Darnell et al 1986). Protein-induced fusion may also have some common factors to those for the lipid membrane fusion: the portion of fusion inducing protein may have a high hydrophobic segment which may penetrate into the membrane hydrophobic core and disturb and mix the membrane molecules of the two interacting cells to result in fusion (Wickner and Lodish 1985, Novick and Hoekstra 1988). Besides those mentioned above, there are a number of other agents to cause membrane fusion: such as proteolytic enzymes (Mizzen et al 1987, Israel et al 1983), large organic cations (Eytan et al 1984) and electric fields (Zimmermann 1982). However, our concept that the membrane surface or molecular surface should attain a sufficient hydrophobicity in order to induce fusion seems to be applicable to all membrane fusion reactions.

ACKNOWLEDGEMENT

This work was supported partly by a grant from the U.S. National Institutes of Health (5R01GM24840).

REFERENCES

Ahkong QF, Cramp FC, Fisher D, Howerll JR, Tampion W, Verrinder M, Lucy JA (1973a) Chemically-induced and thermally-induced cell fusion: lipid-lipid interaction. Nature (London) 242: 215-217

Ahkong QF, Fisher D, Tampion W, Lucy JA (1973b) The fusion of erythrocytes by fatty acids, esters, rentinol and α-tocopherol. Biochem. J 136: 147-155

Arnold K, Herrmann A, Gawrisch K, Pratsch L (1988) Water-mediated effects of PEG on membrane properties and fusion, in Molecular mechanisms of membrane fusion. Ohki S, Doyle D, Flanagan TD, Hui SW, Mayhew E (Eds) Plenum Publ. Co NY pp 255-272

Arnold K, Herrmann A, Pratsch L, Gawrisch K (1985) The dielectric properties of aqueous solution of poly-(ethylene glycol) and their influence on membrane structure. Biochim Biophys Acta 815: 515-518

Arnold K, Pratsch L, Gawrisch K (1983) Effect of poly(ethylene glycol) on phospholipid hydration and polarity of the external phase. Biochim Biophys Acta 728: 121-128

Arnold K, Zschornig O, Gerold W, Barthel D (1987) Effect of PEG on the electrophoretic mobility of liposomes. Mol Cryst Liq Cryst 152: 357-362

Arnold K,Zschornig O (1988) Aggregation of human plasma low density lipoproteins by means of PEG. Biomed Biochim Acta 47: 949-954

Bentz J, Duzgunes N and Nir S (1983) Kinetics of divalent cation induced fusion of phosphatidylserine vesicles: correlation between fusogenic capacities and binding affinities. Biochemistry 22: 3320-3330

Blitz AL, Fine RE (1974) Muscle-like contractile proteins and tubulins in synaptosomes. Proc Natl Acad Sci USA 71: 4472-4476

Blow AMJ, Botham GM, Disher D, Goodall AH, Tilcock CPS, Lucy JA (1978) Water and calcium ions in cell fusion induced by poly(ethylene glycol). FEBS Lett 94: 305-310

Blumenthal R (1987) Membrane fusion, in Current topics in membrane transport, vol 29 Klausner RD, Kempf C (Eds) Academic Press NY pp 203-254

Boni LT, Hah JS, Hui SW, Mukherjee P, Ho JT, Jung CY (1984) Aggregation and fusion of unilamellar vesicles by poly (ethylene glycol). Biochim Biophys Acta 775: 409-418

Boni LT, Stewart TP, Aldenfer JL, Hui SW (1981) Lipid-polyethylene glycol interactions: I.Induction of fusion between liposomes. J Membrane Biol 62: 65-70

Breisblatt W, Ohki S (1976) Fusion in phospholipid sperical membranes. II. Effect of cholesterol, divalent ions and pH. J Membrane Biol 29: 127-146

Cohen FS, Zimmerberg J, Finkelstein A (1980) Fusion of phospholipid vesicles with planar phospholipid bilayer membranes. II. Incorporation of a vesicular membrane marker into the planar membrane. J Gen Physiol 75: 251-270

Creutz CE (1981) Cis-unsaturated fatty acids induce the fusion of chromaffin granules aggregated by synexin. J Cell Biol 91: 247-256

Creutz CE, Pazoles CJ, Pollard HB (1978) Self association of synexin in the presence of calcium: Correlation with synexin-induced membrane fusion and examination of the structure of synexin aggregates. J Biol Chem 254: 552-558

Cullis PR, Hope MJ (1978) Effect of fusogenic agent on membrane structure of erythrocyte ghosts and the mechanism of membrane fusion. Nature (London) 271: 672-674

Darnell J, Lodish H, Baltimore D (1986) Molecular cell biology. Scientific American Books Inc pp 638-640

Davies JR, Rideal EK (1961) Interfacial phenomena. Academic Press NY

Duzgunes N (1985) Membrane fusion. Subcellular Biochem 11: 195-286

Duzgunes N, Bronner F (1988) Current topics in membrane transport. vol 32 Academic Press NY
Duzgunes N, Ohki S (1977) Calcium-induced interaction of phospholipid vesicles and bilayer lipid membranes. Biochim Biophys Acta 467: 301-308
Duzgunes N, Ohki S (1981) Fusion of small unilamellar liposomes with phospholipid planar bilayer membranes and large single-bilayer vesicles. Biochim Biophys Acta 640: 734-747
Duzgunes N, Wilschut J, Fraley R, Papahadjopoulos D (1981) Studies on the mechanism of membrane fusion: Role of head group composition in calcium and magnesium-induced fusion of mixed phospholipid vesicles. Biochim Biophys Acta 642: 182-195
Eckert R, Randall D (1978) Animal physiology. Freeman EH Co, San Francisco
Eisenberg M, Gresalfi T, Riccio T, McLaughlin S (1979) Adsorpiton of monovalent cations to bilayer membranes containing negative phospholipids. Biochemistry 18: 5213-5223
Ellens H, Bentz J, Szoke FC (1986) Fusion of phosphatidylethanolamine containing liposomes and the mechanism of the L $-H_{11}$ phase transition. Biochemistry 25: 4141-4147
Evans E, Needham D (1988) Intrinsic colloidal attraction/repulsion between lipid bilayers and strong attraction induced by non-adsorbing polymers. in Molecular mechanisms of membrane fusion. Ohki S et al (Eds) pp 83-99
Eytan GD, Mary T, Broza R, Shaltin Y (1984) Cationic amphiphile induce fusion of acidic liposomes. Biochim Biophys Acta 778: 38-48
Feigenson CW (1986) On the nature of calcium ion binding between phosphatidylserine lamellae. Biochemistry 25: 5819-5825
Gad AE, Silver BL, Eytan GD (1982) Polycation-induced fusion of negatively-charged vesicles. Biochim BIophys Acta 690: 124-132
Gratzl M, Schudt C, Ekerdt R, Dahl G (1980) Fusion of isolated biological membranes: A tool to investigate basic processes of exocytosis and cell-cell fusion. in Membrane structure and function. vol 3 Bittar EE (Ed) Wiley NY pp 59-92
Hauser H, Finer GE, Darke A (1977) Crystalline anhydrous Ca-phosphatidylserine bilayers. Biochim Biophys Research Comm 76: 267-274
Haywood AM, Boyer BP (1981) Initiation of fusion and disassembly of sendai virus membranes into liposomes. Biochim Biophys Acta 646: 31-35
Heuser JE, Reese TS (1973) Evidence of recycling of synaptic vesicle membrane during transmitter release at the frog neuromuscular junction. J Cell B-ol 57: 315-344
Hiemenz PC (1977) Principles of colloid and surface chemistry. Marcel Dekker NY
Hoekstra D (1982) Role of lipid phase separation and membrane hydration in phospholipid vesicle fusion. Biochemistry 21: 2833-3840.
Hoestra D, Yaron A, Carmel A, Scherphof G (1979) Fusion of phospholipid vesicles containing a trypsin-sensitive fluorogenic substrate and trypsin. FEBS Lett 106: 176-180
Hong K, Duzgunes N, Papahadjopoulos D (1981) Role of synexin in membrane fusion. J Biol Chem 256: 3641-3644
Hong K, Duzgunes N, Papahadjopoulos D (1982) Modulation of membrane fusion by calcium binding proteins. Biophys J 37: 297-305
Houslay MD, Stanley KK (1984) Dynamics of biological membranes. John Wiley & Sons NY
Hui SW, Steward TP, Boni LT, Yeagle PL (1981) Membrane fusion through point defects in bilayers. Science 212: 921-923
Israel S, Ginsberg D, Laster Y, Zakai N, Milner Y, Loyter A (1983) A possible involvement of virus-associated protease in the fusion of sendai virus envelopes with human erythrocytes. Biochim Biophys Acta 732: 337-346
Kantor HL, Prestegard JM (1975) Fusion of fatty acid containing lecithin vesicles. Biochemistry 14: 1790-1795

Kimura Y, Ikegami A (1985) Local dielectric properties around polar region of lipid bilayers. J Membrane Biol 85: 225-231

Knutton S (1979) Studies of membrane fusion. III. Fusion of erythrocytes with polyethylene glycol. J Cell Sci 36: 61-72

LeNeveu DM, Rand RP, Parsegian VA (1976) Measurement of forces between lecithin bilayers. Nature (London) 259: 601-603

Lucy JA (1970) The fusion of biological membranes. Nature (London) 227: 814-817

Lucy JA (1978) Mechanisms of chemically induced cell fusion, in Membrane fusion. Poste G, Nicolson GI (Eds) Elsevier/North Holland Biomedical Press pp 267-304

Lucy JA, Ahkong QF (1986) An osmotic model for the fusion of biological membranes. FEBS 199: 1-11

Lucy JA, Ahkong QF (1988) Osmotic forces and the fusion of biomembranes. in Molecular mechanisms of membrane fusion. Ohki S et al (Eds) pp 163-179

MacDonald RI (1985) Membrane fusion due to dehydration by polyethylene glycol, dextran or sucrose. Biochemistry 24: 4058-4066

Miledi R (1973) Transmitter release induced by injection of calcium ions into nerve terminals. Proc R Soc London Ser B 183: 421-425

Miller C, Arvan P, Telford JN, Racker E (1976) Ca^{2+}-induced fusion of proteoliposomes: Dependence on transmembrane osmotic gradient. J Membrane Biol 30: 271-282

Mizzen L, Daya M, Anderson R (1987) The role of protease-dependent cell membrane fusion in persistent and lytic infections of murine hepatitis virus murine. Adv Exp Med Biol 218: 175-186

Molynex P (1975) Synthetic polymers, in Water, a comprehensive treatise. Franks F (Ed) vol 4 Plenum Publ Co NY pp 569-757

Morgan CG, Williamson H, Fuller S, Hudson B (1983) Melittin induces fusion of unilamellar phospholipid vesicles. Biochim Biophys Acta 732: 668-674

Nir S, Bentz J, Wilschut J, Duzgunes N (1983) Aggregation and fusion of phospholipid vesicles, in Progress in surface science. vol 13 Pergamon Press NY p 54

Novick SL, Hoekstra D (1988) Membrane penetration of sendai virus glycoproteins during the early stages of fusion with liposomes as determined by hydrophobic photoaffinity labelling. Proc Natl Acad Sci USA 85: 7433-7437

Ohki S (1982) A mechanism of divalent ion-induced phosphatidylserine membrane fusion. Biochim Biophys Acta 689: 1-11

Ohki S (1984a) Effects of divalent cations temperature, osmotic pressure gradient and vesicle curvature on phosphatidylserine vesicle fusion. J Membrane Biol 77: 265-275

Ohki S (1984b) Adsorption of local anesthetics on phospholipid membranes. Biochim Biophys Acta 777: 56-66

Ohki S (1985) Membrane fusion: Theory and experiments. Studia Biophysica 10: 95-104

Ohki S (1988a) Surface tension, hydration energy and membrane fusion, in Molecular mechanism of membrane fusion. Ohki S et al (Eds) pp 123-138

Ohki S (1988b) Membrane fusion, hydration energy and hydrophobicity. Studia Biophysica 112: 89-97

Ohki S, Arnold K (1989) Surface dielectric constant, surface hydrophobicity and membrane fusion. Submitted to J Membrane Biol

Ohki S, Doyle D, Flanagan TD, Hui SW, Mayhew E (Eds) (1988) Molecular mechanisms of membrane fusion. Plenum Publ Co NY

Ohki S, Duax J (1986) Interaction of polyamines with phosphatidylserine membranes. Biochim Biophys Acta 861: 177-186

Ohki S, Duzgunes N, Leonards K (1982) Phospholipid vesicle aggregation: Effect of monovalent and divalent ions. Biochemistry 21: 2127-2133

Ohki S, Ohshima H (1984) Divalent cation-induced surface tension increase in acidic phospholipid membranes: Ion binding and membrane fusion.

Biochim Biophys Acta 776: 177-182
Ohki S, Ohshima H (1985) Divalent cation-induced phosphatidic acid membrane fusion. Effect of ion binding and membrane surface tension. Biochim Biophys Acta 812: 147-154
Ohnishi S (1988) Fusion of viral envelopes with cellular membranes, in Current topics in membranes and transport. vol 32 Duzgunes N, Bronner F (Eds) Academic Press NY pp 257-296
Ohnishi S, Ito T (1974) Calcium-induced phase separations in phosphatidylserine-phosphatidylcholine membranes. Biochemistry 13: 881-887
Ohshima H, Ohki S (1985) Effects of divalent cations on the surface tension of a lipid monolayer-coated air/water interface. J Colloid Interface Sci 102: 85-94
Pal R, Petri WA, Barenholz Y, Wagner R (1983) Lipid and protein contributions to the membrane surface potential of vesicular stomatitis virus probed by a fluorescent pH indicator, 4-heptadecyl-7 -hydroxycoumarin Biochim Biophys Acta 729: 185-192
Papahadjopoulos D (1978) Calcium-induced phase changes and fusion in natural and model membranes, in Membrane fusion. Poste G, Cnicolson GL (Eds) Elsevier/North Holland Biomedical Press Amsterdam pp 265-790
Papahadjopoulos D, Meers PR, Hong K, Ernst JD, Goldstein IM, Duzgunes N (1988) Calcium-induced membrane fusion: From liposomes to cellular membranes, in Molecular mechanisms of membrane fusion. Ohki S et al (Eds) pp 1-16
Papahadjopoulos D, Poste G, Schaffer BE, Vail WJ (1974) Membrane fusion and molecular segregation in phospholipid vesicles. Biochim Biophys Acta 352: 10-28
Papahadjopoulos D, Vail WJ, Pangborn WA, Poste G (1976) Studies on membrane fusion. II. Induction of fusion in pure phospholipid membranes by calcium and other divalent cations. Biochim Biophys Acta 465: 579-598
Parente RA, Lentz BR (1986) Rate and extent of poly(ethylene glycol)-induced large vesicle fusion monitored by bilayer and internal content mixing. Biochemistry 25: 6678-6688
Portis A, Newton C, Pangborn W, Papahadjopoulos D (1979) Studies on the mechanism of membrane fusion: Evidence for an intermembrane Ca^{2+}-phospholipid complex synergism with Mg^{2+} and inhibition by spectrin. Biochemistry 18: 780-790
Poste G, Allison AC (1973) Membrane fusion. Biochim Biophys Acta 300: 421-465
Poste G, Nicolson GL (Eds) (1978) Membrane fusion. Elsevier/North Holland Amsterdam
Pratsch L, Donath E (1988) Poly-(ethylene glycol) depletion layers on human red blood cell surfaces measured by electrophoresis. Studia Biophys 123: 101-108
Rasmussen H (1970) Cell communication, calcium ion and cyclic adenosine monophosphate. Science 170: 404-412
Rupert LAM, Engberts, JBFN, Hoekstra D (1988) Effect of poly(ethylene glycol) on the Ca^{2+}-induced fusion of didodecyl phosphate vesicles. Biochemistry 27: 8232-8239
Schullery SE, Schmidt CF, Felgner P, Tillack TW, Thompson TE (1980) Fusion of dipalmitoyl-phosphatidylcholine vesicles. Biochemistry 19: 3919-3923
Siegel DP (1984) Inverted micellar structures in bilayer membranes: Formation rates and half-lives. Biophys J 45: 399-420
Sowers AE (Ed) (1987) A long-lived fusogenic state is induced in erythrocyte ghosts by electric pulses. J Cell Biol 102: 1358-1362
Sowers AE (Ed) (1987) Membrane fusion. Plenum Publ Co NY
Struck DE, Hoekstra D, Pagano RE (1981) Use of resonance energy transfer to monitor membrane fusion. Biochemistry 20: 4093-4099

Sukumaran D, Ohki S (1988) Effect of monovalent cations on polyvalent cation-induced fusion of phosphatidylserine small unilamellar vesicles. Chem Phys Lipids 49: 9-14

Sundler R, Papahadjopoulos D (1982) Control of membrane fusion by phospholipid head groups. I. Phosphatidate/phosphatidyl-inositol specificity. Biochim Biophys Acta 649: 743-750

Szoka F Jr, Papahadjopoulos D (1978) Procedure for preparation of liposomes with large internal aqueous space and high capture by reverse-phase evaporation. Proc Natl Acad Sci USA 75: 4194-4198

Tilcock CPS, Fisher D (1979) Interaction of phospholipid membranes with poly(ethylene glycol)s. Biochim Biophys Acta 577: 53-61

Tilcock CPS, Fisher D (1982) The interaction of phospholipid membranes with poly-(ethylene glycol) vesicle aggregation and lipid exchange. Biochim Biophys Acta 688: 645-652

Uster PS, Deamer DW (1981) Fusion competence of phosphatidylserine-containing liposomes quantitatively measured by a fluorescence resonance energy transfer assay. Arch biochim Biophys 209: 385-395

Vail WJ, Stollery JG (1979) Phase changes of cardiolipin vesicles mediated by divalent cations. Biochim Biophys Acta 551: 74-84

van Oss CJ, Arnold K, Good RJ, Gawrisch K, Ohki S (1989) Interfacial tension and the osmotic pressures of solutions of polar polymers. Submitted to J Macromolecular Sci

van Oss CJ, Chaudhury MK, Good RJ (1987) Monopolar surfaces. Advan Colloid Interface Sci 28: 35-64

van Oss CJ, Chandhury MK, Good RJ (1988) Polar interfacial interactions, hydration pressure and membrane fusion, in Molecular mechanisms of membrane fusion. Ohki S et al (Eds) pp 113-122

Vanderwerf P, Ullman EF (1979) Monitoring of phospholipid vesicle fusion by fluorescence energy transfer between membrane-bound dye labels. Biochim Biophys Acta 596: 302-314

Verkleij AJ, van Echteld CJA, Gerritsen WJ, Cullis PR, DeKruijff B (1980) The lipidic particle as an intermediate structure in membrane fusion processes and bilayer to hexagonal H_{II} transitions. Biochim BIophys Acta 600: 620-624

Waggoner AS, Stryer L (1970) Fluorescent probes of biological membranes. Proc Natl Acad Sci USA 67: 579-589

Wickner WT, Lodish HF (1985) Multiple mechanisms of protein insertion and across membranes. Science 230: 400-407

Wilschut J, Papahadjopoulos D (1979) Ca^{2+}-induced fusion of phospholipid vesicles monitored by mixing of aqueous contents. Nature (London) 281: 690-692

Wilschut J, Duzgunes N, Papahadjopoulos D (1981) Calcium/magnesium specificity in membrane fusion: Kinetics of aggregation and fusion of phosphatidylserine vesicles and the role of bilayer curvature. Biochemistry 20: 3126-3133

Zakai N, Kulka RG, Loyter A (1977) Membrane ultrastructural changes during calcium phosphate-induced fusion of human erythrocyte ghosts. Proc Natl Acad Sci 74: 2417-2421

Zimmermann U (1982) Electric field-mediated fusion and related electrical phenomena. Biochim Biophys Acta 694: 227-277

Membrane fusion and lysis by the amphiphilic peptides induced by charge neutralization: A model study of virus membrane fusion mechanism

Murata,M., Takahashi,S.*, Kanaseki,T.,** and Ohnishi,S.

Department of Biophysics
Faculty of Science, Kyoto University
Kyoto, Japan.
* Institute for Chemical Research
Kyoto University, Uji, Japan.
**Department of Genetics,
Tokyo Metropolitan Institute for Neuroscience,
Tokyo, Japan.

Introduction

Membrane fusions are crûcial phenomena in cellular membrane dynamics such as endocytic and exocytic processing and fertilization. These are mediated by some specific fusion proteins. The understanding of molecular and control mechanisms of membrane fusions is very important and presents attractive problems for elucidating membrane dynamics.

The fusogenic proteins of enveloped viruses have been extensively studied by physicochemical and genetic approaches. Hemaggulutinin (HA) on the influenza viral envelope is a well-known fusogenic protein. This virus can fuse with cell membranes only at acidic pH due to the low pH-induced fusion activity of HA. Therefore, this virus is endocytosed into target cells and it transfers its genome through fusion with the endosome membrane when the intravesicular pH is 5-6 (White et al., 1983; Ohnishi,1988). Most other enveloped viruses fuse in this way. However, paramyxoviruses fuse at neutral pH.

Many fusogenic proteins have a stretch of around 20 hydrophobic amino acids at the N-terminus with a few acidic groups. The hydrophobic segments have been assumed to be involved in the fusion reaction because of the hydrophobicity and the sequence homology between the viral strains (Gething et al.,1987). We think that the protonation of the acidic residues in the hydrophobic segment is the trigger for the low pH-induced fusion (Maeda & Ohnishi,1980). The three-dimensional structure of HA has been determined

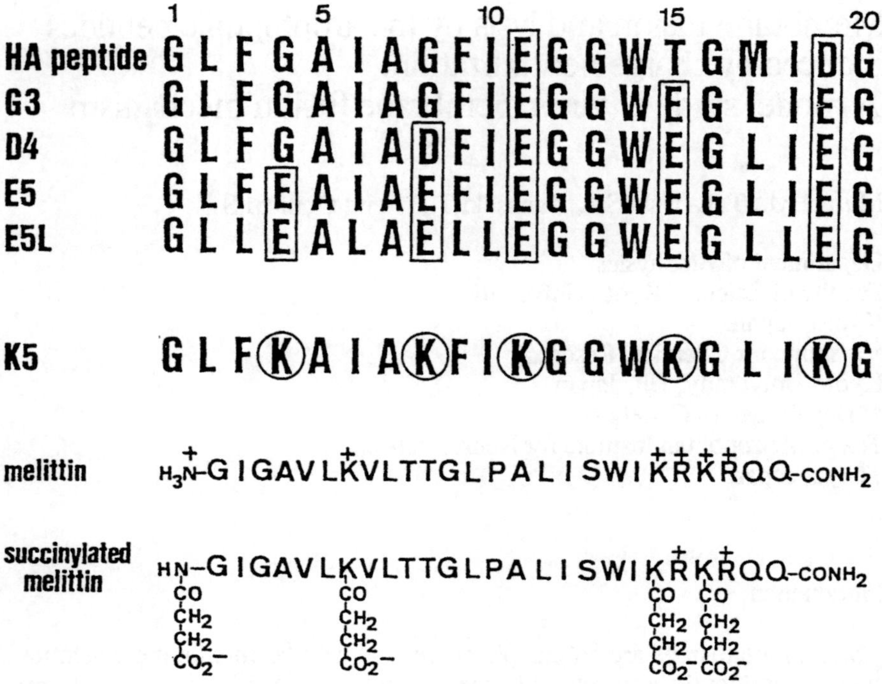

Fig. 1. HA-peptide and its analogues used in the present study. G3, D4, E5 and E5L are the anionic fusogenic peptides containing the 3, 4, 5, and 5 acidic groups, respectively. K5 is the cationic one with the 5 Lys's group. HA-peptide is a 20-amino acid peptide with the same sequence as that of the HA2 N-terminal segment of influenza virus A/PR/8/34. Melittin and succinylated melittin are used for the fusogenic peptides.

(Wilson et al.,1981) and the conformational change induced at acidic pH has been investigated physicochemically (Skehel et al.,1982; Sato et al.,1983). The latter authors suggested that this hydrophobic segment in HA becomes exposed at acidic pH due to the drastic conformational change of HA, and is thus able to interact directly with the target cell membranes. Genetic approaches in the study of the hydrophobic segment in HA include the use of site-specific mutagenesis and have shown that the length of the hydrophobic segment and the substitution of the N-terminus of the HA2 subunit were crucially important for the fusion (Gething et al.,1986). Such hydrophobic segments were also found in other fusogenic proteins and in some proteins which could cause perturbation and lysis of the biological and artificial membranes (Lucy, 1984). These hydrophobic segments showed some different properties from the two well-known hydrophobic segments in membrane proteins, the leader peptide domain and the transmembrane domain (Paterson et al.,1986; Davis et al.,1986).

In this article, we have drawn attention to the putative hydrophobic segment in the virus fusion reactions and synthesized a twenty amino acid peptide (HA-peptide) from the HA2 N-terminal segment of influenza virus A/PR/8/34 (see Fig.1 for the formula). We also made more soluble and amphiphilic peptide analogues from the HA-peptide; four anionic peptides called G3, D4, E5, and E5L and one cationic peptide called K5. All the anionic peptides caused the low pH-induced fusions and the cationic peptide caused the high pH-induced fusion. We also studied fusion by a mixture of the anionic minus E5 and the cationic plus K5 peptides *at neutral pH*. We also studied fusion by succinylated melittin at acidic pH to indicate the essential role of the carboxyl groups for protonation. In all fusion studies, we used sonicated egg PC vesicles as the target membranes, since our attention was on the electrostatic interactions of peptides and not on the interaction of peptides with the vesicle surface charges.

Fusion Induced by the HA-Peptide

HA-peptide caused fusion of egg PC vesicles at acidic pH but not at neutral pH (Fig.2A) (Murata et al.,1987a). The pH profile of fusion activity was similar to that of the parent virus. The light scattering value was the same at neutral pH but greatly increased at acidic pH values, indicating aggregations of vesicles and fusion (Fig.2B). The hydrophobicity of the HA-peptide was measured by Triton X-114 partitioning (Bordier,1981). It increased at pH values lower than 6.0, i.e. closer to the fusion threshold. The protonation of the acidic residues increased the hydrophobicity of peptide which caused aggregation of the peptide-vesicles and interacted more easily with the lipid bilayer membranes.

Lear and DeGrado (1987) also synthesized a 20-amino acid peptide from the HA2 N-terminus of influenza virus B/Lee/40 and showed fusion of DOPC vesicles at neutral pH. This result differs from our results on HA-peptide and other analogues. They also synthesized the 16-amino acid peptide which did not cause fusion. Düzgünes & Gambale (1988) showed that the 17-amino acid peptide from influenza virus X-31 did not cause fusion of small unilamellar vesicles at either pH 7 or 5, although it caused leakage of liposomes. The peptides shorter than 20 amino acids may not contain the fusion activity.

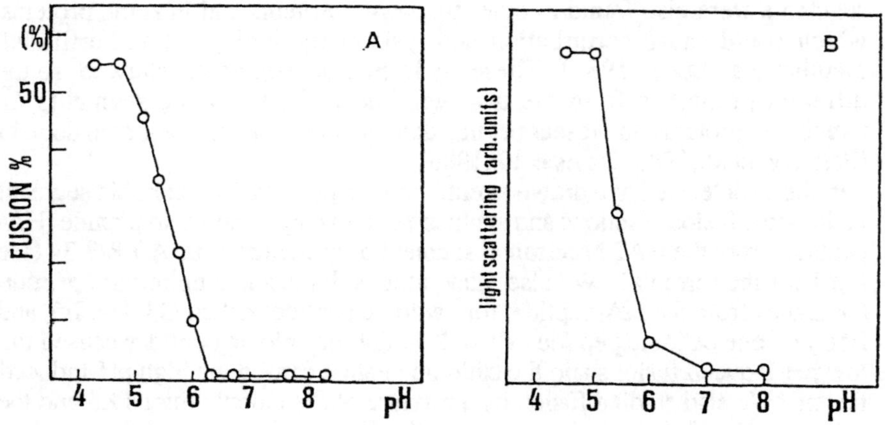

Fig. 2. (A) Membrane fusion activity of HA-peptide. The fusion was assayed at 23°C in a system of the peptide (68 μM), egg PC sonicated vesicles containing 10% spin-labeled PC (0.82 mM), and egg PC (3.3 mM). (B) The light scattering at 400 nm of HA-peptide and egg PC vesicles.

We changed the N-terminus of the HA-peptide by acetylation or succinylation. Both of these derivatives caused a low pH-induced fusion activity but the pH range was shifted to the more acidic side as compared with the parent peptide. The shift was more than 1 pH unit; the threshold was lower than pH 5.2 as compared with pH 6.0 and the fusion still increased at pH 3.8 (Murata et al.,1987a). A mutant of the HA protein with the HA2 N-terminus of Glu for Gly was studied (Gething et al.,1986). It abolished the fusion activity down to pH 4.8. It may be due to a shift of the fusion activity to the more acidic side. Such an inhibitory effect of the modification at the N-terminus is a promising possibility for virus infection.

Protonation of the Carboxyl Groups Triggers Succinylated Melittin-Induced Fusion

Succinylated melittin, which is a derivative of bee venom melittin, contains the hydrophobic segment and is succinylated at the 4 amino groups. It caused low pH-induced fusion, an optimum fusion at pH 5.1 and a threshold lower than pH 5.2. In order to confirm the involvement of protonation of the carboxyl groups, we prepared the ^{13}C-labeled succinylated melittin to study the pK_a value. The pK_a value of the four carboxyl groups was 5.19 in the presence of vesicles. It agreed quite well with the half-maximal pH 5.15 for fusion. This indicates that the protonation of its carboxyl groups causes the fusion (Murata et al.,1987b).

The hydrophobicity of succinylated melittin increased at pH values lower than the threshold fusion. The protonated form of the peptide caused greater hydrophobicity for easier immersion into the hydrocarbon region of the lipid bilayer.

HA-Peptide Analogues and Their Fusion Activity

The CD spectra of the HA-peptide showed an ordered secondary structure. The peptide is amphiphilic with the hydrophilic and the hydrophobic surfaces on the α-helix according to the CPK model(Fig.3). The Gly- and Ala-rich regions are on the hydrophilic surface which make the peptide "less hydrophobic" because of the smaller side chains. We first made the peptide more amphiphilic by introducing the acidic residues on this side, instead of one Glu and one Asp for HA-peptide. Other simplifying and solubilizing substitutions were also made. The peptide solubility was essential for studying the fusion mechanism by physicochemical techniques such as NMR, CD, small angle X-ray scattering, etc.

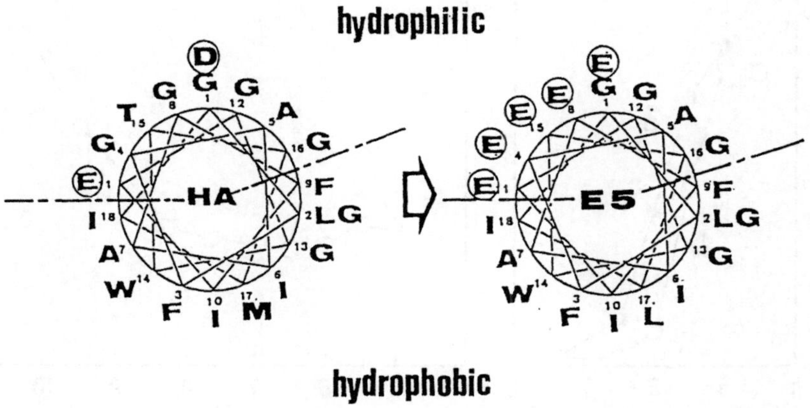

Fig. 3. Schematic helical diagram of HA-peptide and E5 with the hydrophilic and the hydrophobic surfaces.

E5 and E5L have the 5 Glu substitutions for Gly and Ala and more distinct segregation of the hydrophilic and the hydrophobic residues on the opposing helical surfaces (Fig. 3). G3 and D4 have 3 and 4 acidic residues, respectively, and are in between HA-peptide and E5. We have also synthesized one cationic peptide K5 in contrast to the anionic peptides. K5 has the charge-reversed analogue of E5 with the same amino acid sequence except for 5 Lys's instead of 5 Glu's.

All the anionic peptides caused extensive fusion only at acidic pH but not at neutral pH (Fig.4A). The optimum pH was 5.0 for E5, E5L, G3 and 5.8 for D4. The threshold pH was lower than 6.5 for E5, E5L, and 7.5 for G3 and D4. These are compared with the optimum pH 4.8 and the threshold pH lower than 6.2 for HA-peptide. The carboxyl group at the C-terminus for the anionic peptides may not be involved in the low pH-induced fusion. The cationic peptide K5 caused extensive fusion at pH values higher than 8.2 (Fig.4B). It also caused weak fusion at neutral and acidic pH values.

These results indicate that the protonation of acidic residues by H^+ or the deprotonation of Lys residues by OH^- triggered the fusion reaction (Murata et al.,1990a).

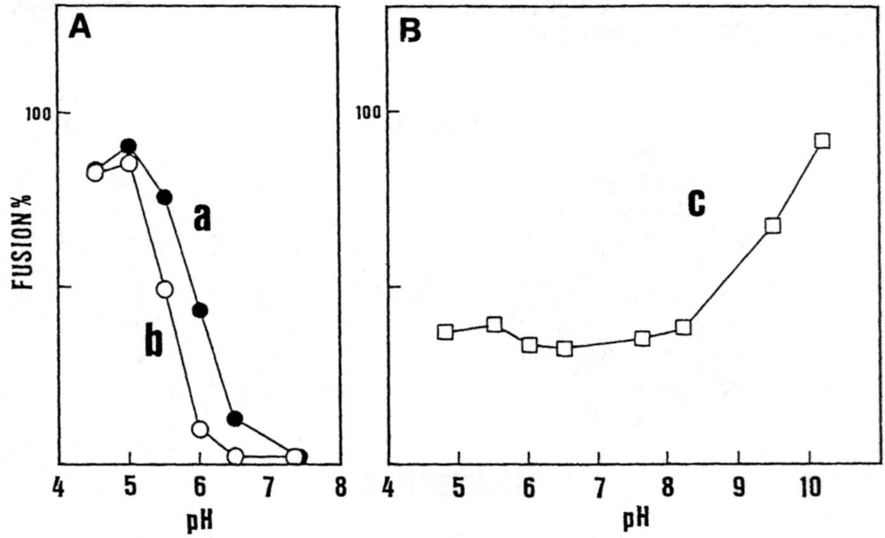

Fig. 4. pH-dependence of fusion of egg PC vesicles induced by (a) G3 and (b) E5 in (A) and by K5 in (B). Fusion was assayed by the spin label method described in Fig. 2. The degree of fusion after 5 min was estimated from the normalized ESR peak height.

Peptide-Vesicle Interactions and CD Spectra

The anionic and cationic peptides bound to egg PC vesicles at various pH values. The fluorescence maximum of Trp in the peptides was at 349 to 355 nm in the absence of vesicles. But it shifted to blue by 10-20 nm within 10 sec after addition of vesicles. The blue shift indicates amino acid residues including Trp associating more to the lipid bilayer hydrocarbon regions.

The light scattering at 400 nm for the anionic peptides in vesicles was nearly the same at neutral pH but it greatly increased at acidic pH. The peptide-vesicles become therefore aggregated at acidic pH and fused. Similar scattering increase was found for the cationic peptide at alkaline pH.

The hydrophobicity of peptides was assayed at various pH values (Fig.5). The anionic peptides became more hydrophobic at pH values lower than the fusion threshold. The cationic peptide also caused an increase in the hydrophobicity higher than the threshold pH value for larger fusions.

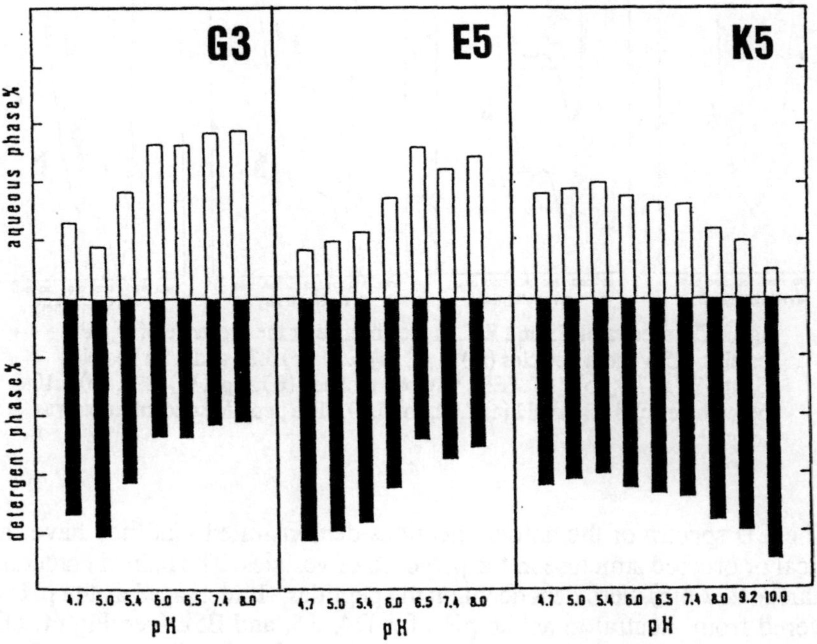

Fig. 5. Hydrophobicity of peptides at various pH values as determined by Triton X-114 partitioning. Peptide (6 μM) was incubated in final 0.5% Triton X-114 solution at the indicated pH values. After separation, the peptide concentration in the detergent and the aqueous phases was determined by the fluorescamine assay.

The oligomerization of peptides was studied by small angle X-ray scattering measurements. For E5 and E5L, the observed radius of aggregation and the oligomerization state increased at acidic pH as compared with those at neutral pH (Kajiwara et al., in preparation). In the presence of vesicles, the protonated peptides may create additional hydrophobic domains within themselves to interact easily with the lipid bilayer.

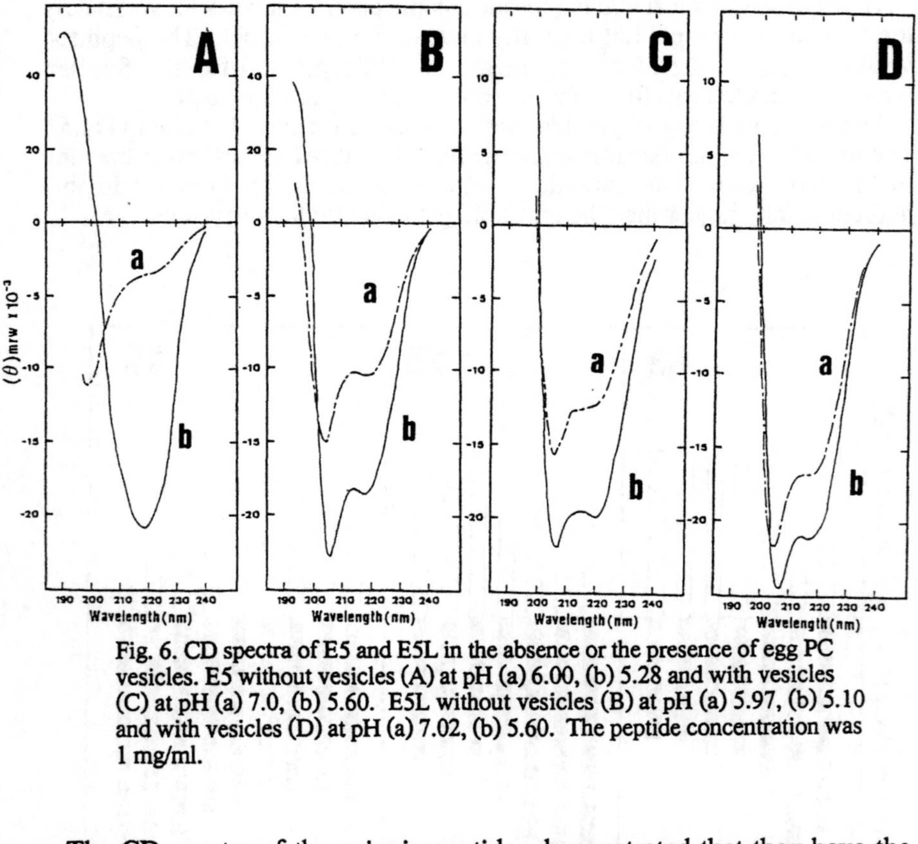

Fig. 6. CD spectra of E5 and E5L in the absence or the presence of egg PC vesicles. E5 without vesicles (A) at pH (a) 6.00, (b) 5.28 and with vesicles (C) at pH (a) 7.0, (b) 5.60. E5L without vesicles (B) at pH (a) 5.97, (b) 5.10 and with vesicles (D) at pH (a) 7.02, (b) 5.60. The peptide concentration was 1 mg/ml.

The CD spectra of the anionic peptides demonstrated that they have the helical or ordered structure in the presence of vesicles. The helical contents (value at 222 nm) were dependent on the peptides. It increased as the pH was lowered from neutral to acidic pH's for D4, E5, and E5L (see Fig.6C,D). HA-peptide was nearly the same and G3 decreased as the pH decreased from neutral to acid. The CD spectra suggest the amphiphilic helical property shown in Fig.3 for HA-peptide and the synthetic peptides on lipid bilayer membranes.

Interestingly, in the absence of vesicles, the CD spectra of E5 showed a random conformation at neutral pH but a helical structure at acidic pH. The CD spectra of E5L showed the typical α-helix at both pH's (Fig.6A,B). The minor variations in their primary sequences affected the secondary structure and their pH-profiles of leakage and hemolytic activity (see Fig.8).

The CD spectra of K5 showed a random coil with some indication of helical structure at neutral pH in the absence of vesicles. However, at alkaline pH, where it caused larger fusions, it showed an ordered structure with a minimum at 222 nm (Murata et al.,1990a).

Anionic E5 and Cationic K5 Cause Fusion *at Neutral pH*

E5 caused fusion at acidic pH but not at neutral pH. K5 caused a large fusion at alkaline pH but weak fusion at neutral pH. The two mixturs caused rapid and extensive fusion at neutral pH. An example is shown in Fig.7. E5 was added to egg PC vesicles but there was no fusion. However, when K5 was added to the reaction mixture after 6 min, it caused an immediate, rapid fusion. More additions of K5 again caused more rapid fusions. Similarly, when E5 was added to the reaction mixture of K5 and egg PC vesicles, it again caused a very rapid fusion.

When equimolar E5 and K5 was subsequently added to vesicles, it caused very efficient fusion. The equimolar mixture gave the most hydrophobic complex. The fusion was greatly dependent on the ionic strength of peptides. It was most rapid and extensive in NaCl from zero to 100 mM and NaP_i buffer at 40 mM and became lower with increasing NaCl concentrations. The fusion was nearly zero at 800 mM NaCl.

The CD spectra of E5 and K5 alone were random coils in the absence of vesicles at neutral pH. Surprisingly, when the two peptides were combined, they showed an ordered structure with a minimum at 222 nm. In the presence of vesicles, K5 was a random coil and E5 now took a helical structure and, when these two were mixed, they formed an ordered structure. This ordered structure caused fusion at neutral pH.

The interaction of two peptides arises mainly from the electrostatic Coulombic force. They can form antiparallel helicals since the two peptides have the same amino acid sequence except for Lys vs Glu and take the ordered secondary structures. The ordered structure may come from the charged interactions. The complex has the charged interactions between the 5 Lys's and 5 Glu's inside and the hydrophobic residues outside, thus forming a hydrophobic moiety. The vesicles containing the two peptides become aggregated by the electrostatic interactions and the hydrophobic moiety on the closely apposed vesicles can interact with the hydrocarbon regions of the bilayers (Murata et al.,1990b).

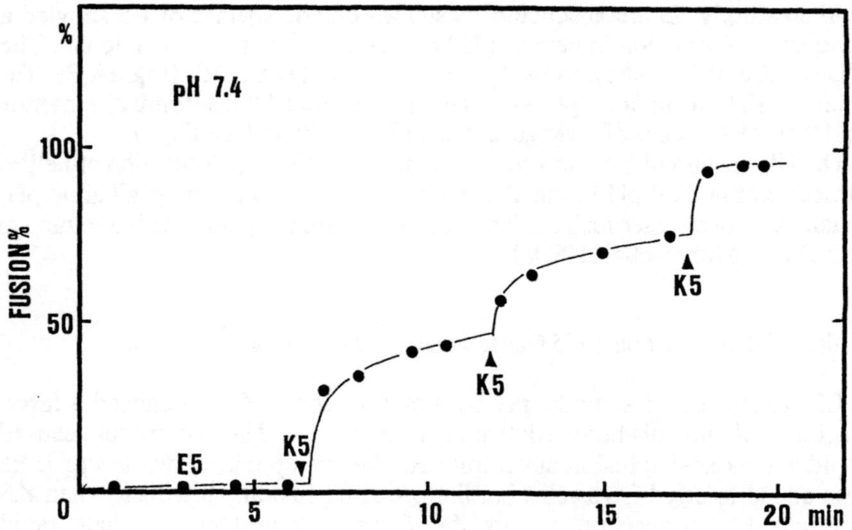

Fig. 7. Effect of further addition of K5 on the reaction mixture of E5 and egg PC vesicles at neutral pH. The degree of fusion was estimated from the normalized ESR peak height.

Lysis and Hemolysis by Peptides

The peptide-induced leakage of low molecular fluorescence marker (ANTS/DPX) from egg PC large unilamellar vesicles was studied by dequenching of the fluorophore ANTS (Ellens et al.,1985). The anionic peptides caused high leakages at acidic pH and also weak leakages at neutral pH, in contrast to fusion only at acidic pH (Fig.8A). E5L was an exception showing only high leakages at acidic pH but no leakages at neutral pH. The cationic peptide K5 caused higher leakages at alkaline pH and weaker ones at neutral pH. Since these peptides bound to vesicles at neutral pH, they may cause some leakage by interacting with the lipid bilayer. At acidic pH, the peptides become hydrophobic due to the charge neutralization and cause large leakages and fusion due to interaction with the hydrocarbon region of bilayers.

Hemolysis of human red blood cells was measured to investigate the peptide activity to biomembranes (Fig.8B). The peptides had more restrictive action on the red blood cells because they first encountered with the membrane surface proteins to reach the lipid bilayers. E5, E5L, and D8 showed hemolysis only at acidic pH but HA-peptide and G3 did not hemolyze. K5 caused greater hemolysis at alkaline pH (Murata et al.,1990a).

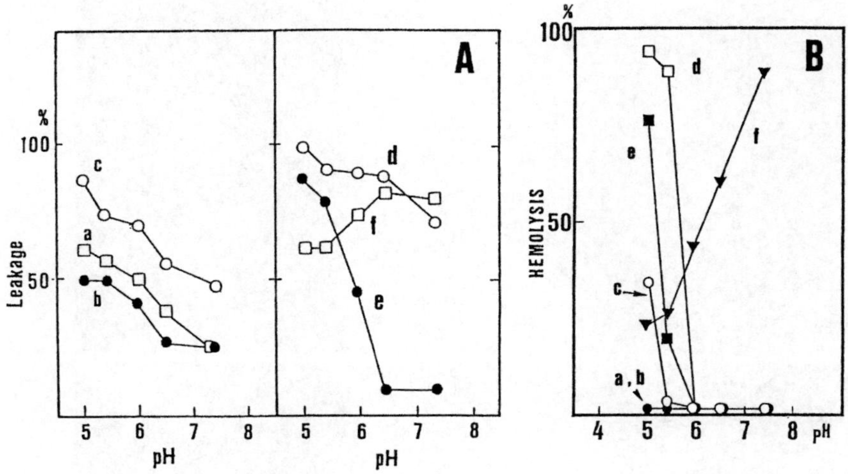

Fig. 8. Leakage of the fluorophore ANTS from egg PC large unilamellar vesicles (A) and hemolysis (B) induced by (a) HA-peptide (b) G3, (c) D4, (d) E5, (e) E5L and (f) K5. The lipid/peptide ratio was 20.

Intermediary Membrane Structure During Fusion

In order to observe the intermediary steps of membrane fusion, we studied the reaction of influenza virus with glycophorin-containing vesicles by quick-frozen-replica electron microscopy. After 30 sec at pH 5.0 at 23°C, about 8 to 10 small protrusions with a diameter of ca. 20 nm arranged in a ring-like manner were observed on the convex membrane faces of the vesicles. Each protrusion has a complimentary pit on the opposite concave face (Fig.9). On favorable fracture planes, penetrating viral spikes were observed. Such protrusions and pits were not observed after incubation at neutral pH. The protrusions and pits may be involved as the micro-domains for membrane fusion (Kawasaki, K., 1987; in preparation). In the highly curved regions, the phospholipid packing of the outer layer may be loose and that of the inner layer may be compressed more tightly. The lipid bilayers become too unstable to be fused.

The peptides, after becoming hydrophobic, may interact with the lipid bilayers to create such highly curved areas.

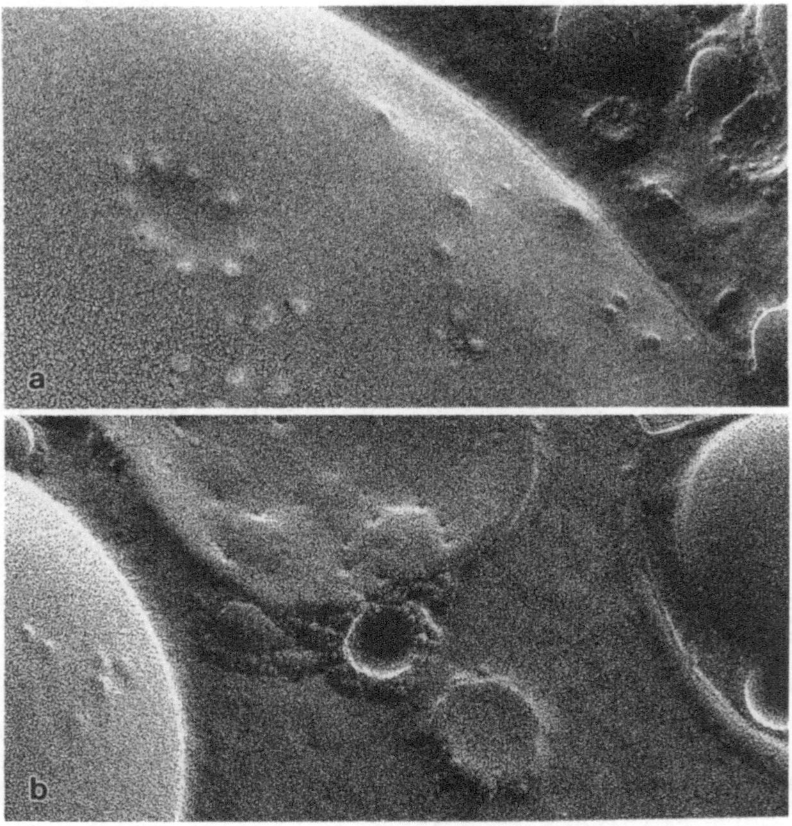

Fig. 9. Quick-frozen electron microscopy of influenza virus with glycophorin-containing egg PC-cholesterol (2:1) vesicles after 30 sec at pH 5.0 at 23°C. Magnification; x 156,600 (upper) and x 163,800 (lower).

Fusion Mechanism

The anionic peptides cause fusion at acidic pH probably because of protonation of the acidic groups. The amphiphilic peptides have affinity to egg PC vesicles at neutral pH. They become more hydrophobic due to the charge neutralization at acidic pH. They can change the oligomeric state to become more hydrophobic. The hydrophobic peptides on vesicles cause their aggregations and may interact with the bilayer hydrocarbon regions of closely apposed vesicles (Fig.10A). In the case of the cationic peptide, the deprotonation of the lysine amino groups at alkaline pH causes the peptide to neutralize, to become more hydrophobic, and to interact with the lipid bilayers in a similar manner to the anionic peptides (Fig.10B). The anionic E5 and the cationic K5 peptides caused fusion at neutral pH. The charge neutralization occurs through the specific Coulombic interaction between the plus and the minus groups of the peptides. Then, the two peptides become hydrophobic and the peptide-vesicles aggregate. The hydrophobic peptides now interact with the lipid bilayers as stated above (Fig.10C).

The peptides used in the present studies have an ordered secondary structure under the fusion conditions. The E5 and K5 mixture forms the ordered structure through charge interactions. The hydrophobic helical peptides may be effective to interact with the hydrocarbon regions of the bilayers. The helical structure was observed for melittin, succinylated melittin, and the 20-amino acid peptide by Lear and DeGrado (1987). No helical structure was observed for the 16-amino acid peptide, which cannot cause fusion. The HA2 N-terminal segment in intact HA did not have the helical conformation at neutral pH. However, it adopted an amphiphilic helical structure on binding to membranes (Harter et al.,1989).

All the peptides become neutralized and hydrophobic for the anionic peptides by protonation, the cationic peptide by deprotonation, and the peptide-peptide by specific charge interaction. These neutralized peptide interactions may be important for cellular fusions. Several studies have shown such charge neutralization between the plus charges in polypeptides and protein and the minus charge on the phosphatidylserine-containing vesicles. An acrosomal protein lysin has both the hydrophobic and the polycationic regions in its sequence and the former interacts with the hydrocarbon regions of the lipid bilayer and the latter facilitates the close apposition of two membranes (Hong & Vacquier,1986). In polylysine-induced (Walter et al.,1986) or polyhistidine-induced (Uster et al.,1985) fusions, the peptide side plus charges and the serine minus charges on vesicle surfaces were important.

The hydrophobic peptides may attack the hydrocarbon regions of the bilayers to create highly curved areas as shown for influenza virus and vesicles. Fusion of peptides with vesicles was dependent on the curvature, and was more effective with sonicated egg PC vesicles of ca. 20 nm diameter than with large unilamellar vesicles of ca. 50 nm diameter. This may indicate the high curvature in inducing fusion.

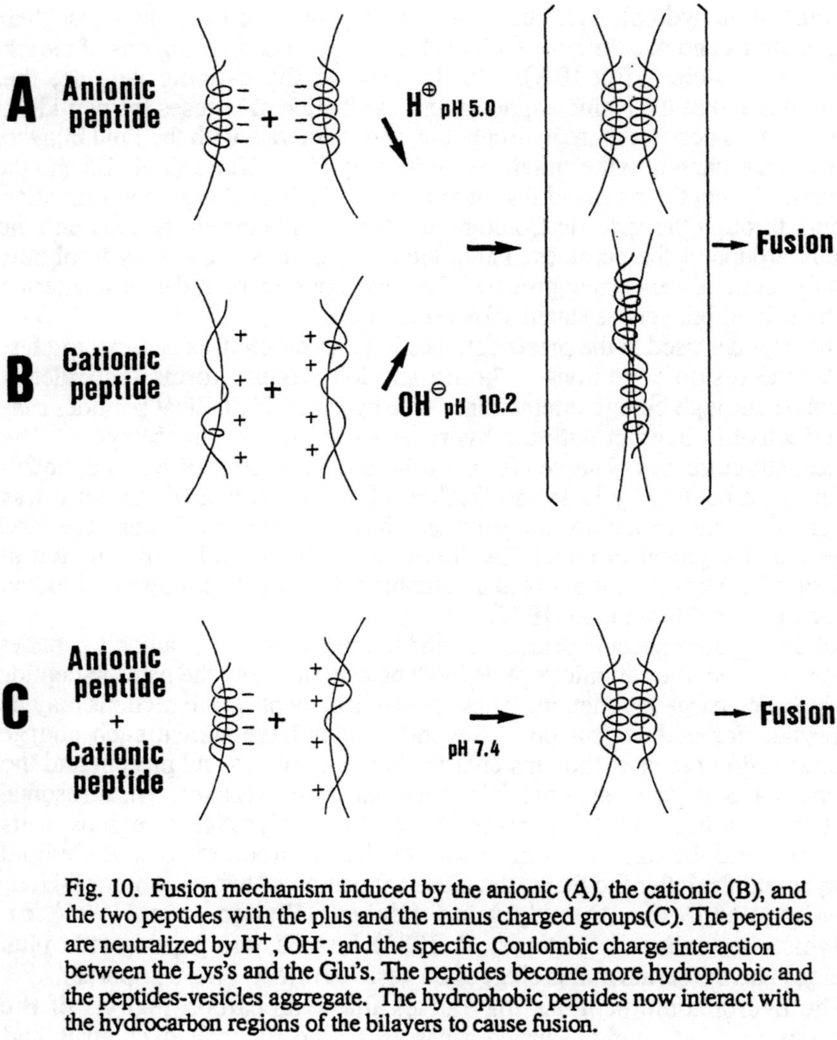

Fig. 10. Fusion mechanism induced by the anionic (A), the cationic (B), and the two peptides with the plus and the minus charged groups(C). The peptides are neutralized by H^+, OH^-, and the specific Coulombic charge interaction between the Lys's and the Glu's. The peptides become more hydrophobic and the peptides-vesicles aggregate. The hydrophobic peptides now interact with the hydrocarbon regions of the bilayers to cause fusion.

REFERENCES

Bordier,C.(1981) Phase separation of integral membrane protein in TritonX-114 solution. J. Biol.Chem. 256,1604-1607.

Davis,N.G.,& Hsu,M.-C.(1986) The fusion -related hydrophobic domain of Sendai F protein can be moved through the cytoplasmic membrane of Escherichia coli., Proc.Natl.Acad.Sci.U.S.A.83,5091-5095.

Düzgünes,N.,& Gambale,F.(1988) Membrane action of synthetic N-terminal peptides of influenza virus hemagglutinin and its mutants. FEBS Lett. 227, 110-114.

Ellens,H., Bentz,J.,& Szoka,F.C.(1985) H^+-and Ca^{2+}-induced fusion and destabilization of liposomes. Biochemistry 24,3099-3106.

Gething,M.-J., White,J.M.,& Waterfield,M.D.(1978) Purification of the fusion protein of Sendai virus:analysis of the NH -terminal sequence generated during precursor activation. Proc.Natl.Acad.Sci.U.S.A. 75 2737-2740.

Gething,M.-J.,Doms,R.W.,York,D.,& White,J. (1986) Studies on the mechanism of membrane fusion: site-specific mutagenesis of the hemagglutinin of influenza virus. J.Cell Biol. 102,11-23.

Harter,C.,James,P.,Bächi,T.,Semenza,G.,& Brunner,J.(1989) Hydrophobic binding of the ectodomain of influenza hemagglutinin to membranes occurs through the "fusion peptide" J.Biol.Chem. 264,6459-6464.

Hong,K.,& Vacquier,V.D.(1986) Fusion of liposomes induced by a cationic protein from the acrosome granule of abalone spermatozoa. Biochemistry 25,543-549.

Kawasaki,K.,Murata,M.,Ohnishi,S.,Ikeuchi,Y.,& Kanaseki,T. Membrane fusion of influenza virus with liposomes studied by quick-frozen-replica technics. Abstract:(1987) Cell Structure and Function 12,No. 6, 717.

Lear,J.D.,& DeGrado,W.F. (1987) Membrane binding and conformational properties of peptides representing the NH_2terminus of influenza HA-2. J.Biol.Chem. 262,6500-6505.

Lucy,J.,A. (1984) Do hydrophobic sequences cleaved from cellular polypeptides induce membrane reaction in vivo? FEBS Lett. 166,223-231.

Maeda,T.,& Ohnishi,S. (1980) Activation of influenza virus by acidic media causes hemolysis and fusion of erythrocytes. FEBS Lett. 122,283-287.

Murata,M.,Sugahara,Y.,Takahashi,S.,& Ohnishi,S.(1987a) pH-dependent membrane fusion activity of a synthetic twenty amino acid peptide with the same sequence as that of the hydrophobic segment of influenza virus hemagglutinin. J.Biochem. 102,957-962.

Murata,M.,Nagayama,K.,& Ohnishi,S. (1987b) Membrane fusion activity of succinylated melittin is triggered by protonation of its carboxyl groups. Biochemistry 26,1046-4062.

Murata,M.,Takahashi,S.,Kagiwada,S.,& Ohnishi,S.(1990a) pH-dependent membrane fusion activity of neutral phospholipid vesicles induced by synthetic amphiphilic peptides. Biochemistry submitted.

Murata,M.,Kagiwada,S.,Takahashi,S.,& Ohnishi,S.(1990b) Membrane fusion induced by mutual interaction of the two charge-reversed synthetic peptides. Biochemistry submitted.

Ohnishi,S.(1988) Fusion of viral envelopes with cellular membranes . In " Current Topics in Membrane and Transport".32,257-296.

Paterson,R.G.,& Lamb,R.A.(1987) Ability of the hydrophobic fusion - related external domain of a paramyxovirus F protein to act as a membrane anchor. Cell 48,441-452.

Sato,S.B.,Kawasaki,K.,& Ohnishi,S.(1983) Hemolytic activity of influenza virus hemagglutinin glycoproteins activated in mildly acidic environments. Proc.Natl.Acad.Sci.U.S.A. 80,3153-3157.

Skehel,J.J.,Bayley,P.M.,Brown,E.B.,Martin,S.R.,Waterfield,M., White,J.,Wilson,I.A.,& Wiley,D.C.(1982) Changes in the conformation of influenza virus hemagglutinin at the pH optimum of virus-mediated membrane fusion . Proc.Natl. Acad.Sci.U.S.A. 79,968-972.

Walter,A.,Steer,C.J.,& Blumenthal,R. Polylysine induces pH-dependent fusion of acidic phospholipid vesicles:a model for polycation-induced fusion. (1986) Biochim.Biophys.Acta 861,319-330.

White,J.,Kielian,M.,& Helenius,A.(1983) Membrane fusion proteins of enveloped animal viruses. Q.Rev.Biophys.16,151-195.

Wilson,I.A.,Skehel,J.J.,& Wiley,D.C.(1981) Structure of the haemagglutinin membrane glycoprotein of influenza virus at 3Å resolution. Nature(Lond.) 289,366-373.

Uster,P.S.,& Deamer,D.W.(1985) pH-dependent fusion of liposomes using titratable polycations. Biochemistry 24,1-8.

Significance of hydrophobic interactions in membrane fusion of enveloped viruses

A comparison with model membranes

Steven L. Novick and Dick Hoekstra
Laboratory of Physiological Chemistry
University of Groningen
Bloemsingel 10
9712 KZ Groningen
The Netherlands

I. INTRODUCTION

An essential prerequisite for a virus to infect a cell is its ability to deposit the nucleocapsid into the cytoplasm of that 'host'-cell. In the case of *enveloped* viruses, the nucleocapsid is surrounded by a membrane, consisting of a lipid bilayer that resembles the plasma membrane composition of the host-cell in which the virus was grown, and a unique set of virus-specific membrane proteins. To get access to the cytoplasm, the nucleocapsid must therefore traverse two membrane barriers. To accomplish this crossing, a virus uses a most efficient mechanism, namely membrane fusion. The cellular site where this fusion process occurs depends on the family to which the virus belongs. Some virions (for example, those belonging to the paramyxoviruses such as Sendai virus, Newcastle disease virus and measles virus) fuse directly with the plasma membrane at neutral pH. Most viruses, however, are first internalized by receptor-mediated endocytosis and subsequently fuse in the endosomal compartment 'from within' (Fig. 1). It has been shown that these viruses [such as orthomyxoviruses (influenza), toga viruses (Semliki Forest virus, SFV), and rhabdoviruses (vesicular stomatitus virus, VSV)] require a mild acidic pH in order to acquire fusogenic capacity (White et al., 1983).

Specific viral membrane proteins are intimately involved in the overall process that leads to fusion. For most viral membranes the protein composition is relatively

----

S.L. Novick's present address is: Division of Chemistry and Chemical Engineering, California Institute of Technology, Pasadena, CA 91125, USA.

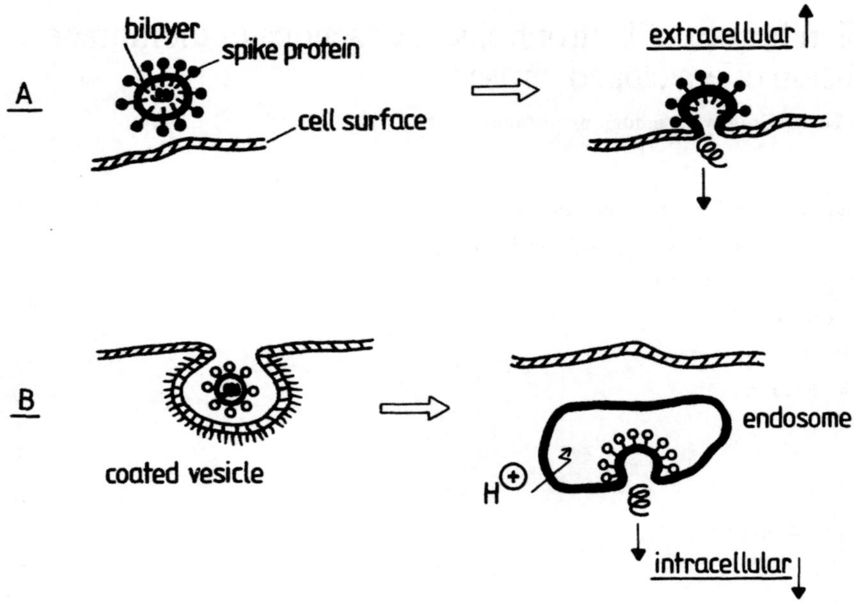

FIGURE 1. ENTRY PATHWAYS OF ENVELOPED VIRUSES. *The entry mechanism is family-dependent. Some virions fuse directly with the plasma membrane at neutral pH (A). Others are first internalized by receptor-mediated endocytosis. Subsequently, fusion is triggered when the endosomal compartment acidifies (B).*

simple, as they may contain as few as one or two types of proteins. In many instances, one of these proteins governs both binding to and fusion with a cellular membrane; in some cases, however, one protein contains only binding activity whereas a second protein contains the fusion activity. It is also possible, as might be the case for entry of Herpes simplex virus (Spear et al., 1989), that the overall fusion process is the result of a cooperative sequential chain of events requiring several (i.e., more than two) different viral envelope proteins.

Many of the aforementioned proteins have been identified; however, much of their structural and functional features remain to be elucidated. In this regard, the major influenza envelope protein, HA (hemagglutinin), represents an exception. To date, this protein is the best-characterized viral protein with respect to its synthesis, structure, and function (Wiley & Skehel, 1987; Doms & Helenius, 1988). The HA protein possesses a homotrimeric structure, the assembly of which occurs in the endoplasmic reticulum. Post-translational modifications take place after transfer to the Golgi complex (carbohydrate addition) and during subsequent trafficking to the cell surface. The latter involves a proteolytic cleavage, at or near the cell surface, which is essential for converting the inactive precursor HA_0 into a fusion-active HA, consisting of two subunits HA_1 and HA_2. An analogous phenomenon is seen for the fusion protein, F, of paramyxoviruses. In this case the inactive precursor F_0 is converted into the active form F, which also consists of two polypeptides, F_1 and F_2 (Choppin & Scheid, 1980). Human immunodeficiency virus (HIV) displays a similar feature as the predominant envelope protein gp 160 also needs to be activated by endoproteolytic cleavage in order to acquire fusion activity. As a result, a heterodimer consisting of two polypeptides, gp 120 and gp 41, is formed (McCune et al., 1988). However, not all viral proteins share this property. In the case of the fusion proteins of SFV (E_1) and VSV (G) no precursors have been identified (White et al., 1983).

Common to many viral fusion proteins appears to be the presence of a stretch of amino acids which is particularly hydrophobic, or acquires this property under appropriate conditions. This has led to the proposal that these regions are related to the ability of a virus to gain fusion activity. In the present paper, we will discuss the experimental evidence that supports this concept. In addition, guided by extensive work carried out with simple artificial membrane systems, we will briefly describe parameters that govern membrane fusion of liposomes and discuss whether similar parameters may govern virus fusion as well, emphasizing the role of hydrophobic interactions. In order to obtain insight into the role of various parameters in virus fusion, the development of sensitive techniques to monitor fusion continuously and directly, has proven to be of significant importance. We will therefore briefly outline the principles of these procedures, before discussing experimental details.

II. TECHNIQUES FOR MEASURING VIRUS FUSION

Relevant to a mechanistic understanding of viral fusion is the quantitative detection and characterization of the fusion process. This entails that fusion should be monitored with direct techniques, i.e. procedures that monitor the fusion of a virus as such, as opposed to more indirect approaches such as virus-induced hemolysis or cell-cell fusion. Evidently, although quite useful to qualitatively determine the fusogenic potency of a virus preparation, the latter procedurės register the consequences of a virus-target membrane interaction rather than the occurence of virus membrane fusion *per se*. Moreover, the lack of hemolysis does not necessarily imply that the virus does not fuse. Whether or not hemolysis occurs depends on the viral membrane structure, which in turn is affected by the age of the virus (Homma et al., 1976) or by the conditions of storage ("freezing and thawing") of the virus. For example, Sendai virus, harvested 24 h after infection, does fuse with erythrocytes without the induction of significant hemolysis. The extent of hemolysis greatly increases, however, when such virions undergo a freezing and thawing cycle. Similarly, when reconstituting the Sendai virus envelope, using the nonionic detergent octyl-glucoside, no detectable hemolysis is observed unless the reconstituted particles are subjected to freezing and thawing. Irrespective of such treatments, the fusion activity, monitored by a quantitative assay, is essentially the same (Harmsen et al., 1985). These examples thus emphasize the relevance of determining virus fusion in a direct manner.

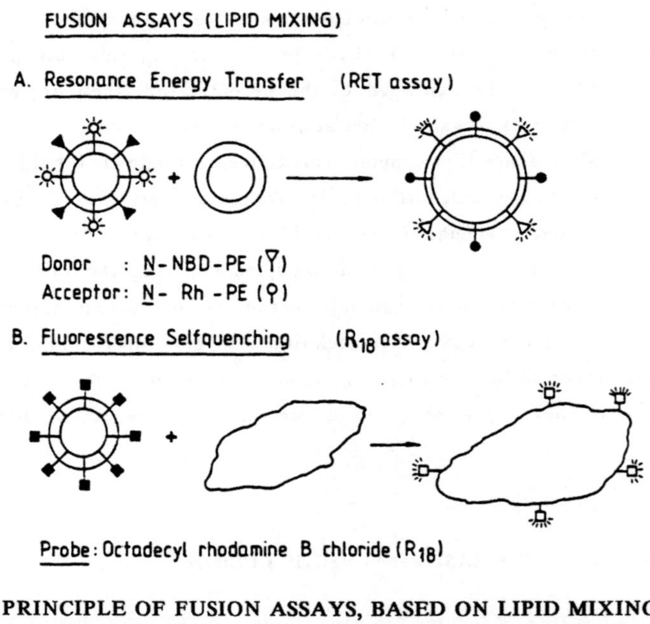

FIGURE 2. **PRINCIPLE OF FUSION ASSAYS, BASED ON LIPID MIXING.**

A. RET assay. The increase in donor fluorescence is usually monitored. This will occur when the energy transfer efficiency decreases as a result of probe dilution when a labeled membrane fuses with an unlabeled counterpart.

B. R_{18} assay. In this assay the relief of fluorescence selfquenching is monitored, which occurs when one membrane population, labelled with a selfquenching concentration of the probe, fuses with unlabelled membranes. For details and references, see text.

The most common method of monitoring virus-membrane fusion is based on lipid mixing, using either resonance energy transfer (RET) between fluorescently labelled membrane components or relief of selfquenching of a fluorescent lipid-like derivative (Fig. 2). In this way, fusion of complex biological membranes can be measured continuously and quantitatively. In conjunction with a kinetic mass action model, such an approach allows detailed kinetic analysis of the fusion reaction, and evaluation of the physical and chemical determinants of membrane fusion.

The fluorescent derivative octadecyl Rhodamine B chloride (R_{18}) can be readily inserted into intact membranes such as viral membrane envelopes (Hoekstra et al., 1984). As the density of the probe increases, its fluorescence, being sensitive to intermolecular distance, becomes increasingly selfquenched. Thus when a population of labelled virus particles fuses with a population of unlabelled membranes, lipid intermixing between the two membranes allows lateral diffusion of the probe to greater intermolecular distances, with the observable result of a relief of self-quenching. This assay ("R_{18}-assay") overcomes some of the difficulties encountered in the addition of fluorescently labelled phospholipids to intact membranes such as in the RET assay, using phosphatidylethanolamine molecules bearing headgroup fluorophores (N-NBD-PE and N-Rh-PE; Struck et al., 1981). These lipid derivatives have been used successfully in measurement of fusion of viruses and their reconstituted envelopes with both liposomes and biological membranes, as the probes can be readily incorporated into the liposomal membranes or viral envelopes during their preparation (Eidelman et al., 1984; Harmsen et al., 1985; Stegmann et al., 1985; Chejanovsky & Loyter, 1985).

Mixing of aqueous contents is also a useful assay of fusion, since true fusion consists of both lipid and contents mixing. Contents mixing assays are based on interactions between aqueous fluorophores, such as Tb/DPA (terbium/dipicolinic acid; Wilschut & Papahadjopoulos, 1979) and ANTS/DPX (aminonaphthalenetrisulfonic acid and p-xylylene bis(pyridinium) bromide; Ellens et al., 1985). Particularly in studies of fusion between artificial membranes, the use of both approaches, i.e., membrane and contents mixing, have been highly profitable in analyzing fusion events, as membrane mixing *per se* may be accompanied by lipid exchange processes (Hoekstra & Wilschut, 1989). Unfortunately, such a direct approach cannot be undertaken with intact virus particles, given the limitation on encapsulating aqueous content-markers into intact virions. In principle, encapsulation is possible when reconstituting viral envelopes, but thusfar such efforts have been limited (Loyter et al., 1988). It is important, therefore, to include extensive control experiments that exclude lipid exchange when monitoring virus fusion by membrane mixing alone.

Based upon kinetic measurements of Sendai virus fusion with both liposomes

(Klappe et al., 1986) and erythrocyte ghosts (Hoekstra et al., 1985; Hoekstra & Klappe, 1986), the application of a mass action kinetic model has allowed distinction between the kinetics of binding and fusion and evaluation of the corresponding rate constants (Nir et al., 1986a, 1986c). It should be noted that with the fluorescence assays, the overall fusion rate (including adherence of virions and the fusion reaction itself), is measured when the virus particles are added to the target membrane at 37°C. By kinetic simulation, employing the mass action kinetic model, insight into the parameters that govern binding and fusion *per se* can thus be obtained (Nir et al, 1986a,b,c). The model views the overall fusion reaction as a sequence of a second order process of virus-target membrane adhesion, which is partially reversible, followed by the first-order fusion reaction itself.

Apart from fusion of Sendai virus, fusion of other viruses has also been characterized with the fluorescence assays decribed above, including influenza (Stegmann et al., 1986; Bundo-Morita et al., 1987), vesicular stomatitis virus (VSV, Blumenthal et al., 1987; Puri et al., 1988) and Newcastle disease virus (Lorge et al., 1986). Although the N-NBD-PE/N-Rh-PE energy transfer pair is most commonly used in cases in which lipid vesicles are fusion targets, other energy transfer donor/acceptor pairs (Vanderwerf & Ullman, 1980; Uster & Deaner, 1981) may also be used for monitoring virus fusion.

Other direct methods of measuring virus fusion activity are available such as those based on using radiolabelled virus (Haywood & Boyer, 1982) or virions labelled with paramagnetic probes (Maeda et al., 1975; 1981a,b; Yamada & Ohnishi, 1986). Although useful and more informative than indirect methods such as infectivity and polykaryon formation (White et al., 1983; Stein et al., 1987), these procedures are generally more time-consuming. Evidently, the spectroscopic assays have the advantage of rapid and sensitive data acquisition, as well as continuous measurement. It thus becomes possible to monitor the occurrence of membrane fusion under a wide variety of conditions, providing the possibility to correlate membrane and protein physical properties with their roles in viral entry and membrane fusion. In spite of the availability of quantitative assays for membrane fusion, many details of the mechanism of fusion remain to be elucidated. A major limitation stems from the fact that virus and cell membranes are complex systems, making it difficult to isolate structural membrane components and their associated physical contributions to fusion. Artificial membranes are therefore commonly used as models in membrane fusion (Düzgüneş, 1985; Wilschut & Hoekstra, 1986). Although these model systems oversimplify the biological membranes involved in fusion, they provide an important and useful means of identifying basic structural elements involved in fusion (Hoekstra, 1988; Hoekstra & Wilschut, 1989). Model studies may thus provide fundamental insight into the more complex fusion of biological membranes, although, as noted before (cf. Klappe et al., 1986; Stegmann et al., 1986; Hoekstra & Wilschut, 1989), a direct extrapolation to pure biological systems is not always warranted.

III. PARAMETERS AFFECTING MEMBRANE FUSION

In artificial systems, membrane fusion is dependent on disruption of the combination of hydration and hydrophobic stabilizing interactions, governing membrane self-assembly and stability (Israelachvili et al., 1980; Tanford, 1980; Rand, 1981; Rand & Parsegian, 1984). The close apposition of bilayers necessary for fusion requires overcoming the repulsive hydration layers surrounding each membrane. The repulsive regime of hydration forces is significant at interbilayer distances of less than 30Å. Thus, disruption of the hydration shells is a prerequisite for fusion to occur. It should be noted that the work required for establishing fusion-susceptible contact between membranes is enormous. It has been estimated that an external pressure of 10^3-10^4 atm is required to overcome the hydration barrier (Leikin et al., 1987), thus illustrating the extreme potency of fusogenic agents. Since the product of fusion is a newly formed membrane, the hydrophobic interactions involved in stabilization of both the fusing and product membranes must also be involved in the thermodynamic driving force for fusion (Ohki, 1985; Hoekstra et al., 1988; Hoekstra & Wilschut, 1989).

In addition to the repulsive forces between membranes at short distances, Rand et al. (1988) have recently described an alternative model of intermembrane forces. In this model, the hydration and fusion behavior of membranes is related to a combination of attractive and repulsive forces, where the attractive forces result from intermembrane networks of hydrogen-bonded water. Although awaiting further experimental verification, this model is consistent with some observations of intermembrane interactions.

Artificial membranes have been especially useful in the analysis of membrane hydration and its role in fusion. The most prominent example in this respect is the cation-induced fusion of bilayers consisting of negatively charged phospholipids, such as phosphatidylserine (PS; for references, see Hoekstra & Wilschut, 1989). Upon addition of Ca^{2+} to such vesicles, the gel/liquid-crystalline phase transition temperature of the lipid increases dramatically, consistent with removal of water from the bilayer interface. A variety of physical techniques (X-ray, NMR and Raman scattering) have provided evidence that Ca^{2+}/PS binding results in an essentially complete dehydration of intermembrane contact sites. By contrast, upon addition of Mg^{2+}, which, as opposed to Ca^{2+}, does not induce fusion of (large) PS vesicles, the Mg^{2+}/PS complexes formed still contain substantial amounts of water. Consequently, with Mg^{2+} the apposed bilayers display larger interbilayer separations than with Ca^{2+}, and, therefore, fusion will not take place. Synthetic non-phospholipid amphiphiles, which exhibit polymorphic phase behavior similar to that of phospholipids, have also been useful in relating membrane hydration to fusion (Rupert et al., 1985, 1986, 1987a,b, 1988a,b; Hoekstra et al., 1988).

In addition to the hydration barrier, there is also a hydrophobic barrier to fusion. The thermodynamic bilayer stabilization must be disrupted to induce formation of a fusion intermediate, regardless of whether this intermediate will display a defined structure. Synthetic phospholipid vesicles have also been used for investigation of the structure of fusion intermediates (Prestegard & O'Brien. 1987; Siegel, 1986; Siegel et al., 1989), leading to the notion that nonbilayer intermediates are formed during fusion of liposomes (Prestegard & O'Brien, 1987; Verkleij et al., 1984). Proposed intermediates in fusion include monolayer stalks (Leikin et al., 1987; Markin et al., 1984) and "inverted micellar intermediates" (IMI, Siegel 1986). Recent efforts haven been made to identify the latter by cryo-transmission electron microscopy (Siegel et al, 1989).

The extensive work carried out with simple model systems has substantially contributed to an understanding of virus fusion, providing the current model of viral glycoprotein function in fusion, namely to bring viral and target membranes into close proximity to overcome the repulsive hydration shells and to induce formation of a nonbilayer fusion intermediate in the target membrane via hydrophobic interactions (White et al., 1983; Choppin & Scheid, 1980; Asano & Asano, 1984; Loyter & Volsky, 1982). It should be noted, however, that viral protein-induced formation of any type of nonbilayer fusion intermediate has not been reported thus far.

IV. PROPERTIES OF VIRAL PROTEINS

Virus entry into cells is accomplished by fusion of the viral lipid envelope with a membrane of the target cell. The viral glycoproteins (Fig.3) have the roles of binding to cell surface receptors and mediating the membrane fusion event (Choppin & Scheid, 1980; Loyter & Volsky, 1982; White et al., 1983).

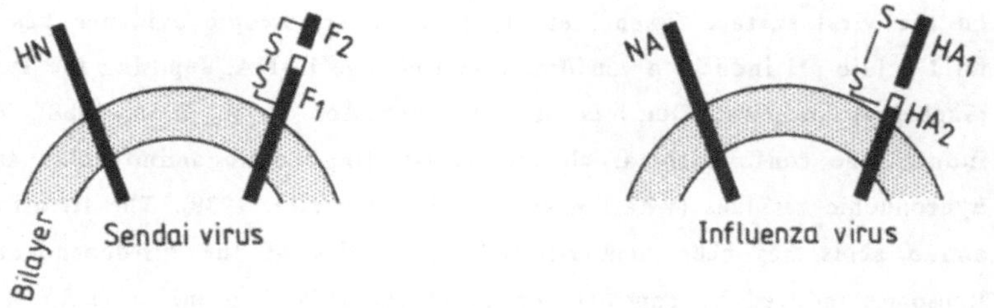

FIGURE 3. SCHEMATIC REPRESENTATION OF SOME VIRAL MEMBRANE PROTEINS
The viral membrane glycoproteins comprise spike-like projections on the surface of the virus. For Sendai virus, HN is necessary for binding while fusion is mediated by the F protein. From the inactive precursor, F_o, the fusion active protein is generated, which consists of two polypeptides, F_1 and F_2, linked by a disulfide bridge. The hydrophobic segment resides in F_1 (open region). For influenza virus, only HA is relevant to binding (contained in HA_1) and fusion (contained in HA_2). The hydrophobic N-terminus of HA_2 (open segment) becomes exposed when the virus encounters a mild acid pH environment. The NA glycoprotein contains neuraminidase activity; its precise function is unknown, but this protein likely plays a role in viral assembly.

Many viruses have a glycoprotein which carries out both of these functions. Examples are the influenza hemagglutinin (HA) and the VSV G protein (White et al., 1983; Pal et al., 1987). Paramyxoviruses such as Sendai virus have separate binding (hemagglutinin/neuraminidase, HN) and fusion (F) proteins (Choppin & Scheid, 1980). Virus-cell binding takes place without any triggering event, but fusion activity requires some type of activation. In the case of influenza HA and VSV G proteins, a low pH activation is necessary to induce virus-cell fusion (White et al., 1983; Pal et al., 1987). In contrast, Sendai virus enters the cell by fusion with the plasma membrane at neutral pH (Choppin & Scheid, 1980; Scheid & Choppin, 1977; White et al., 1983).

A wide range of viruses have fusion proteins that contain a stretch of primarily hydrophobic amino acids in their primary sequence (Choppin & Scheid, 1980; White et al., 1983; White et al., 1983; Asano & Asano, 1984; Ohnishi, 1988). These hydrophobic sequences are usually twenty or more amino acids in length, and often occur at the N-terminus of the proteins. The Sendai and influenza virus hydrophobic sequences are located at the N-termini of the F_1 and HA_2 subunits, respectively (Blumberg et al., 1985; Wiley & Skehel, 1987; Ohnishi, 1988). X-ray crystallography of the bromelain-cleaved influenza HA_2 ectodomain (BHA_2) revealed that the fusion peptide is located near the viral membrane, hidden within the hemagglutinin trimer on the viral surface (Skehel et al., 1982). Spectroscopic evidence has shown that mild acidic pH induces a conformational change in HA, exposing the fusion peptide (Skehel et al., 1982; Doms et al., 1985, Wharton et al., 1986, 1988). The low pH induces the conformational change by titrating acidic amino acids and exposing hydrophobic residues (Wharton et al., 1988; Ohnishi, 1988). The titration of acidic amino acids has been demonstrated by studies of the pH-dependent fusion of liposomes induced by peptides similar to the HA_2 N-terminus (Lear & De Grado, 1987). This mechanism is probably also active in fusion of other viruses, such as VSV, at low pH (Pal et al., 1987), as well as in fusion of liposomes induced by proteins such as lysozyme, clathrin, and α-lactalbumin (Arvinte et al., 1986; Hong et al., 1985; Kim & Kim, 1986; Maezawa et al., 1989). The ability of these proteins to induce fusion at low pH reflects the large negative free energy of insertion of hydrophobic peptides into liposomal membranes at low pH (Von Heijne, 1981; Ohnishi, 1988).

In addition to indirect evidence of hydrophobic interactions between fusion-inducing proteins and membranes, some attempts have been made to directly observe these protein-lipid interactions by photoaffinity labelling. In these experiments, a lipid-bound label or a hydrophobic label partitioning into the bilayer is photoactivated to covalently label proteins in contact with the hydrophobic core of the membrane. The label's fate is generally followed by using radioactive photoaffinity probes.

Hydrophobic photoaffinity labelling has proven a useful technique for identification of transmembrane segments of proteins, and protein subunits associated with membranes (Brunner, 1981; Sigrist & Zahler, 1985). Both lipid soluble probes (Brunner & Semenza, 1981; Richards & Brunner, 1980; Gitler & Bercovici, 1980) and lipid-bound probes (Chakrabarti & Khorana, 1975; Radhakrishan et al., 1980; Brunner et al., 1983; Schroit et al., 1987; Harter et al., 1988) have been successfully employed.

In hydrophobic photolabelling studies of influenza BHA, Harter et al. (1988) demonstrated the specific low pH-induced labelling of the BHA_2 subunit after incubation with liposomes containing photoreactive lipids. Similar results were observed by Boulay et al. (1986) using the membrane partitioning label 3-(trifluoromethyl)-3-(m-[^{125}I]iodophenyl)diazirine (TID).

Membrane organization of the Sendai virus F protein and the VSV G protein have been studied by photolabelling with lipid-bound and membrane soluble probes (Moscufo et al., 1987; Zakowski & Wagner, 1980). Viral protein-protein interactions for these two viruses were also examined by chemical crosslinking (Markwell & Fox, 1980; Zakowski & Wagner, 1980). It is clear from these studies that protein-protein interaction is important in viral assembly and function. Supporting evidence suggesting the importance of protein-protein interaction in viral assembly was reported by Vaux et al. (1988), demonstrating interactions between the cytoplasmic tail of SFV E_2 glycoprotein and the capsid protein, while trimerization of HA represents an essential feature in the biogenesis of influenza virus (Copeland et al., 1988). In addition to a possible role in viral assembly, protein-protein interactions might also be important in stabilizing (fusion) active glycoprotein conformations.

V. HYDROPHOBIC INTERACTION, A KEY EVENT IN VIRUS FUSION

Hydration and hydrophobic interactions are responsible for the stabilization of membrane bilayers (see section III). As far as a virus is concerned, it is very likely therefore that its ability to undergo fusion is related to the physical properties of the target membranes, such as membrane packing, hydration, and phase. In this section several of these aspects will be discussed.

A. Penetration of viral proteins

During the past several years, substantial evidence of the requirement for fusion proteins to initiate virus-membrane fusion has been reported (see Section IV). Based on the observation of hydrophobic amino acid sequences in a wide variety of fusion proteins, it was postulated that viral proteins initiate fusion by hydrophobic interactions with the target membrane (Choppin & Scheid, 1980; Loyter & Volsky, 1982; White et al., 1983; Asano & Asano, 1984). In this mechanism, insertion of the hydrophobic peptide into the target membrane might destabilize the bilayer structure and lead to formation of a fusion intermediate. This type of mechanism is supported by reports of liposomal fusion induced by hydrophobic peptides at low pH (Hong et al., 1985; Kim & Kim, 1986; Maezawa et al., 1989). Titration of acidic amino acid side chains and exposure of hydrophobic surfaces for interaction with the membrane are necessary for promoting fusion.

Recently, several investigators have shown that the application of photoaffinity labels could be a promising tool in elucidating structural features of viral proteins (section IV), including their interaction sites with target membranes. Photolabelling experiments are generally conducted by photolysis of a mixture of virus or viral protein with liposome or cell membranes containing a photoreactive label, after an incubation period for equilibration. These experiments provide important information about membrane organization of viral proteins, but do not necessarily give insight into

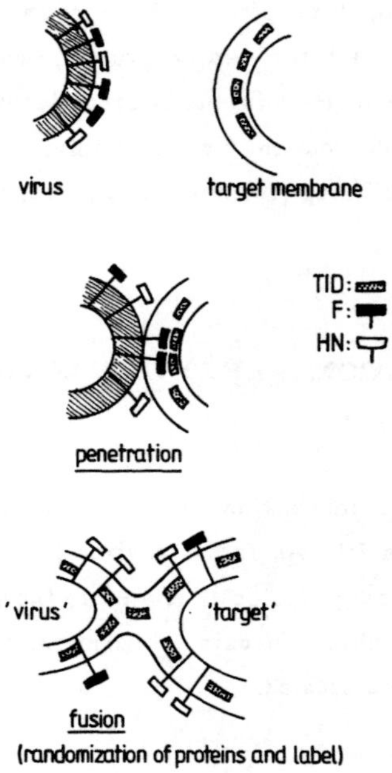

FIGURE 4. PRINCIPLE AND PROBLEMS OF USING PHOTOAFFINITY LABELLING TO DETECT PROTEIN PENETRATION
Target membranes are labelled with the hydrophobic probe TID. A careful knowledge of the kinetics of fusion is essential in order to "catch" the penetration step and to avoid at random labelling that occurs when fusion proceeds to total membrane mixing. In the latter case transmembrane segments of all viral membrane proteins will become labelled. In addition, some probe may diffuse spontaneously from the target membrane, causing non-specific labelling of the viral proteins. Appropiate controls are therefore required. The scheme depicts the fusion of Sendai virus, which requires HN for binding and F for fusion activity.

the mechanism of protein-induced membrane fusion. The incubation period prior to photolysis may result in protein labelling after membrane fusion, i.e. under equilibration conditions. Hence, rather than reflecting the capacity of a putatively fusogenic peptide to induce fusion, it would reflect exposure of protein to the probe after reorganization of the membrane components accompanying fusion. In order to obtain mechanistic information about the role of hydrophobic interactions between a viral fusion peptide and a membrane involved in fusion, it is necessary to carry out the photoactivation *during* the early stages of the fusion reaction and before randomization of membrane components at stages secondary to fusion (Fig.4). This type of time-dependent photolabelling of a viral protein was recently reported (Novick

& Hoekstra, 1988), during Sendai virus fusion with liposomes. In these experiments, the hydrophobic photoaffinity label 3-(trifluoromethyl)-3-(m-[^{125}I]iodophenyl)-diazirine(TID) was used. This molecule partitions effectively into membranes (Brunner et al., 1983). Fusion between Sendai virus and reconstituted Sendai virus envelopes (RSVE) with both liposomes (Novick & Hoekstra, 1988) and erythrocyte membranes (unpublished observations) was examined. The photoaffinity label was incorporated into the target membranes (large unilamellar vesicles composed of cardiolipin (CL) or PS, or erythrocyte membranes) by exogenous addition to the membrane preparations (for details, cf. Novick & Hoekstra, 1988). Virus and target membranes are mixed for various reaction times, and the samples are then photolyzed. After photolysis, the proteins are precipitated and separated by reducing SDS-PAGE and the fate of the TID probe is analyzed by gamma counting of slices of the polyacrylamide gel.

The TID probe, as a trifluoromethyldiazirine, preferentially generates a singlet carbene upon photolysis at temperatures in the range of 0 to 37°C (Turro, 1978; Richards & Brunner, 1980). The singlet carbene shows zwitterionic reactivity and engages readily in "insertion" reactions. It displays a preference for addition to double bonds and insertion into polar single bonds, rather than a preference for formal C-H and C-C insertion that would be expected from a triplet (Turro, 1978; Kanakarajan et al., 1988). Phenyl azides commonly used in photolabelling usually generate a triplet nitrene. The result of the singlet carbene formation upon TID photolysis is that there is a preferential reaction with polar and double bonds present in proteins over the C-C and C-H bonds in the bilayer core. This effect permits substantial labelling of a transiently inserting protein in an excess of lipid, constituted by the membrane core. Normalization of the protein labelling is also possible as a result of the singlet reactivity, by comparison of the viral glycoprotein labelling (F/HN) under various conditions. Normalization of the data is important in viral work, due to the variability in viral activity with passage and handling.

The use of a lipid-soluble probe may result in some nonspecific labelling by diffusion out of the membrane prior to photolysis, causing background labelling of proteins not specifically involved in fusion. The extent to which such background labelling interferes with labelling during fusion can be readily evaluated by analyzing virus labelling with the free probe. When doing so for Sendai virus, it was excluded that diffusional labelling was responsive to conditions of the reaction, as labelling with the "free" probe was very dissimilar compared to labelling seen during fusion (Novick & Hoekstra, 1988).

As described above, the fusion protein of Sendai virus (F) consists of two subunits, F_1 and F_2 which are generated upon proteolytic cleavage of the inactive precursor, F_0. As a result, a new N-terminal region is generated in F_1 and it is this

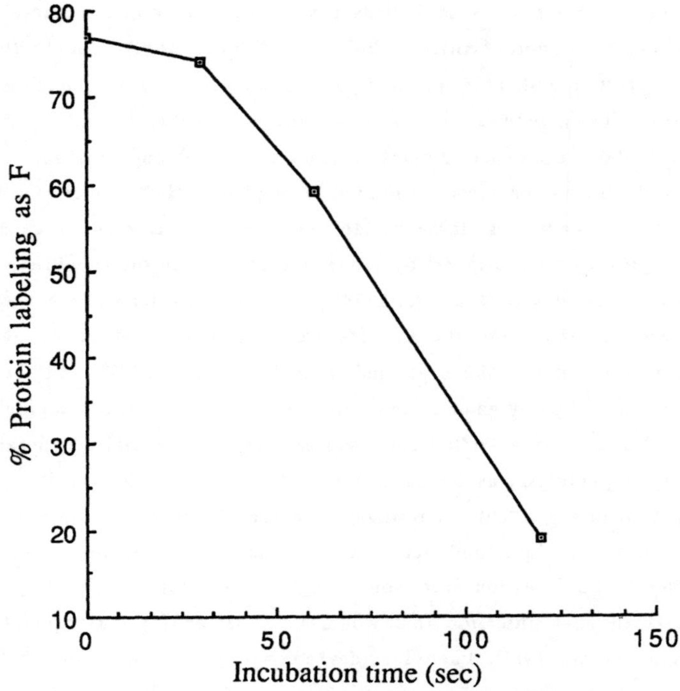

FIGURE 5. PENETRATION OF SENDAI VIRUS F PROTEIN DURING VIRUS-LIPOSOME FUSION
Sendai virus was incubated with ^{125}I-TID-labelled CL vesicles for various incubation times, followed by photolysis of the mixture for 30 secs. Afterwards, the viral proteins were separated by gelelectrophoresis, slices of the gel were counted and the labelling of F_1, as a percentage of total protein labelled, was calculated. Note that (i) F_1 and F_2 are separated on SDS-gels (cf. Fig.3) and that (ii) F_1 is almost exclusively labelled when irradiation is started immediately upon mixing virus and liposomes (t=o). After longer incubation times, much of the label is associated with other viral proteins as well (cf. Fig.4). Data are from Novick & Hoekstra, 1988).

strongly hydrophobic region (about 20 amino acids) which is thought to penetrate into the target membrane. Using TID-labelled liposomes, consisting of CL or PS, as target membranes and incubating such labelled membranes with Sendai virus, direct evidence was obtained, supporting this concept. During the early interaction events between virus and target membrane, a preferential labelling of F_1 is observed (Fig. 5), consistent with the proposed hydrophobic insertion of this peptide into a target membrane. Furthermore, the extent of labelling increases substantially (about 2.5-3-fold) when the incubation temperature increases from 2 to 37°C. The increase

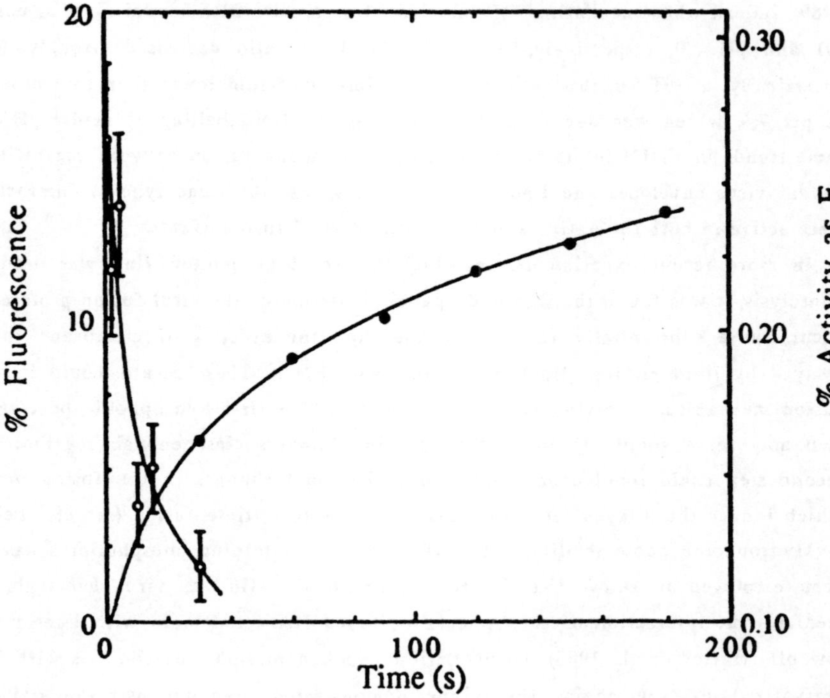

FIGURE 6. KINETICS OF SENDAI VIRUS F_1 PENETRATION VERSUS KINETICS OF FUSION WITH PHOSPHOLIPID VESICLES

Sendai virus was incubated with TID-labelled CL vesicles and F_1 labelling was determined as a function of the incubation time. The amount of radioactivity associated with F1 (open circles) is expressed as a percentage of the total liposome-associated activity. In this experiment photolysis was conducted with a high photon flux, using a 1000W Xe arc lamp. In a parallel experiment, the kinetics of fusion was assayed, using the R_{18} assay (closed circles). For details, see text.

in labelling parallels the increase in fusion activity of the virus, as monitored by a kinetic assay, as described in section II.

Fusion of viruses with CL liposomes do not in every respect follow the fusion characteristics seen when a biological membrane is used as the target (Klappe et al., 1986; Stegmann et al., 1986). One peculiar aspect involves the ability of Sendai virus, normally fusing in the neutral pH region, to fuse with acidic phospholipid vesicles at acidic pH as well. This low pH-induced fusion has been proposed to result from an involvement of HN in fusion at those conditions (Klappe et al., 1986; Amselem et al., 1986) i.e., HN becomes an ordinary fusion protein, analogous to a variety of other, normally non-fusogenic, proteins that acquire an increased hydrophobicity when the pH is lowered (for references, see Hoekstra & Wilschut,

1989). Indeed when incubating TID-labelled CL liposomes with Sendai virus at neutral pH and pH 5.0, respectively, the F/HN labelling ratio decreased strongly. More specifically, at pH 5.0, this ratio was approximately 5-fold lower than that observed at pH 7.4, which was due to a strong increase in HN-labelling at acidic pH. The same trends in F/HN interactions could be seen during fusion between reconstituted Sendai virus envelopes and liposomes, signifying that the same type of interactions were active in both the native and reconstituted viral fusion events.

In more recent experiments, in which a very high photon flux was used for photolysis, it was found that the hydrophobic insertion of the viral fusion protein (F_1) occurs on a substantially faster timescale than the kinetics of membrane fusion, assayed by lipid mixing (Hoekstra et al., 1984; Fig.6). These results could imply a fusion mechanism involving more than one step. The first hydrophobic penetration step may serve simply to bring the apposing bilayers close enough together. The second step could involve additional conformational changes in the fusion protein which induce the bilayer destabilization necessary to initiate fusion (see also below).

Hydrophobic photolabelling with TID and TID-containing phospholipids has also been employed to show that the HA_2 subunit of influenza virus hemagglutinin mediates the interaction of the bromelain-cleaved HA ectodomain with liposomes at low pH (Harter et al., 1988). Using tritium labelled phosphatidylcholines with TID-derivatized *sn*-2 acyl chains, the authors demonstrated unambiguously that only the HA_2 subunit interacts hydrophobically with the membrane under fusogenic conditions (low pH). These data agree with the model of influenza virus fusion at low pH, in which a conformational change exposes the HA_2 hydrophobic N-terminus for interaction with the target membrane (Doms et al., 1985). In labelling experiments with lipid-soluble TID, some residual labelling of BHA_1 was observed. This is probably due to the presence of hydrophobic pockets on the protein in close proximity with, but not penetrating into, the membrane. Using water soluble proteins, the authors showed that TID labels hydrophobic pockets of proteins which are not in contact with the membrane. Labelling of these proteins by one of the photoreactive lipids may indicate that water soluble proteins do interact hydrophobically with membranes at low pH.

Other investigators have also examined lipid interactions of viral fusion proteins using TID. Asano and Asano (1984) reported the hydrophobic photolabelling of the intact Sendai virus F protein by TID, partioned into the plasma membrane of cells. The F protein was labelled when virus was added to cells at 4°C, indicating that a close association of the protein with the target cell membrane occurs during the binding step, since fusion does not occur at temperatures below ca. 20°C.

The use of equilibrated systems for photolabelling must be avoided in mechanistic studies, however, in that the labelling under these conditions reflects interactions taking place after much of the fusion reaction has occurred. Since one would expect the hydrophobic interaction to occur at an early point during the fusion reaction, it will be necessary to photolyze the samples at early timepoints. This is often inconvenient and expensive, as the reaction must be carefully timed, and little probe will have accumulated in the fusion protein. This necessitates use of a larger amount of probe, which is an expensive proposition.

B. Effect of target membrane packing

Given the need for hydrophobic penetration of a viral fusion protein to trigger the fusion reaction, one could intuitively envision that penetration as such might be affected by the packing density of the target membrane. To investigate this parameter, liposomes are evidently a most convenient tool as biological membranes are not readily amenable to extensive changes in lipid composition. Various studies in which liposomes of different compositions have been used to investigate (viral) protein-induced fusion or protein-membrane interactions *per se* indicate that packing of the hydrophobic acyl chains can indeed affect fusion. In general, poorer packing promotes protein insertion (cf. Boggs et al., 1980) and fusion. The VSV G protein shows greater activity in the presence of cis-unsaturated lipids (Yamada & Ohnishi, 1986), while clathrin induces fusion of DOPC vesicles, but not of egg PC vesicles (Hong et al., 1985). Furthermore, Sendai virus fuses more readily with CL liposomes than with PS liposomes (Klappe et al., 1986). In fact, kinetic simulation of those data revealed that the fusion rate constant of virus-CL vesicle fusion was enhanced relative to that of virus-PS vesicle fusion, without a significant change in the aggregation rate constants (Nir et al., 1986c). The amounts of TID labelling seen during virus fusion with CL vesicles, as discussed above, were, moreover, consistently higher than those during virus-PS vesicle fusion (about 2-3-fold), which is consistent with the higher extent and kinetics of fusion in the former case.

C. Role of membrane hydration in virus fusion

In studies of fusion of liposomes, it has become apparent that hydration forces govern to a major extent the ability of apposed membranes to fuse (Rand, 1981; Düzgüneş, 1985; Wilschut & Hoekstra, 1986; Hoekstra & Wilschut, 1989). As described above (section III), Ca^{2+} appears to form a strongly dehydrated complex with PS, acting as a nucleation site in Ca^{2+}-induced fusion of PS vesicles. Inclusion of a strongly hydrated lipid such as PC in PS vesicles, inhibits Ca^{2+}-induced fusion

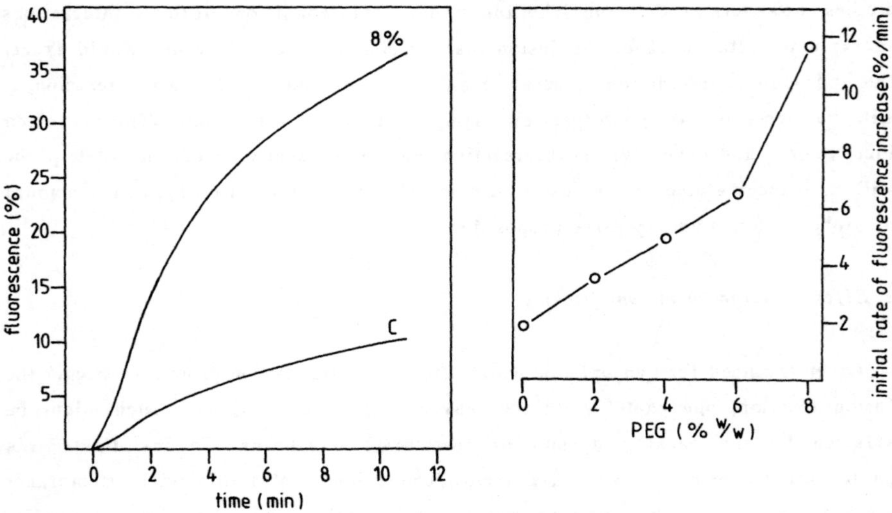

FIGURE 7. PEG FACILITATES SENDAI VIRUS-ERYTHROCYTE MEMBRANE FUSION
Sendai virus was labelled with a selfquenching concentration of R_{18} and fusion with erythrocyte ghosts was monitored as depicted in Fig.2. In the left panel, fluorescence tracings are shown for fusion taking place in the absence (C) or presence (8 wt%) of PEG 8K. The increase in the initial rate is dependent of the PEG concentration (right panel). The final extent of fusion, as determined after long-term incubations, does not increase.

whereas the fusion event is sustained when PE, a weakly hydrated phospholipid, is incorporated in negatively charged bilayers (Düzgünes, 1985). In fusion studies involving the use of synthetic amphiphiles (Rupert et al., 1985,1986,1987a), the overall process of vesicle aggregation and fusion itself is triggered at different divalent cation concentrations. This control over both processes allows the detailed analysis of each step separately. In the case of the negatively charged amphiphile didodecylphosphate (DDP), it could be shown that fusion is triggered by formation of an anhydrous trans-Ca^{2+}/DDP complex (Rupert et al., 1987a,b), whose formation is dependent on membrane surface dehydration.

The dehydrating polymer poly(ethylene glycol) (PEG) has proven a useful tool in the study of hydration effects in membrane fusion. Binding of water molecules by PEG decreases the polarity of the solvent phase near the membrane surface, destabilizing the membrane (Arnold et al., 1988; Hoekstra et al., 1988). This dehydrating capacity limits water available for binding to lipid headgroups. By varying the state of surface hydration, low concentrations of PEG substantially increase the Ca^{2+}-dependent rate of fusion of PS vesicles (Hoekstra, 1982). Thus, membrane dehydration can have a profound effect on fusion of artificial membranes.

To determine whether membrane hydration is also central to fusion of biological membranes, similar approaches have been used as those described for artificial systems. When studying Sendai virus-liposome interactions it was observed that, analogous to Ca^{2+}-induced fusion of liposomes, the presence of PC causes inhibition of fusion between the virus and negatively charged bilayers, with increasing mole ratios of PC. By contrast, the fusion reaction was sustained when similar concentrations of PE were included. Using a mass action kinetic model, it was shown that PC caused a two-fold decrease in the fusion rate constant, while barely affecting the rate constant for aggregation (Nir et al., 1986a,c). In contrast, small amounts (0-8%) of PEG 8K caused the fusion rate constant to triple with little effect on the aggregation rate constant. Both of these results suggest that hydration of the membrane surface plays a critical role in regulating virus-liposome fusion.

As noted above, virus-liposome fusion does not necessarily reflect the fusion conditions as they occur in a physiological environment. That membrane hydration may be critical to biological membrane fusion as well, was suggested by experiments which demonstrated that low concentrations of PEG (which do not cause fusion by themselves) substantially increase the kinetics of virus-erythrocyte membrane fusion (Fig.7; Hoekstra et al., 1989). It was excluded that this increase was due to a non-specific fusion process, since the limited number of the virus particles that fuse with one erythrocyte (approx. 100-200 out of 1000 active virus particles that bind) was the same, regardless of the presence of the polymer. Kinetic analysis showed that the aggregation rate constant doubled and the fusion rate constant increased by nearly an order of magnitude in the presence of PEG. These results indicate that viral membrane fusion with a purely biological target is also strongly dependent on membrane hydration. The details of the effect of PEG have been further examined by determining the extent to which a separate preincubation of viruses or target membranes in PEG-containing media modulated the fusion reaction. It was observed (Hoekstra et al., 1989) that the rate of fusion doubled when the virus was preincubated in a PEG-containing medium. By contrast, no change was seen when the target membranes were preincubated. This implies that the virus itself is highly sensitive towards the polarity of its environment and becomes more fusogenic in a more hydrophobic environment.

The observation that PEG also slightly increases the aggregation rate constant implies that the interaction of the binding protein of Sendai virus, HN, is not soleley governed by electrostatic interactions with sialic acid-containing receptors on the cell surface. Indeed, previous results (Hoekstra & Klappe, 1986) indicated that structural requirements have to be met as well for the binding protein to interact properly with the receptor. It could thus be suggested that structural integrity of HN is necessary to allow for hydrophobic interactions, accompanying the initial electrostatic interactions. It should be noted in this regard, that at this initial attachment the interbilayer distance between virus and target membrane can be on the order of ca. 100 Å, a distance at which hydration forces are of minor significance. With liposomes, changes in the aggregation rate constants were less pronounced than those observed with erythrocyte membranes (Nir et al., 1986c). This suggests differences in the mechanism of attachment of virions in the process of fusion with liposomes as compared to that during virus-erythrocyte membrane fusion. Indeed, except in the case of PS liposomes, PEG induced an enhancement of the final levels of fusion between Sendai virus and cardiolipin and cardiolipin/ phosphatidylcholine vesicles (Nir et al., 1986c). It would appear, therefore, that fusion of a virus with a biological target membrane is more strictly controlled than the fusion of a virus with a liposomal membrane.

The effect of PEG on virus-erythrocyte membrane fusion suggests that the hydrophobic environment can modulate the fusion efficiency of the virus. After penetration of F_1 into the target membrane an interbilayer distance (between virus and target membrane) of about 100 Å still exists and this gap obviously has to be overcome before merging of both bilayers occurs. One would anticipate that additional structural alterations have to take place in the F protein to "bridge" this cap. As a preincubation of virus alone in PEG causes a doubling of the virus fusion rate, relative to that seen for fusion (in a PEG-containing medium) without preincubation, it is possible that hydrophobic dehydration may trigger such a conformational change. If this scenario were true, the *overall* activation of the fusion protein of viruses fusing at neutral pH would entail a two step activation mechanism. The first step would involve proteolytic cleavage of the inactive fusion-protein precursor, occurring during virus biogenesis, while the second step would represent a structural change, which accompanies the penetration step. This sequence of events would bear some similarity to that seen for viruses fusing at acidic pH. For viruses penetrating via receptor-mediated endocytosis, mild acidic pH induces a major conformational change in the protein structure, responsible for fusion, before the (buried) hydrophobic sequence becomes exposed (Doms et al., 1985; Wiley & Skehel, 1987).

In summary, it is evident that many details of the mechanimsm of virus fusion remain to be eludiated. It is also clear, that differences exist in virus fusion mechanisms when comparing artificial and biological target membranes. However, in developing basic strategies to unravel the details relevant for fusion at physiological conditions, liposomes can provide a valuable tool.

ACKNOWLEDGMENTS

Personal work cited in this paper was supported by grants from the National Institutes of Health (AI 25534 and RR07003), Monsanto and Amersham. S.L.N. is the recipient of an NSF Graduate Fellowship. The expert secretarial assistance of Mrs Rinske Kuperus and Mrs Lineke Klap is gratefully acknowledged.

REFERENCES

Amselem S, Barenholz Y, Loyter A, Nir S, Lichtenberg D (1986) Fusion of Sendai virus with negatively charged liposomes as studied by pyrene-labelled phospholipid liposomes. Biochim Biophys Acta 860: 301-313

Arnold K, Hermann A, Gawrisch K, Pratsch L (1988) Water-mediated effects of PEG on membrane properties and fusion in: Molecular mechanisms of membrane fusion, Ohki S, Doyle D, Flanagan TD, Hui SW, Mayhew E (eds) Plenum Press, New York: 255-272

Arvinte T, Hildenbrand K, Wahl P, Nicolau C (1986) Lysozyme-induced fusion of liposomes with erythrocyte ghosts at acidic pH. Proc Natl Acad Sci USA 83: 962-966

Asano A, Asano K (1984) Molecular mechanism of virus entry to target cells. Tumor Res 19: 1-20

Blumberg BH, Giorgi C, Rose K, Kolakofsky D (1985) Sequence determination of the Sendai virus fusion protein. J Gen Virol 66: 317-331

Blumenthal R, Bali-Puri A, Walter A, Covell D, Eidelman O (1987) pH-dependent fusion of vesicular stomatitis virus with Vero cells. J Biol Chem 262: 13614-13619

Blumenthal R, Seth P, Willingham MC, Pastan I (1986) pH-dependent lysis of liposomes by adenovirus. Biochemistry 25: 2231-2237

Boggs JM, Stollery JG, Moscarello MA (1980) Effect of lipid environment on the motion of a spin-label covalently bound to myelin basic protein. Biochemistry 19: 1226-1234

Boulay F, Doms RW, Helenius A (1986) Photolabelling of the influenza hemagglutinin with a hydrophobic probe in: Positive strand RNA viruses, Brinton MA, Rueckert RR (eds) Alan R Liss, New York: 103-112

Brunner J (1981) Labelling the hydrophobic core of membranes. Trends Biochem Sci 6: 44-46

Brunner J, Semenza G (1982) Selective labelling of the hydrophobic core of membranes with 3-(trifluoromethyl)-3-(m-[^{125}I]iodophenyl)diazirine, a carbene-generating reagent. Biochemistry 20: 7174-7182

Brunner J, Spiess M, Aggeler R, Huber P, Semenza G (1983) Hydrophobic labelling of a single leaflet of the human erythrocyte membrane. Biochemistry 22: 3812-3820

Bundo-Morita K, Gibson S, Lenard J (1987) Estimation of radiation inactivation of the size of functional units governing Sendai and influenza virus fusion. Biochemistry 26: 6223-6227

Chakrabarti P, Khorana HG (1975) A new approach to the study of phospholipid-protein interactions in biological membranes. Synthesis of fatty acids and phospholipids containing photosensitive groups. Biochemistry 14: 5021-5033

Chejanovsky N, Loyter A (1985) Fusion between Sendai virus envelopes and biological membranes. The use for quantitative estimation of virus-membrane fusion. J Biol

Chem 260: 7911-7918

Choppin PW, Scheid A (1980) The role of viral glycoproteins in adsorption, penetration and pathogenicity of viruses. Rev Infect Dis 2: 40-61

Copeland CS, Zimmer KP, Wagner KR, Healy GA, Mellman I, Helenius A (1988) Folding, trimerization and transport are sequential events in the biogenesis of the influenza virus hemagglutinin. Cell 53: 197-209

Doms RW, Helenius A, White JM (1985) Membrane fusion activity of the influenza virus hemagglutinin: The low pH-induced conformational change. J Biol Chem 260: 2973-2981

Doms RW, Helenius A (1985) Properties of a viral fusion protein in: Molecular mechanisms of membrane fusion, Ohki S, Doyle D, Flanagan TD, Hui SW, Mayhew E (eds) Plenum Press NY: 385-398

Düzgüneş N (1985) Membrane fusion. Subcell Biochem 11: 195-286

Eidelmann O, Schlegel R, Tralka TS, Blumenthal R (1984) pH-dependent fusion induced by vesicular stomatitus virus glycoprotein reconstituted into phospholipid vesicles. J Biol Chem 259: 4622-4628

Ellens H, Benz J, Szoka FC (1985) H^+- and Ca^{2+}-induced fusion and destabilization of liposomes. Biochemistry 24: 3099-3106

Gitler C, Bercovici T (1980) Use of lipophilic photoactivatable reagents to identify the lipid-embedded domains of membrane proteins in: Applications of Photochemistry in Probing Biological Targets, Tometsko AM, Richards FM (eds) NY Acad Sci: 199-211

Harmsen MC, Wilschut J, Scherphof G, Hulstaert C, Hoekstra D (1985) Reconstitution and fusogenic properties of Sendai virus envelopes. Eur J Biochem 149: 591-599

Harter C, Bachi T, Semenza G, Brunner J (1988) Hydrophobic photolabelling identifies BHA_2 as the subunit mediating the interaction of bromelain-solubilized influenza virus hemagglutinin with liposomes at low pH. Biochemistry 27: 1856-1864

Haywood AM, Boyer BP (1982) Sendai virus membrane fusion: Time course and effect of temperature, pH, calcium and receptor concentration. Biochemistry 21: 6041-6046

Hoekstra D (1982) Role of lipid phase separations and membrane hydration in phospholipid vesicle fusion. Biochemistry 21: 2833-2840

Hoekstra D (1988) Glycolipids, glycoproteins and membrane fusion. Ind J Biochem Biophys 25: 76-84

Hoekstra D, De Boer T, Klappe K, Wilschut J (1984) Fluorescence method for measuring the kinetics of fusion between biological membranes. Biochemistry 23: 5675-5681

Hoekstra D, Klappe K (1986) Sendai virus-erythrocyte membrane interaction: quantitative and kinetic analysis of viral binding, dissociation, and fusion. J Virol 58: 87-95

Hoekstra D, Klappe K, De Boer T, Wilschut J (1985) Characterization of the fusogenic properties of Sendai virus: kinetics of fusion with erythrocyte membranes. Biochemistry 24: 4739-4745

Hoekstra D, Rupert LAM, Engberts JBFN, Nir S, Hoff H, Klappe K, Novick SL (1988) On the role of hydrophobic interactions in membrane fusion: effects of poly(ethylene glycol). Studia Biophys 127: 105-112

Hoekstra D, Wilschut J (1989) Membrane fusion of artificial and biological membranes. Role of local membrane dehydration. In: Water Transport in Biological Membranes (Benga G, ed) CRC Press, Boca Raton, FL: 143-176

Hoekstra D, Klappe K, Hoff, H, Nir S (1989) Mechanism of fusion of Sendai virus: Role of hydrophobic interactions and mobility constraints of viral membrane proteins. Effects of poly(ethylene glycol). J Biol Chem, in press

Homma M, Shimizu YK, Ishida D (1976) On the study of Sendai virus hemolysis. I. Complete Sendai virus lacking in hemolytic activity. Virology 71: 41-47

Hong K, Yoshimura T, Papahadjopoulos D (1985) Interaction of clathrin with liposomes: pH-dependent fusion of phospholipid membranes induced by clathrin. FEBS Lett 191: 17-23

Israelachvili JN, Marcelja S, Horn RG (1980) Physical principles of membrane organization. Q Rev Biophys 13: 121-200

Kanakarajan G, Goodrich R, Young MJT, Soundararajan S, Platz MS (1988) EPR spectroscopy of triplet aryl nitrenes covalently bound to α-chymotrypsin. The application of low temperature methods to photoaffinity labelling. J Am Chem Soc 110: 6536-6541

Kim J, Kim H (1986) Fusion of phospholipid vesicles induced by α-lactalbumin at acidic pH. Biochemistry 25: 7867-7874

Klappe K, Wilschut J, Nir S, Hoekstra D (1986) Parameters affecting fusion between Sendai virus and liposomes. Role of viral proteins, liposome composition, and pH. Biochemistry 25: 8252-8260

Lear JD, De Grado WF (1987) Membrane binding and conformational properties of peptides representing the NH_2 terminus of influenza HA_2. J Biol Chem 262: 6500-6505

Leikin SL, Kozlov MM, Chernomordik LV, Markin VS, Chizmadzhev YA (1987) Membrane fusion: overcoming of the hydration barrier and local restructuring. J Theor Biol 129: 411-425

Lorge P, Cabiaux V, Long L, Ruysschaert JM (1986) Fusion of Newcastle disease virus with liposomes; role of the lipid composition of liposomes. Biochim Biophys Acta 858; 312-316

Loyter A, Citovsky V, Blumenthal R (1988) The use of fluorescence dequenching measurements to follow viral membrane fusion events. Meth Biochem Anal 33: 129-164

Loyter A, Volsky DJ (1982) Reconstituted Sendai virus envelopes as carriers for the introduction of biological material into animal cells. Cell Surf Rev 8: 215-266

Maeda T, Asano A, Ohki Y, Ohnishi S (1975) A spin-label study on fusion of red blood cells induced by hemagglutinating virus of Japan. Biochemistry 14: 3736-3741

Maeda T, Kawasaki K, Ohnishi S (1981a) Interaction of influenza virus hemagglutinin with target membrane lipids is a key step in virus-induced hemolysis and fusion at pH 5.2. Proc Natl Acad Sci USA 78: 4133-4137

Maeda T, Kuroda K, Toyama S, Ohnishi S (1981b) Interaction of hemagglutinating virus of Japan with erythrocytes as studied by release of a spin probe from virus. Biochemistry 20: 5340-5345

Maezawa S, Yoshimura T, Hong K, Duzgunes N, Papahadjopoulos D (1989) Mechanism of protein-induced membrane fusion: Fusion of phospholipid vesicles by clathrin associated with its membrane binding and conformational change. Biochemistry 28: 1422-1428

Markin VS, Kozlov MM, Borovjagin VL (1984) On the theory of membrane fusion. The stalk mechanism. Gen Physiol Biophys 2: 361-377

McCune JM, Rabin LB, Feinberg MB, Lieberman M, Kosek JC, Reyes GR, Weissmann JL (1988) Endoproteolytic cleavage of gp160 is required for the activation of human immunodeficiency virus. Cell 53: 55-67

Markwell MAK, Fox CF (1980) Protein-protein interactions within paramyxoviruses identified by native disulfide bonding or reversible chemical cross-linking. J Virol 33: 152-166

Moscufo N, Gallina A, Schiavo G, Montecucco C, Tomasi M (1987) Multiple lipid interactions of the Sendai virus fusogenic protein. J Biol Chem 262: 11490-11496

Nir S, Klappe K, Hoekstra D (1986a) Kinetics and extent of fusion between Sendai virus and erythrocyte ghosts: Application of a mass action kinetic model. Biochemistry 25: 2155-2161

Nir S, Stegmann T, Wilschut J (1986b) Fusion of influenza virus with cardiolipin liposomes at low pH: Mass action analysis of kinetics and extent. Biochemistry 25: 257-266

Nir S, Klappe K, Hoekstra D (1986c) Mass action analysis of kinetics and extent of fusion between Sendai virus and phospholipid vesicles. Biochemistry 25: 8261-8266

Novick SL, Hoekstra D (1988) Membrane penetration of Sendai virus glycoproteins during the early stages of fusion with liposomes as determined by hydrophobic photoaffinity labelling. Proc Natl Acad Sci USA 85: 7433-7437

Ohki S (1985) Membrane fusion: Theory and experiments. Studia Biophys 110: 95-104

Ohnishi S (1988) Fusion of viral envelopes with cellular membranes. Current Topics in Membranes and Transport 32: 257-296

Pal R, Barenholz Y, Wagner RR (1987) Vesicular stomatitis virus membrane proteins and their interactions with lipid bilayers. Biochim Biophys Acta 906: 175-193

Prestegard JH, O'Brien MP (1987) Membrane and vesicle fusion. Ann Rev Phys Chem 38: 383-411

Puri A, Winick J, Lowy RJ, Covell D, Eidelman O, Walter A, Blumenthal R (1988) Activation of vesicular stomatitis virus fusion with cells by pretreatment at low pH. J Biol Chem 263: 4749-4753

Radhakrishnan R, Gupta CM, Erni B, Robson RJ, Curatolo W, Majumdar A, Ross AH, Takagaki Y, Khorana HG (1980) Phospholipids containing photoactivatable groups in studies of biological membranes. In: Applications of Photochemistry in Probing Biological Targets, Tometsko AM, Richards FM (eds) NY Acad Sci: 165-198

Rand RP (1981) Interacting phospholipid bilayers; Measured forces and structural changes. An Rev Biophys Bioeng 10: 277-314

Rand RP, Parsegian VA (1984) Variation in hydration forces between neutral phospholipid bilayers: Evidence for hydration attraction. Can J Biochem Cell Biol 62: 752-759

Rand RP, Fuller N, Parsegian VA and Rau DC (1988) Physical force considerations in model and biological membranes. Biochemistry 27: 7711-7722

Richards FM, Brunner J (1980) General labelling of membrane proteins. In: Applications of Photochemistry in Probing Biological Targets, Tometsko AM, Richards FM (eds) NY Acad Sci: 144-164

Rupert LAM, Engberts JBFN, Hoekstra D (1986) Role of membrane hydration and membrane fluidity in the mechanism of anion-induced fusion of didodecyldimethylammonium bromide vesicles. J Am Chem Soc 108: 3920-3925

Rupert LAM, Engberts JBFN, Hoekstra D (1988a) Effect of poly(ethylene glycol) on the Ca^{2+}-induced fusion of didodecylphosphate vesicles. Biochemistry 27: 8232-8239

Rupert LAM, Hoekstra D, Engberts JBFN (1985) Fusogenic behavior of didodecyldimethylammonium bromide bilayer vesicles. J Am Chem Soc 107: 2628-2631

Rupert LAM, Hoekstra D, Engberts JBFN (1987b) Ca2+-mediated fusion of didodecylphosphate vesicles. J Colloid Interface Sci 120: 125-134

Rupert LAM, Van Breemen JFC, Hoekstra D, Engberts JBFN (1988b) pH-Dependent fusion of didodecylphosphate vesicles. Role of hydrogen-bond formation and membrane fluidity. J Phys Chem 92: 4416-4420

Rupert LAM, Van Breemen JFC, Van Bruggen EFJ, Engberts JBFN, Hoekstra D (1987a) Calcium-induced fusion of didodecylphosphate vesicles: The lamellar to hexagonal II (H_{II}) phase transition. J Membr Biol 95: 255-263

Scheid A, Choppin PW (1977) Two disulfide-linked polypeptide chains constitute the active F protein of paramyxoviruses. Virology 80: 54-66

Schroit AJ, Madsen J, Ruoho A (1987) Radiodinated, photoactivatable phosphatidylcholine and phosphatidylserine: Transfer properties and differential photoreactive interaction with human erythrocyte membrane proteins. Biochemistry 26: 1812-1819

Siegel DP (1986) Inverted micellar intermediates and the transitions between lamellar, cubic, and inverted hexagonal lipid phases. II. Implications for membrane-membrane interactions and membrane fusion. Biophys J 49: 1171-1183

Siegel DP, Burns JL, Chestnut MH, Talmon Y (1989) Intermediates in membrane fusion and $L_\alpha/I_{II}/H_{II}$ phase transitions imaged by time-resolved cryotransmission electron microscopy. Biophys J 55: 342a

Sigrist H, Zahler PI (1985) Selective covalent modification of membrane components. In: The Enzymes of Biological Membranes (Martonosi AN, ed) Plenum NY: 333-370

Skehel JJ, Bayley PM, Brown EB, Martin SR, Waterfield MD, White JM, Wilson IA, Wiley DC (1982) Changes in the conformation of influenza virus hemagglutinin at the pH optimum of virus-mediated membrane fusion. Proc Natl Acad Sci USA 79: 968-972

Spear PG, Wittels M, Fuller AD, WuDunn D, Johnson R (1989) Herpes simplex virus: pathway of entry into cells in: Cell biology of virus entry, replication and pathogenesis, Compans R, Helenius A, Oldstone MBA (eds) Alan R Liss Inc NY:163-175

Stegmann T, Hoekstra D, Scherphof G, Wilschut J (1985) Kinetics of pH-dependent fusion between influenza virus and liposomes. Biochemistry 24: 3107-3113

Stegmann T, Hoekstra D, Scherphof G, Wilschut J (1986) Fusion activity of influenza virus: A comparison between biological and artificial target membrane vesicles. J Biol Chem 261: 10966-10969

Stein BS, Gowda SD, Lifson JD, Penhallow RC, Bensch KG, Engleman EG (1987) pH-Independent HIV entry into CD4-positive T-cells via virus envelope fusion to the plasma membrane. Cell 49: 659-668

Struck DK, Hoekstra D, Pagano RE (1981) Use of resonance energy transfer to monitor membrane fusion. Biochemistry 20: 4093-4099

Tanford D (1980) The hydrophobic effect. Wiley-Interscience NY

Turro J (1978) Modern molecular photochemistry. Benjamin/Cummings, Menlo Park, CA: 551-557

Uster PS, Deamer DW (1981) Fusion competence of phosphatidylserine-containing liposomes quantitatively measured by a fluorescence resonance energy transfer assay. Arch Biochem Biophys 209: 385-395

Vanderwerf P, Ullman EF (1980) Monitoring of phospholipid vesicle fusion by fluorescence energy transfer between membrane-bound dye labels. Biochim Biophys Acta 596: 302-314

Vaux DJT, Helenius A, Mellman I (1988) Spike-nucleocapsid interaction in Semliki Forest virus reconstructed using network antibodies. Nature 336: 36-42

Verkleij AJ, Leunissen-Bijvelt J, De Kruijff B, Hope M, Cullis PR (1984) In: Cell Fusion, Evered D, Whelan J (eds) Pitman, London: 45-59

Von Heijne G (1981(On the hydrophobic nature of the signal sequence. Eur J Biochem 116: 419-422

Wharton SA, Ruigrok RWH, Martin SR, Skehel JJ, Bayley PM, Weis W, Wiley DC (1988) Conformational aspects of the acid-induced fusion mechanism of influenza virus hemagglutinin; Circular dichroism and fluorescence studies. J Biol Chem 263: 4474-4480

Wharton SA, Skehel JJ, Wiley DC (1986) Studies of influenza hemagglutinin-mediated membrane fusion. Virology 149: 27-35

Wiley DC, Skehel JJ (1987) The structure and function of the hemagglutinin membrane glycoprotein of influenza virus. Ann Rev Biochem 56: 365-394

White J, Helenius A (1980) pH-Dependent fusion between the Semliki Forst virus membrane and liposomes. Proc Natl Acad Sci USA 77: 3273-3277

White J, Kielian M, Helenius A (1983) Membrane fusion proteins of enveloped animal viruses. Q Rev Biophys 16: 151-195

Wilschut J, Hoekstra D (1986) Membrane fusion; Lipid vesicles as a model system. Chem Phys Lipids 40: 145-166

Wilschut J, Papahadjopoulos D (1979) Ca^{2+}-induced fusion of phospholipid vesicles monitored by mixing of aqueous contents. Nature 281: 690-692

Yamada S, Ohnishi S (1986) Vesicular stomatitis virus binds and fuses with phospholipid domain in target cell membranes. Biochemistry 25: 3703-3708

Zakowski JJ, Wagner RR (1980) Localization of membrane-associated proteins in vesicular stomatitis virus by use of hydrophobic membrane probes and cross-linking reagents. J Virol 36: 93-102

Cell contact with solid surfaces

David Gingell
Department of Anatomy and Developmental Biology
University College and Middlesex School of Medicine
Gower St., London W1

Introduction

It is my intention to outline the main biophysical studies on cell-to-substratum contact which have been done in this laboratory during the past few years. A brief explanation of the way in which the topics to be discussed are connected may be helpful. In the dark ages, when it was suspected that cell adhesion could be understood solely as a long-range balance of intermolecular attractive and repulsive forces, without direct molecular contact, I joined Dr.V.A.Parsegian to learn how to calculate the size of these forces (Parsegian & Gingell, 1973) and went on to devise several ways of measuring them for cell contacts. Since the idea of a long-range force balance implies a gap between the cell membrane and the substratum, it was natural to try to visualize it optically and measure it if possible. This led to an interest in interferometry which eventually culminated in a quantitative theory of interference reflection microscopy, IRM, (Gingell & Todd, 1979; Gingell et al., 1982). Cell-to-substratum water gaps were indeed detected in certain situations for aldehyde-fixed red blood cells (Gingell & Todd, 1980) and for free living amoebae (Gingell & Vince, 1982b). The theory helped us to understand why IRM images could sometimes be very hard to interpret, and our attention turned to total internal reflection fluorescence as an alternative and superior method of visualizing and measuring cell-to substratum contacts. The success of this technique has exceeded our most optimistic predictions.

The rationale behind our earlier methods of assessing cell-to-substratum attraction forces was to increase the electrostatic repulsion force to a point where adhesion failed to take place. From this (and ideally a measurement of the width of the water gap between the cell and the substratum, at which cells just managed to attach) the attraction could be estimated. However, we finally convinced ourselves beyond any doubt that the adhesion of unmodified red blood cells, even though showing a visible water gap, involved bridging macromolecules (Trommler et al., 1985). This by no means disproved the action of van der Waals and electrostatic forces, but indicated that the overall effect of molecular forces is to pull cells into direct contact with the substratum.

Our attention therefore turned to ways of measuring the strength of direct molecular contacts. With Dr.L.Fisher and Dr.G.Francis I devised an exceedingly sensitive method of pulling single attached cells, using a flexible micropipette (Francis et al., 1987). These very encouraging results, as well as those from the application of viscous shear to cells in a conventional laminar flow chamber, have underlined the care needed in distinguishing between forces of adhesion and forces of removal. It is also clear that contact is not necessarily synonymous with adhesion.

Our studies of adhesion forces have entailed measuring cell adhesion to well-defined substrata. More recently these have concentrated on Langmuir-Blodgett molecular films, and we have begun to address the question of the nature of the chemical groups to which cells can adhere. Our studies on defined surfaces led us to describe a remarkable and unexpected motile response, seen in free-living amoebae as well as mononuclear leucocytes. This fired our interest in transductive events which seem to be triggered by contact with solid surfaces.

Our attempts to prevent adhesion by maximizing electrostatic repulsive forces, and the realization that repulsive hydration forces can greatly exceed them (Le Neveu et al., 1977; Parsegian et al., 1979; see the chapter by Gawrisch et al. in this volume), led us into the topical area of anti-adhesive surface treatments, and we have begun to study the conditions under which certain flexible uncharged polymers successfully prevent cell adhesion.

Adhesion implies molecular contact

A suspicion that cells adhere to other cells or to solid surfaces, despite being apparently separated by a gap of 10-20 nm, came from early electron microscope studies of cell contacts. Improved staining methods capable of showing the glycocalyx, and the appreciation of the fact that it consists of extracellular carbohydrate-linked segments of integral membrane glycoproteins, more or less killed off the water gap idea. The contact zone came to be viewed as an aqueous compartment spanned by the polycationic segments of membrane macromolecules. The early notion of adhesion across a water gap led to the suggestion (Curtis, 1962) that it might be a secondary minimum energy balance - a concept borrowed from colloid chemistry where attractive van der Waals forces between particles are thought to be balanced by repulsive elestrostatic forces at a few tens of nanometers. Even when it was realized that the water "gap" contained membrane glycoproteins, the nagging question remained as to whether adhesions could nevertheless exist in the reversible secondary minimum mode.

Measurements on the electrostatic repulsive forces needed to prevent adhesion of aldehyde-fixed erythrocytes to an oil/water interface (Gingell et al., 1977; Todd & Gingell, 1980) and to a polarizable electrode (Gingell & Fornes, 1976) were quantitatively consistent with the operation of a long range intermolecular attraction with a possible secondary minimum balance (Parsegian & Gingell, 1980). However, our more recent observations on unfixed erythrocytes have shown clearly that secondary minimum adhesions with a glass surface cannot be demonstrated (Trommler et al., 1985). Our experiments were based on the centrifugal removal of adherent cells in salt solutions over a wide range of salt concentrations. The basic idea was that lowering the salt concentration would cause a concomitant increase in the electrostatic force of repulsion between cells and the glass surface, thereby reducing the force needed to pull cells out of a secondary minimum balance. The crucial observations were the following. Cells were allowed to settle at a given salt concentration, then subjected to a tiny removal force of unit gravity. This sequence was repeated at several dilutions down to 1.5mM. The response curve showed that fewer cells stuck at lower salt concentrations. Subsequent centrifugation showed that those which did stick could be removed, but that the force needed was independent of the salt concentration. In the next part of the experiment, cells were allowed to settle at a given salt concentration, the solution was carefully diluted and the number of cells remaining attached was counted. This sequence was repeated, allowing cells to settle at various salt concentrations before carrying out further dilution. Despite the fact that dilution greatly increases the Debye electrostatic repulsion, we found that it did not remove a single cell. These results show without any doubt that increasing the electrostatic repulsion reduces the probability of initial adhesion at low salt concentrations, but once adhesions have formed they cannot be broken by lowering the salt concentration. This is irreconcilable with adhesion in a secondary minimum energy configuration, which must always be reversible. It follows that direct molecular connections are established between the cell surface, presumably the glycocalyx, and the glass substratum.

Adhesion and cell breakage

Since we know that cells in direct molecular contact with glass can be removed by centrifugation, the next question to arise concerns the site of separation. Does this correspond with the site of initial adhesive contact (the glycocalyx/glass junction) or does breakage take place within the glycocalyx, the membrane, or the cell itself? This is a very important question, since as L.Weiss pointed out many years ago, measurements of removal forces will give only minimal estimates of adhesion forces, if breakage occurs at some site other than that of initial adhesion. It should be emphasized that although a study of removal forces enabled us to reject the notion of secondary minimum adhesion of erythrocytes in the case considered, it is far from clear that the mechanism of molecular contact could be deduced from the adhesion force or energy, even if it could be convincingly measured.

To study the question of cell breakage, we examined erythrocyte detachment from glass surfaces in a laminar flow chamber, using interference reflection microscopy (IRM) and scanning electron microscopy of cells which were chemically fixed during flow. Technical difficulties prevented direct observation of the detachment process during centrifugation. We found that detachment under shear stress on both highly hydrophobic glass (covalently derivatized with octadecyl dimethyl chlorosilane) and normal coverslip glass in physiological saline left numerous broken dendritic cytoplasmic fragments on the solid surface (fig.7 Owens et al., 1988b). Extensive dendritic fragments were also left behind when cells which had settled on normal glass in 157mM NaCl were subjected to flow in 1.5mM NaCl (Gingell & Trommler, unpublished). Others have described membranous debris left behind when adherent red cells are removed from surfaces

(Borenstein & Brash, 1986). In our experiments we noticed the discrepancy between the broad dark areas of intimate cell-glass apposition seen under IRM before starting the liquid flow, and the multiple punctate adhesions at the tips of the dendritic tubes drawn out from the cell bodies during flow. This alerted us to the need for caution in attempting to relate either the closeness or the area of contact to the strength of adhesion.

The fact that removal of erythrocytes from solid surfaces can sometimes involve cell breakage suggests an explanation for our observations on the effects of shear stress on cells attached to deposited monolayer films (see fig.6 in Owens et al., 1988b). The curves for cell removal from octadecyl derivatized glass and the same surface coated with a coherent hydrophilic monolayer film of docosanol ($C_{22}H_{45}$-OH) are almost identical. This result is difficult to explain unless the adhesions to both surfaces exceed the tensile strength of the cytoplasmic tubes (tethers) which connect the cell to its adhesion points. They are formed as the cell is pushed by the flowing liquid.

We anticipate that the uncertainties associated with attempting to calculate cell detachment forces from measured wall shear stresses in a laminar flow chamber will be obviated by using a novel micropipette suction technique (Francis et al., 1987). This also makes it possible to examine the contact zone for cell debris during the process of detachment, using high power IRM. A finely drawn, thin walled, flexible capillary with an upturned orifice is brought up to an adherent cell, whose area of contact is in the field of an interference reflection microscope. Suction is applied via the capillary in order to grab the cell, then the capillary is moved a few micrometers further from the glass, with the result that the cell is smoothly unpeeled from the glass and drawn into the capillary. Interference reflection images showing changes in cell contact with the glass are recorded on videotape and a continuous record of the force applied to the cell is obtained from opto-electronic recording of the bending of the micropipette. Using this equipment, which has now been much refined, it is possible to measure detachment forces with a sensitivity exceeding 10^{-10}N, which is much less than the force needed to break a C-C bond! Our observations showed that detachment occurs by centripetal peeling. Recent measurements on erythrocytes stuck to coverslip glass in physiological saline give a mean detachment force of about 6 x 10^{-9} N (Francis, personal communication) with a slightly lower value for cells on glass which had been hydrophobed with trimethyl chlorosilane. On normal glass, cell breakage is seen in about 50% of detachments, while it is less frequent on trimethyl glass. However, for reasons that are not understood, cells attached to glass hydrophobed with dimethyl dichlorosilane almost always rupture on detachment. A particularly interesting result is that if a cell which had been allowed to settle on clean glass are partially unpeeled and then allowed to re-attach, a second attempt at removal gives a force at least as great as the first (G.Francis, personal communication). Thus it would appear that in this example separation takes place at the plane of the original adhesion. Since detailed calculations giving the work of separation per unit area are not yet published (Francis, in preparation) it is not possible to comment on the relative forces and energies of cell interaction with hydrophilic versus hydrophobic glass. In view of the hydrodynamic shear experiments, a vital question is whether the similar <u>removal</u> forces on hydrophobic and hydrophilic glass imply cell or membrane breakage. Similar <u>adhesion</u> forces (per unit area) on such different materials would be suprising. However, it is clear that cell rupture on detachment sometimes occurs on both, but there is evidence that clean separation can also take place on both. If clean detachment really does happen, it will be possible to learn about the forces of adhesion from the mechanics of detachment, even where adhesion involves molecular contact. To demonstrate the legitimacy of this, and to know when it is legitimate, would represent a significant advance in the interpretation of adhesion data.

Imaging cell contacts

Until relatively recently, IRM provided the only useful way of imaging the contacts between living cells and their substrata. The potential of this extremely valuable technique was first fully appreciated by Izzard & Lochner (1976) who correctly identified focal contacts in fibroblasts (review, Verschueren, 1985). The monopoly hitherto enjoyed by interferometry has recently been challenged by the introduction of a fluorescence method based on total internal reflection. This gives images of the contacts of living cells (which the antibody technique of Neyfakh et al.,(1983) does not) and can actually give novel information about the degree of molecular penetration beneath cells, a major source of uncertainty in all methods employing labelled antibodies to putative contact site molecules (Avnur & Geiger, 1981; see Grinnell, 1986).

The essential idea behind the new fluorescence method has long been appreciated by physical chemists investigating protein adsorption onto glass surfaces. If a laser beam passes through a glass block and it is totally internally reflected at its upper face, which is in contact with a liquid of lower refractive index (ie. cell culture medium), a very thin zone of evanescent light is set up at the glass/liquid interface. This light dies off exponentially as it penetrates into the aqueous medium, and it has a characteristic thickness or penetration depth, (the distance in which its energy falls to 1/e of its value at the interface) of around 50-150nm. This distance falls if the refractive index step across the interface is increased and also falls as the angle of incidence of the laser beam increases. The latter feature is very convenient, since the angle can be changed during an experiment to adjust the penetration depth. In early applications to cell biology, Axelrod (1981) and Hafeman et al., (1981) used the evanescent wave to stimulate fluorescent labels attached to the surface of cells which had spread on a glass substratum. This technique, called TIRF (Total Internal Reflection Fluorescence), had the effect of illuminating only the contact points, since the short "probing" depth of the evanescent wave is insufficent to reach any of the more distant parts of the cell surface, which may be hundreds or thousands of namometers from the glass. This method requires that the cell surface carries a fluorescent label, for example a lipid analoque or a fluorescent labelled antibody.

Our modification of the TIRF method gives very satisfactory images and is based on the far more convenient use of non-adsorbing fluorescent markers in the aqueous volume (Gingell et al., 1985). The temptation of coining our own acronym eventually proved irresistable and we later called our method TIRAF (A=aqueous). This showed up contacts in explant cultures of chick fibroblasts in high contrast as dark spots and streaks which co-rresponded with focal contrast seen under IRM. These characterize the less mobile and stationing cells, whereas the more motile peripheral cells lack clear focal contacts. The generally dark images of these cells showed that the lower surfaces are extensively and closely applied to the glass. The 4000 m.wt. fluorescent labelled dextran which was used to mark the external volume provided topographic maps of the cellular contacts which were indistinguishable from those given by fluorescein alone (400 m.wt.). In contrast, a much smaller available volume was found beneath well spread locomoting cells when probed by 50,000 m.wt. dextran carrying the same label, and the cell contacts consequently looked very dark. If the high molecular weight material was removed and replaced with low, the contact pattern typical of the latter was re-established (see Gingell et al., 1985; fig.3;i,j,k). This kind of experiment, together with arguments based on diffusion rates, shows that there are substantial areas beneath adherent cells which are inaccessible to large molecules, but accessible to relatively small ones. Large ones cannot diffuse in, given any amount of time. A reasonable interpretation of these results is that in such zones the external segments of transmembrane glycoproteins span an aqueous region between the lipid bilayer and the substratum, providing adhesion and acting as a kind of molecular sieve to the diffusion of aqueous molecules. The zones which do not possess such discrimination, admitting both large and small dextran molecules, are probably too widely separated for glycoproteins to span the gap. TIRAF can therefore give information, which cannot be obtained by any other existing methods, about the physical state of the glycocalyx and associated macromolecules in living cells.

To be confident of interpreting the images seen under TIRAF, we developed an exact electromagnetic theory of totally internally reflected waves in multilayered media, modelling the cell contact region as parallel dielectric films (Gingell et al.,1987). This brought out several very encouraging features: principally that TIRAF is far more sensitive than IRM for assessing small cell-substratum gaps; that structural inhomogeneities in the cytoplasm have almost no effect on the image, and that in distinction to IRM, it should be possible to image even the thinnest cytoplasmic lamellae (around 0.1um) without the intrusion of optical information originating at the upper surface of the cell, or beyond it. We carried out several tests of the theory. The most persuasive involved measuring the thickness of deposited magnesium fluoride films by means of TIRAF in media of various refractive indices (Mellor et al., 1988). Film thicknesses were independently measured by a crystal oscillator; the TIRAF values subsequently obtained over the entire experimental range from 2-160 nm agreed with the expected values to within a few percent. Since this encompasses the expected dimensions of cell to substratum contacts, we are now in a position to measure them by TIRAF.

Perhaps the most impressive result to come from TIRAF so far concerns the contacts of amoebae and mononuclear leucocytes. Both types of cells are capable of producing extensive ultrathin cytoplamic lamellae of 100 nm or less, and I pointed out several years ago on theoretical grounds that these would produce potentially confusing IRM images (Gingell 1981). Our investigations showed that even under optimal conditions for IRM, such complex images can be generated. We found that Dictyostelium amoebae can have a fried-egg shape, the outer (thin) part generating a dark image and the inner (thick) zone appearing lighter. The conventional interpretation of dark areas in IRM images is that they represent intimate contacts with the substratum. That this can happen is well established in the case of focal contacts, but the situation is otherwise in the amoeboid cells. TIRAF observations on coverslip glass showed a very striking pattern, which was a contrast reversal of the interference reflection pattern (Todd et al., 1988). The central zone appeared dark and the outer part showed up as a pale penumbra. This is precisely the TIRAF image that would be expected of a cell with a shape like a fried-egg and making a uniform contact over its entire under surface, when viewed in conditions where the evanescent wave can penetrate the thin peripheral cytoplasmic sheet and stimulate fluorescence in the medium on the side of the cell. Our interpretation was shown to be correct by performing TIRAF on cells under conditions where the penetration depth of the evanescent wave was much less than the expected thickness of the peripheral cytoplasmic lamella. A sufficiently low penetration depth of 34 nm was achieved using high refractive index glass (n=1.83) together with a large angle of incidence of the laser beam. Under these conditions the cell looked uniformly dark by TIRAF. This shows that the volume available for the fluorescent marker is constant over all parts of the contact. It is consistent with the interpretation that the glycocalyx spans a uniform aqueous gap between the lipid bilayer and the glass, forming a penetrable meshwork of approximately constant thickness. These observations were made on cells which had spread but were not locomoting and it remains to be seen whether actively moving cells have a similar contact pattern.

Contact versus adhesion

While it is one thing to map the contact topography (the dimensions of the aqueous compartment beneath spread cells) it is quite another matter to define the pattern of adhesions between a cell and the substratum. In general terms it is not possible to tell whether a given contact is adhesive from optical data alone, since a stress must be applied to the cell. This is brought out in the case of fibroblasts which can show a complex pattern of "close" contacts which have a rather wide spectrum of aqueous separations (Bailey & Gingell, 1988), but it is not at all clear what (if any) adhesive function is exercised over this range of appositions. Even in the situation where there is a relatively uniform gap between the bilayer membrane and the substratum, there is not necessarily uniform adhesion. This can be demonstrated by applying a mechanical stress which tends to lift the cell. One way of doing this is to dilute the saline medium while maintaining a constant osmotic pressure: the greatly increased electrostatic repulsive forces in the polyanionic glycocalyx cause it to swell and push the cell membrane several tens of nanometres further from the glass. When this was done with amoebae a speckled dark pattern of adhesions interspersed with bright wide gaps was displayed under IRM (Gingell & Vince, 1982b). When the diluted medium was gently exchanged for normal medium, the uniform dark contact pattern was re-established. Cells could be taken through several such cycles. Similarly, if amoebae or red blood cells which were stuck to glass with essentially uniform contacts were subjected to a gentle detaching force using viscous flow, the positions of adhesions could often be distinguished from mere appositions under IRM (unpublished observations).

Cells on molecular films

We have studied cells attached to relatively well-defined surfaces in order to try to understand more about the physical chemistry of adhesion and molecular recognition. Our recent efforts have concentrated on molecular films deposited on glass surfaces. These were obtained by the Langmuir-Blodgett method of dipping a hydrophobed glass coverslip in a controlled way through an air/water interface at which is spread an insoluble monolayer. Hydrophobic glass was prepared by covalent derivatization with octadecyl dimethyl chlorosilane. The monolayer becomes transfered to the coverslip, with the polar headgroups oriented towards the water and the hydrocarbon chains associated with the hydrophobic glass. This configuration is stable as long as the preparation is kept under water. The advantages of deposited monolayers are that they provide an optical surface for studying cell behaviour and they are chemically well-defined with a known number of headgroups per unit area. We have studied the spreading and locomotion of Dictyostelium amoebae on deposited monolayers of synthetic glycolipids including C_{22} galactose, C_{18} glucose and C_{18} mannose in order to find out whether the amoebae can distinguish between the monosaccharide headgroups. We found no indication that the amoebae can do so, though they adhered to and spread on all the films (Owens & Gingell, in preparation). Adhesion was assessed by quantitative hydrodynamic shearing in a parallel plate flow chamber. Spreading occured with active peripheral ruffling, but no locomotion was evident. The apparent lack of saccharide specific recognition was suprising in view of reports to the contrary (Chipowsky et al., 1973). We can only surmise that a degree of projection and flexibility of the headgroups which may be necessary for stereospecific ligand-receptor fitting is not available in the glycolipids that we used. The fact that rhodamine-ConA fails to label any of the saccharide films supports this.

The spreading and adhesive behaviour of amoebae on monolayer films of synthetic $C_{22}H_{45}$-OH at a density of 5 groups per 1.0nm^2 was found to be the same as on the saccharide monolayers (Owens et al., 1988b). The fact that adhesion occurs on these films is of considerable interest, because the major difference between bacteriological polystyrene and tissue culture polystyrene culture dishes is thought to be due to the introduction of -OH groups into the former (Curtis et al., 1983). Our result represents the only example of cell adhesion to a surface containing nothing but alcoholic -OH groups. Indeed, it appears to be the only example of adhesion of living cells to a solid interface that contains but a single chemical group. Under the conditions of our experiment there is almost certainly no significant protein contamination of the interface, thus ruling out the possibility that cells are really attached to an adsorbed protein film. The monolayer films were shown to be highly coherent by using rhodamine labelled ConA, which adsorbs avidly to the octadecyl glass base on which the films are deposited, but fails to stain the films. It is also extremely unlikely that the cells are able to make any direct contact with the octadecyl chains beneath the films, or indeed with the paraffin chains of the monolayers themselves, since cell behaviour on a paraffinic surface is most distinctive.

Contact transduction

When the amoebae spread on an adjacent control area of hydrophobed glass they were seen under IRM to form dark ultrathin peripheral lamellae by a process of centripetal retraction. This was usually followed by locomotion (Gingell & Vince, 1982a). In order to try to understand what triggers this intriguing cellular response, we utilized our earlier finding that it is also stimulated by polycations adsorbed to glass (Gingell & Vince, op. cit.) by preparing glass with covalently linked amino groups (Owens et al., 1988a). The point of this is that the NH_3^+ charge density can be controlled by adjusting the pH. The zeta potentials and hence the charge densities of the surfaces were determined from electrokinetic streaming potentials. Underivatized glass and hydrophobic octadecyl glass were used as control surfaces. We also investigated the influences of ionic strength and ion binding to the aminated surface. The conclusion from our analysis was that the trigging of the lamellar response could not be correlated with the degree of hydrophobicity of the interface, nor with its overall charge or with the density of ionogenic -N== groups. The same response is triggered by strongly hydrophobic surfaces with a negative potential, and by hydrophilic surfaces with a surface potential less negative than -30mV. To both types of surface, the amoebae stick strongly. Our tentative hypothesis is that the contact-mediated trigger is somehow related to the energetics of adhesion, a matter that we hope to test.

Since this cytoplasmic response is not limited to humble soil amoebae, but is shown by human mononuclear cells (Todd et al., 1988) it is anticipated that the subject of contact trigged cellular responses will receive wider recognition as an important subject area of membrane transductive process. The way in which the transducer might operate can be inferred from related systems. Mechanisms by which the cross-linking of cell surface receptors might initiate signal transduction have been discussed by Metzger (1983). He envisages two classes, "intrinsic mechanisms" in which very close contact between receptors trigger conformational changes in the receptor complexes themselves, and "extrinsic mechanisms" whereby a sufficiently high local concentration of receptors facilitates low energy interactions between the cytoplasmic parts of the receptor complexes and molecules in the peripheral cytoplasm which are responsible for the next stage in the signalling concatenation. Kane et al. (1988) examined the cross-linking of IgE receptors by rigid bivalent antigens and found that triggering occured even when the rigid links exceeded 20 nm in length. This suggests an intrinsic triggering mechanism and is consistent with the idea that the cytoplasmic ends of the receptor complexes are interacting with multiple sites on the cytoskeleton. A recent report by Chasis et al., (1988) shows that antibody binding to glycophorin A induces stiffening of the cytoskeleton in intact red cells. There is a correlation between the increase in rigidity and the ammount of glycophorin that becomes associated with the cytoskeleton. Similar results have been reported for the interaction of platelet (Painter & Ginsberg, 1982) and leukocyte glycoproteins (Sheterline & Hopkins, 1981) with the cytoskeleton. Feltkamp et al. (1987) also showed that capping of H2 antigens of leukaemia cells induced mechanical linkage to the detergent insoluble cytoskeleton.

It therefore seems reasonable to guess that any external events that sufficiently restrict the lateral diffusion of certain classes of membrane macromolecules, effectively trapping and concentrating them, will transductively initiate changes in the peripheral cytoskeleton, and that these in turn influence cell motility. If so, the mere act of adhesive contact with a solid surface (perhaps in a very nonspecific way) will fulfil this requirement and lead to characteristic contact-mediated responses, including cell spreading and phagocytosis. In the latter case it is known that interaction with surface receptors can induce structural organization in the cytoplasm (Sheterline et al., 1984; Wang et al., 1984). Evidence relating to the concentration of membrane molecules at contact sites is discussed in the chapter by Andre & Bongrand in the present volume.

Anti-adhesion

The majority of cells will stick to most non-cellular surfaces, to a greater or lesser extent. This has serious implications for the practical objective of designing non-adhesive surfaces, particularly blood compatible interfaces. On the whole this has proved a most elusive goal (see Brash, 1983; Ruckenstein & Gourisankar, 1984). Merely increasing the negative charge density on a surface in media of physiological ionic strength seems unable to provide repulsion of sufficiently long range to veto adhesion. An imaginative approach to the problem has involved the use of phospholipids. Margolis et al., (1978) claimed that fibroblasts cannot spread on thick hydrated layers of phospholipids that are in a liquid crystalline state, whereas cells were able to spread when the lipids were converted into the gel ("solid") phase. However, certain aspects of this work involving phosphatidyl choline are hard to reconcile with studies from McConnell's laboratory (see discussion in Owens et al., 1988). In a related approach, Durrani et al., (1986) have pioneered the use of phospholipid headgroups directly attached to solid surfaces, and have drawn particular attention to the use of phosphoryl choline, whose parent molecule phosphatidyl choline appears to be non-adhesive towards leucocytes (Hafeman et al., 1981; McConnell et al., 1986). The mechanism of such action is unclear, but it is not simply a matter of negative charge since the choline headgroup is electrically neutral overall. Another encouraging approach has employed flexible synthetic polymers (see Andrade, 1985; Norde et al., 1986) particularly those containing polyethoxy groups (Whicher & Brash, 1978; Sa da Costa et al., 1980, 1981; Mori et al., 1982). These certainly provide repulsion, and Klein & Luckham (1982) and Luckham & Klein (1985) succeeded in directly measuring the repulsive force between mica plates bearing adsorbed polyethylene oxide molecules. Recently we found that a block copolymer comprising segments of hydrophilic and of relatively hydrophobic character could adsorb to octadecyl glass and veto the adhesion of bacteria, erythrocytes and amoebae (Owens et al., 1987). Although the way in which this remarkable action is achieved is presently unknown, it is clearly not electrostatic, and it would seem that entropic repulsion associated with chain motion (Mori et al., 1982), or hydration repulsion due to water structuring (see the chapter by Gawrisch et al. in this volume) are involved. We are continuing these studies and are hopeful that they may lead to materials of practical utility.

Summary

1) Detailed theoretical and experimental studies have shown that even the most rigorous IRM analyses of cell contacts can be difficult to interpret and that serious ambiguities can arise that cannot be satisfactorily disentangled using the technique. We have proved that TIRAF is a far more sensitive tool for examining cell contacts. It avoids all the pitfalls of the interferometric method and it can give a completely different picture

of the contact zone in certain situations.

2) The secondary minimum mode of adhesion cannot be demonstrated for red blood cells in contact with glass in simple salt solutions. Increasing electrostatic repulsion reduces the likelihood of adhesion, but all adhesions involve direct molecular connection. These results do not prove that secondary minimum adhesion can never play a role in cell adhesion phenomena, but it does seem unlikely.

3) Strong molecular adhesions are indicated by the fact that mechanical removal of cells can lead to cell breakage. Trying to infer mechanisms of adhesion from cell removal is therefore frought with difficulties, and it is always imperative to follow the process of cell removal with great care to ascertain whether cell components are left behind. Recent results with a highly sensitive cell puller indicate that in certain cases adhesion force and energy can be obtained from the measured removal force.

4) Deposited lipid films can be used as substrata for examining the chemical requirements of adhesion. We have been able to show that monolayers of several synthetic glycolipids as well as docosanol ($C_{22}H_{45}$-OH) can support cell adhesion. The common feature is -OH groups. The absence of any other chemical groups on the docosanol surface clearly implicates -OH in adhesion.

5) Cell behaviour may be governed not only by the cell surface content of adhesive molecules but by anti-adhesive elements. We have shown that cell adhesion can be prevented by certain macromolecules with a polyoxyethylene backbone structure. This may have important implications for the design of anti-adhesive surfaces.

6) We have described a striking motile response to contact with a solid surface which appears to be a transductively mediated cytoplasmic process triggered by a sufficient adhesive energy. This may be a member of a wider class of transductive processes initiated by cell contact.

Acknowledgements I would like to thank the Science and Engineering Research Council and the Wellcome Trust for their support.

References

Andrade, J.D. (1985) Surface and interfacial aspects of biomedical polymers. vol.1 Surface chemistry and physics. Editor, J.D.Andrade. Plenum Press, New York and London.

Avnur,Z. & Geiger,B. (1981) The removal of extracellular fibronectin from areas of cell-substrate contact. Cell 25: 121-132

Axelrod,D. (1981) Cell substrate contacts illuminated by total internal reflection fluorescence. J.Cell Biol., 89: 141-145

Bailey,J. & Gingell,D. (1988) Contacts of chick fibroblasts on glass: results and limitations of quantitative interferometry. J.Cell Sci., 90: 215-224

Boresnstein,N. & Brash, J.L. (1986) Red blood cells deposit membrane components on contacting surfaces. J.Biomed. Mater. Res., 20: 723-730

Brash,J.L. (1983) Hydrophobic polymer surfaces and their interaction with blood. Ann. N.Y. Acad. Sci., 283: 356-371

Chasis,J.A., Reid,M.E., Jensen,R.H. & Mohandas,N. (1988) Signal transduction by glycophorin A: role of extracellular and cytoplasmic domains in a modulatable process. J.Cell Biol., 107: 1351-1357

Chipowsky,S., Lee,Y.C. & Roseman,S. (1973) Adhesion of cultured fibroblasts to insoluble analoques of cell-surface carbohydrates. Proc. Natl. Acad. Sci. USA 70: 2309-2312

Curtis,A.S.G. (1962) Cell contact and adhesion. Biol.Rev., 37: 82-129

Curtis,A.S.G., Forrester,J.V., McInnes,C. & Lawrie,F. (1983) Adhesion of cells to polystyrene surfaces. J.Cell Biol., 97: 1500-1506

Durrani,A.A., Hayward,J.A. & Chapman,D. (1986) Biomembranes as models for polymer surfaces. 11. The synthesis of reactive species for covalent coupling of phosphorylcholine to polymer surfaces. Biomaterials, 7: 121-125

Feltkamp,C.A., Spiele,H. & Roos,E. (1987) Interaction of plasma membrane-associated filaments and H2 histocompatibility antigens before and after induced patching and capping. J.Cell Sci., 88: 313-325

Francis,G.W., Fisher,L.R., Gamble,R.A. & Gingell,D. (1987) Direct measurement of cell detachment force on single cells using a new electromechanical method. J.Cell Sci., 87: 519-523

Gingell,D. (1981) The interpretation of interference reflection images of spread cells: significant contributions from thin peripheral cytoplasm. J.Cell Sci., 49: 237-247

Gingell,D. & Fornes,J.A. (1976) Interaction of red blood cells with a polarized electrode: evidence of long-range intermolecular forces. Biophys.J., 16: 1131-1135

Gingell,D., Heavens,O.S. & Mellor,J.S. (1987) General electromagnetic theory of total internal reflection fluorescence: the quantitative basis for mapping cell-substratum topography. J.Cell Sci., 87: 677-693

Gingell,D. & Todd,I. (1979) Interference reflection microscopy. A quantitative theory for image interpretation and its application to cell-substratum separation measurement. Biophys.J., 26: 507-526

Gingell,D. & Todd,I. (1980) Red blood adhesion. 11. Interferometric examination of the interaction with hydrocarbon and glass. J.Cell Sci., 41: 135-149

Gingell,D., Todd,I. & Bailey,J. (1985) Topography of cell-glass apposition revealed by total internal reflection fluorescence of volume markers. J.Cell Biol., 100: 1334-1338

Gingell,D., Todd,I. & Heaven,O.S. (1982) Quantitative interference microscopy: effect of microscope aperture. Optica Acta, 29: 901-908

Gingell,D., Todd,I. & Parsegian,V.A. (1977) Long-range attaction between red cells and a hydrocarbon surface. Nature, 268: 767-769

Gingell,D. & Vince,S. (1982a) Substratum wettability and charge influence the spreading of Dictyostelium amoebae and the formation of ultrathin cytoplasmic lamellae. J.Cell Sci., 54: 255-285

Gingell,D. & Vince,S. (1982b) Cell-glass separation depends on salt concentration and valency: measurements on Dictyostelium amoebae by finite aperture interferometry. J.Cell Sci., 54: 299-310

Grinnell,F. (1986) Focal adhesion sites and the removal of substratum-bound fibronectin. J.Cell Biol., 103: 2697-2706

Hafeman,D.D., von Tscharner,V. & McConnell,H.M. (1981) Specific antibody-dependent interaction between macrophages and lipid haptens in polar lipid monolayers. Proc. Natl. Acad. Sci. U.S.A., 78: 4552-4556

Izzard,C.S. & Lochner,L.R. (1976) Cell-to-substrate contacts in living fibroblasts: an interference reflexion study with an evaluation of the technique. J.Cell Sci., 21: 129-159

Kane,P.M., Holowka,D. & Baird,B. (1988) Cross-linking of IgE-receptor complexes by rigid bivalent antigens >200A in length triggers cellular degranulation. J.Cell Biol. 107: 969-980

Klein,J. & Luckham,P. (1982) Forces between two adsorbed polyethylene oxide layers immersed in a good aqueous solvent. Nature, 300: 429-431

Le Neveu,D.M., Rand,R.P., Parsegian,V.A. & Gingell,D. (1977) Measurement and modification of forces between lecithin bilayers. Biophys.J., 18: 209-230

Luckham,P.F. & Klein,J.K. (1985) Forces between mica surfaces bearing adsorbed polyelectrolytes. J.Chem. Soc. Faraday Trans. 1. 80:865-878

Margolis,L.B., Dyatlovitskaya,E.V. & Bergelson,L.D. (1978) Cell-lipid interactions. Cell attachment to lipid substrates. Expl. Cell Res., 111: 454-457

McConnell,H.M., Watts,T.H., Weis,R.M. & Brian,A.A. (1986) Supported planar membranes in studies of cell-cell recognition in the immune system. Biochim. Biophys. Acta, 864: 95-106

Mellor,J.S., Gingell,D. & Heavens,O.S. (1988) Measurement of the thickness of deposited magnesium fluoride films by evanescent wave fluorescence: a critical test of the general TIRAF theory. J.Modern Optics, 35: 623-628

Metzger,H. (1983) The receptor on mast cells and related cells with high affinity for IgE. Contemp. Top. Mol. Immunol., 9: 115-145

Mori,Y., Nagaska,S., Takiuchi,H., Kikuchi,T., Noguchi,N., Tanzawa,H. & Noishiki,Y. (1982) A new antithrombogenic material with long polyethyleneoxide chains. Trans. Am. Soc. Artif. Int. Organs, 28: 459-462

Neyfakh,A.A., Tint,I.S., Svitkina,A.D., Bershadsky,A.D. & Gelfand,V.I. (1983) Visualization of cellular focal contacts using a monoclonal antiboby to 80 kD serum protein adsorbed on the substratum. Exp. Cell Res., 149: 387-396

Norde,W., MacRitchie,F., Nowicka,G. & Lyklema,J. (1986) Protein adsorption at solid-liquid interfaces: reversibility and conformational aspects. J.Coll. Interface Sci., 112: 447-456

Owens,N.F., Gingell,D. & Bailey,J. (1988a) Contact-mediated triggering of lamella formation by Dictyostelium amoebae on solid surfaces. J.Cell Sci., 91: 367-377

Owens,N.F., Gingell,D. & Rutter,P.R. (1987) Inhibition of cell adhesion by a synthetic polymer adsorbed to glass shown under defined hydrodynamic stress. J.Cell Sci., 87: 667-675

Owens,N.F., Gingell,D. & Trommler,A. (1988b) Cell adhesion to hydroxyl groups of a monolayer film. J.Cell Sci., 91: 269-279

Painter,R.G. & Ginsberg,M. (1982) Concanavalin A induces interactions between surface glycoproteins and the platelet cytoskeleton. J.Cell Biol., 92: 565-573

Parsegian,V.A. & Gingell,D. (1973) A physical force model of biological membrane interaction. In: Recent advances in adhesion. Editor, L.H.Lee. Gordon & Breach Science Publishers, Inc., New York. pp 153-192

Parsegian,V.A. & Gingell,D. (1980) Red blood cell adhesion. 111. Analysis of forces. J.Cell Sci., 41: 151-157

Parsegian,V.A., Rand, R.P., Fuller, N. & Rau,D.C. (1979) Osmotic stress for the direct measurement of intermolecular forces. Methods in Enzymology, 127: 400-416

Ruckenstein,E. & Gourisankar,S.V. (1984) A surface energetic criterion of blood compatibility to foreign surfaces. J.Coll. Interface. Sci., 101: 436-451

Sa da Costa,V., Brier-Russell,D., Trudel,G., Waugh,D.F., Salzman,E.W. & Merrill,E.W. (1980) Polyether-polyurethane surfaces: thrombin adsorption, platelet adsorption and ESCA scanning. J.Coll. Interface Sci. 76: 594-596

Sa da Costa,V., Brier-Russell,D., Salzman,E.W. & Merrill,E.W. (1981) ESCA studies on polyurethanes: blood platelet activation in relation to surface composition. J.Coll. Interface Sci., 80: 445-452

Sheterline,P. & Hopkins,C.R. (1981) Transmembrane linkage between surface glycoproteins and components of the cytoplasm in neutrophil leukocytes. J.Cell Biol., 90: 743-754

Sheterline,P., Rickard,J.E. & Richards,R.C. (1984) F_c receptor-directed phagocytic stimuli induce transient actin assembly at an early stage of phagocytosis in neutrophil leukocytes. Eur.J Cell Biol., 34: 80-87

Todd,I. & Gingell,D. (1980) Red blood cell adhesion. 1. Determination of the ionic conditions for adhesion to an oil-water interface. J.Cell Sci., 41: 125-133

Todd,I., Mellor,J.S. & Gingell,D. (1988) Mapping cell-glass contacts of Dictyostelium amoebae by total internal reflection aqueous fluorescence overcomes a basic ambiguity of interference reflection microscopy. J.Cell Sci., 89: 107-114

Trommler,A., Gingell,D. & Woolf,H. (1985) Red blood cells experience electrostatic repulsion but make molecular adhesions with glass. Biophys.J., 48: 835-841

Wang,E., Michl,J., Pffeffer,L.M., Silverstein,S.C. & Tamm,I. (1984) Interferon suppresses pinocytosis but stimulates phagocytosis in mouse peritoneal macrophages: related changes in cytoskeletal organization. J.Cell Biol., 98: 1328-1341

Whicher,S.J. & Brash,J.L. (1978) Platelet-foreign surface interactions: release of granule constituents from adherent platelets. J.Biomed. Mat. Res., 12: 181-201

Verschueren,H. (1985) Interference reflection microscopy in cell biology: methodology and applications. J.Cell Sci., 75: 279-301

Cell-cell contacts

Pascale André & Pierre Bongrand

Laboratoire d'Immunologie, Hôpital de Sainte-Marguerite, BP 29, 13277 MARSEILLE Cedex 09, FRANCE.

1 - INTRODUCTION.

Cell-cell or cell-substrate adhesion is involved in many important biological processes such as cell division (1), migration (2) or differentiation (3). During the last years, many authors characterized cell surface molecules with various adhesive specificities, such as fibronectin (4), LFA-1 (5) or N-CAM (6). Also, low affinity interactions were demonstrated between molecules currently found on the cell periphery, particularly carbohydrates (7). It might thus be tempting to consider adhesion as a straightforward process, resulting from expected associations between complementary sites born by adjacent surfaces. However, several lines of evidence suggest that cell attachment is a complex event requiring active cellular functions.

First, binding is a time- and concentration- dependent phenomenon : indeed, when rat thymocytes were incubated with concanavalin A, a multivalent lectin that binds mannosyl derivatives, the minimal ligand density required to induce substantial agglutination was about 500 lectin molecules/μm^2 when cells were subjected to continuous agitation, resulting in transient intercellular contacts with a duration on the order of 1/10 second. By contrast, when cells were incubated in stationary tubes to provide longer contacts, about tenfold lower concanavalin A densities could initiate cell agglutination. Also, when Adams and colleagues studied the adhesion of tumor cells to macrophage monolayers by determining the centrifugation speed required to obtain substantial detachment (9), they found that adhesive strengths increased from about 1.6 x 10^{-10} newton/cell to 24 x 10^{-10} N/cell during the first 60 minutes following binding initiation, and this increase was abolished by metabolic inhibitors.

Secondly, intercellular binding may require that some repulsive intercellular force be overcome to allow a close contact between adhesive sites. Thus, Grinnell (10) reported that centrifugating cells on a plàne substrate (700 g) made adhesion insensitive to cold or metabolic inhibitors, whereas adhesion obtained after 1-g sedimentation was prevented under similar conditions.

Thirdly, many experimental data suggest that adhesion may occur in absence of any specific intermolecular bond. For example, phagocytic cells such as tissue macrophages or blood granulocytes readily adhere to a variety of artificial polymers (e. g. latex beads) that share some hydrophobic properties and are unlikely to be recognized by specific cell receptors (11, 12).

Fourthly, binding involves and probably requires marked structural changes of adhering cells, with bulk membrane deformation allowing geometrical adaptation of interacting surfaces (13) and lateral movement of membrane molecules resulting in concentration of specialized molecules in binding regions as was repeatedly reported (14). Indeed, freezing rat thymocytes by glutaraldehyde fixation resulted in marked decrease of their ability to be agglutinated by concanavalin A without any significant alteration of lectin binding sites.

The present report is aimed at providing a general framework for a quantitative understanding of cell adhesion. In order to make our point clearer, we shall first describe a tentative scheme of the adhesive process, then we shall propose quantitative estimates for the different parameters involved in the sequential steps we have defined.

2 - QUALITATIVE DESCRIPTION OF CELL-CELL ADHESION.

Different steps of cell adhesion are shown on figure 1 :

i) Cell-to-cell approach :

Two main situations may be considered : cells may be driven together by fluid flow, as occurs in blood, lymph, airways or the digestive and urinary tracts (figure 1.i.a). Alternatively, contact may result from the active sending of a lamellipodium by a substrate-bound cell (figure 1.i.b).

ii) Initial repulsion :

Clearly, it is necessary that some kind of repulsive interaction stabilize natural cell suspensions to prevent spontaneous collapse (15). This should involve the most external membrane structures, i.e. the glycocalyx. Also, during the latest stages of a rapid intercellular approach,the presence of a fluid film between membranes may result in repulsive hydrodynamic interactions (16).

iii) Formation of a first bond at the contact region.

The formation of a first bond (or bond cluster) will transiently stabilize an intercellular contact. This may require the egress of repulsive molecules from contact zones, depending on the depth at which binding molecules are buried in the cell coat. This event is likely to occur on the tip of a microvillus or the margin of a lamellipodium, as suggested by the mere observation of scanning electron micrographs displaying the structure of cell surfaces.

iv) Structural changes of binding areas and extension of contact zones.

This may be the most complex step of cell adhesion. The formation of an increasing number of intermolecular bonds requires the following events :

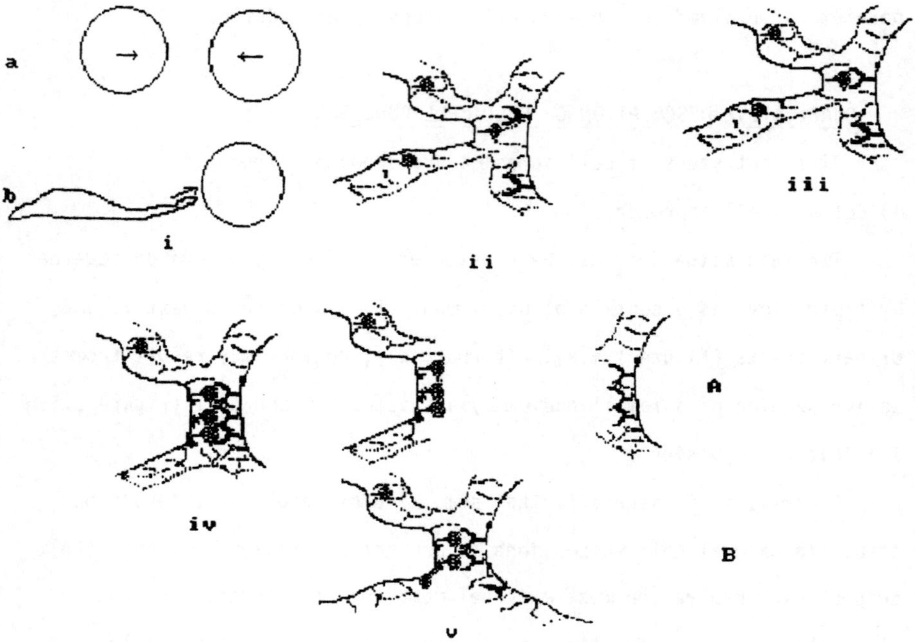

<u>FIGURE 1</u> : STAGES OF CELL ADHESION
i) Cell-cell passive (A) or active (B) approach.
ii) Initial repulsion between glycocalices.
iii) Formation of a first bond.
iv) Formation of numerous bonds involving lateral diffusion of membrane molecules and deformation of the cell membranes.
v) Cell-cell separation by simultaneous (A) or sequential (B) disruption of intercellular bonds.

- Ligand molecules must undergo lateral diffusion in order to allow close contacts between complementary molecular sites.
- Molecular crowding in the contact zones (17) may render necessary the egress of "bystander molecules" to make more room available for interacting structures.
- Repulsive interactions may be decreased by the sorting out of the

molecular species involved in repulsive forces.

- Finally, it must be reminded that the lateral movements of surface molecules are often modulated by parallel reorganization of submembranar cytoplasmic structures such as cytoskeletal elements (18). This may clearly be dependent on the cell metabolic status.

v) Maintenance of adhesion.

The last step of adhesion may be cell-cell separation. It is thus of interest to understand how binding ends. Clearly, bonds may be broken simultaneously (figure 1.v.A) or sequentially (figure 1.v.B) depending on the membrane mechanical properties and the strength of disruptive forces. Indeed, a low separative force will induce slow cell deformation with progressive bond breakage, whereas higher forces would trigger rapid cell separation with possible membrane rupture (19) but minimal deformation due to the high cell surface viscosity.

3 - QUANTITATIVE FEATURES OF CELL STRUCTURE AND MOVEMENTS.

3.1 - A working model for the cell surface.

The most important prerequisite for a biophysical description of cell behaviour is the elaboration of a realistic model of cell structure. This must be done at the nanometer and the micrometer level.

Molecular model of the cell surface.

The basic structure of the cell plasma membrane is a phospholipid bilayer where intrinsic glycoproteins are embedded (figure 2). The protein/lipid ratio (weight/weight) is on the order of unity (20). Assuming that the bilayer is mainly composed of phosphatidylcholine (or similar molecules) and cholesterol, we consider a phospholipid/cholesterol ratio of 1/0.72 corresponding to a molecular area of 0.77 nm^2 for about 1,000 g lipids (21). Cell surface proteins may be modelled as globular units of about 50,000 molecular weight and 4 nm diameter. This figure is consistent with biochemical data (20) and the observation of intercalated particles on frozen-etched samples (22). The area occupied by a protein molecule is thus about 12.5 nm^2. The fraction of the membrane area occupied by protein molecules is therefore on the order of 25 % (i.e. 12.5/(12.5+50x0.77)). This is consistent with recent estimates on "molecular crowding on the cell surface" (17).

Now, cell surface carbohydrates may be modelled as follows : it is widely accepted that the fraction of the plasma membrane dry weight accounted for by carbohydrates is on the order of 10 % (20, 23). Most of these molecules are oligosaccharide chains shorter than about 20 residues (24) and linked to cell surface proteins or lipids. Since the length of a monosaccharide unit is about 0.5 nm (25), these chains are likely to remain concentrated in a region of about 10 nm thickness external to the phospholipid bilayer (region B shown on figure 2).

Further, it has been found that most -if not all - eukaryotic cells were surrounded by a polysaccharide-rich "fuzzy coat" or "glycocalyx" revealed by suitable staining, such as ruthenium red. Estimates for the thickness of this structure range between 50 and 500 nm (26, 27). The glycocalyx is mainly composed of glycosaminoglycans and proteoglycans and may be viewed as a network of polysaccharide chains between several hundreds and several thousands of residues long (28). It is difficult to obtain a quantitative estimate of the glycocalyx mass, since there may exist wide variations between the tightness of fixation of different molecules to the cell surface. Also, proteoglycans are synthetised and released by most cell types (28). It is therefore difficult to discriminate between loosely bound and free molecules. However, it is likely that the

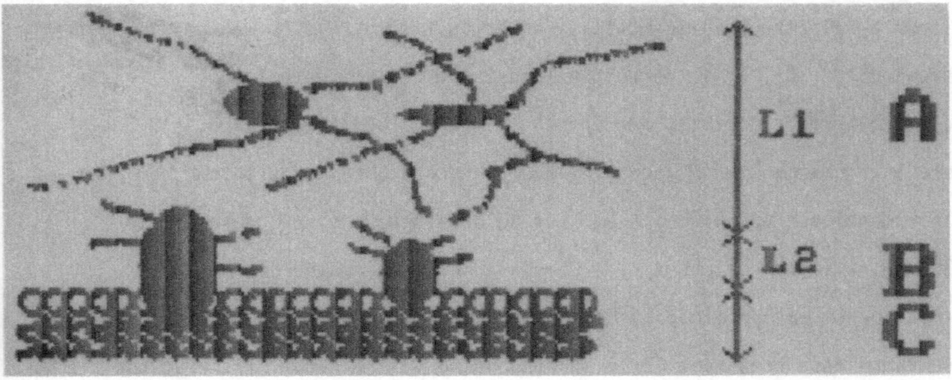

FIGURE 2 : MODEL OF THE CELL SURFACE
A : low density external part of the glycocalyx.
B : higher density proximal part of the cell coat.
C : Lipid bilayer.

polysaccharide chains of cell surface glycosaminoglycans and proteoglycans account for a small fraction of the total carbohydrate content (24). Hence, we shall tentatively represent the external part of the cell coat as a region of 100 nm width containing about 10 % of the cell surface sugars.

The above assumptions lead to the following quantitative description of cell surface carbohydrates :

- The 10 nm thick region external to the phospholipid bilayer (region B, see figure 2) should contain about 2 sugar residue per nm^2, or 0.2 residue/nm^3.
- The 90 nm thick outer region may contain about 0.2 residue per nm^2, i.e. 0.002 residue/nm^3.

The volume fraction of the glycocalyx may be derived by assuming that a saccharide residue is sixfold larger than a solvent molecule, which yields 3.6 % and 0.036 % for the volume fraction of carbohydrates in regions B and A of figure 2 respectively.

Electric charge.

It is well known that most living cells bear a net negative charge, as shown by electrophoretic mobility determinations (29). As was pointed out previously (30), only the most external charges are detected by electrophoretic mobility measurements. Hence, the value of about 10^7 electric charges per lymphocyte or erythrocyte obtained with the aforementioned techniques (29) must be considered as an underestimate. Indeed, a figure of about 2.4×10^7 sialic acid molecules per human erythrocyte was obtained by biochemical titration of molecules released by neuraminidase treatment (31), corresponding to a minimum of 1 electron charge per 6.8 nm^2. We shall retain this estimate for the charge density in the glycocalyx, wich amounts to assume that about one sugar residue out of 14 bears a net negative charge.

Description of the cell surface at the micrometer level.

Electron microscopic studies made it clear that nearly all cells (excepted erythrocytes) are studded with cylindrical or lamellar protrusions with linear size ranging between 0.1 and a few micrometers. The actual (or electron microscopic) membrane area is higher that the apparent (or light microscpical) area by about 50 % (32, 33). When a cell adheres to a substrate and spreads, its apparent area increases while its surface becomes smoother at the electron microscopic level. Morphometric data are consistent with the view that cells deform at constant volume and surface area (33, 34). Further, the observation that most cells are spherical when they are maintained in suspension, and the monitoring of the length of protrusions induced by prolonged aspiration of blood leukocytes with micropipettes led Evans to model these cells as viscous liquid droplets surrounded by an elastic membrane with an initial tension of about 10^{-5} N/m (34). The time required by a cell to recover a spherical shape after extensive deformation was on the order of several tens of seconds, in accordance with previous in vivo observations (35). Interestingly, these results were fairly consistent with previous studies made by Schmid-Schonbein et al. on leukocyte response to small deformations. When cells were modelled as homogeneous viscoelastic spheres, the elasticity and viscosity coefficients were on the order of 50 N/m^2 and 10 $N.second/m^2$ respectively. The authors noticed that a given cell yielded fairly similar material coefficients when sucked at different places, but large differences were found between cells. However, as pointed out by Evans, despite a fairly good agreement between theoretical and experimental deformation curves, the validity of the hypothesis of cell homogeneity was not proved since only limited deformations were studied, thus probing only the cortical cell region.

Now, the main problem remains to estimate the mechanical properties of cell surface asperities, since they are expected to play a major role in adhesion. Clearly, the aforementioned data are equally consistent with two different models :

- i) cell surface microvilli might be mainly responsible for mechanical resistance. The material constants derived from micropipette aspiration data might thus be used to estimate microvillus deformation during the earliest stages of adhesion. This model was used in a recent attempt at deriving adhesive interaction energies between rough biological surfaces (37).

ii) Alternatively, the cytoskeletal network parallel to the plasma membrane might account for the essential of cell surface mechanical properties et microvilli might be much less stiff than expected from model (i) (see figure 3).

At the present moment, no conclusive evidence can discriminate between both models. The key experiment would be to monitor local microvillus deformation after subjecting cells to a known force applied on a limited area of their surface.

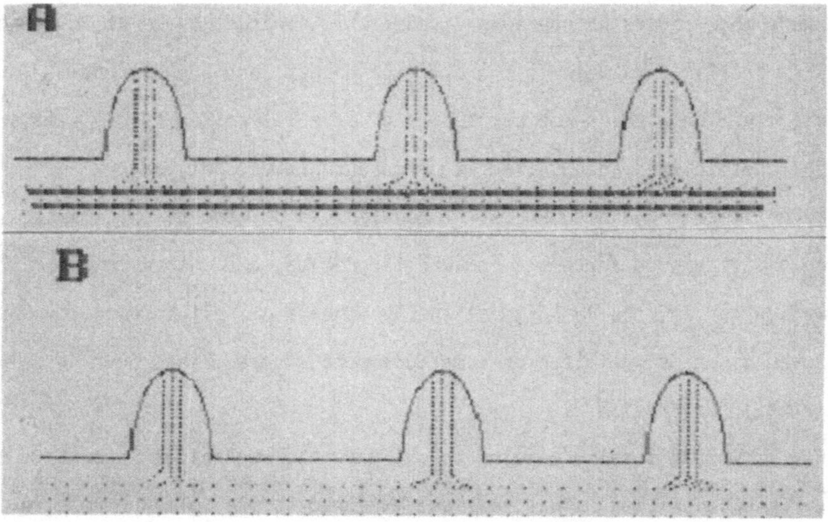

FIGURE 3 : MODELS FOR THE MECHANICAL PROPERTIES OF CELL SURFACE MICROVILLI
A : Main resistance due to cytoskeletal elements parallel to the plasma membrane.
B : Microvilli as stiff as the remaining part of the plasma membrane.

Another important point is that subjecting cells to mechanical stress was found to induce active movements within a few minutes (34, 38). Hence, it should be reminded that mechanical parameters derived from short term experiments (say less than one minute) may not be used to model long term responses to adhesive or mechanical stimulation, such as cell spreading.

The forces exerted by cells during active deformation were studied in different models : fibroblasts deposited on an elastic substrate exerted a traction of about 0.001 dyne per micrometer of "advanced margin" (39). Using a double chamber technique, Inouye and Takeushi found that the motive force of a dictyostelium slug was proportional to its volume with a coefficient of 5.9×10^7 newton/m^3 (40). The pressure gradient generated by a cylindrical pseudopod of a motile amoeba was $5x10^4$ N/m^3 for a pseudopod of 100 µm diameter and 0.2 cm length (41). The motive force generated at the ectoplasm/endoplasm boundary in a plant cell was about 0.16 N/m^2 (42). Unfortunately, it is not possible to extract a universal force coefficient from these data, due to the heterogeneity of cellular sizes and probably force generating mechanisms.

3.2 - Conditions of cell-cell encounter.

The importance of intercellular forces cannot be assessed unless we know which stress is exerted on biological surfaces under physiological conditions. Two limiting situations may be considered.

Flow conditions.

We shall only onsider blood flow, although shearing forces may occur in different tisues. It may be useful to recall some basic results from fluid mechanics : when a laminar shear flow is established in a cylindrical pipe, the fluid velocity is zero at the duct wall and maximum on the axis. The shear rate G is the derivative of the velocity with respect to the radial coordinate (i.e. the distance to the duct axis). G is expressed in $second^{-1}$. the wall shear stress is the (tangential) viscous force exerted on the wall by the fluid per unit area. It is equal to the product of the shear rate and the fluid viscosity (this is 10^{-3} poiseuille in pure water). Now, when a cell is in close contact with the wall, it is expected to roll with an angular velocity on the order of G (see e.g. 43 for more details).

The wall shear stress is on the order of 0.2 N/m^2 (G 1000 s^{-1}) in human arteries (44). In post-capillary venules, the velocity of rolling granulocytes is on the order of 100-500 µm/s, and the wall shear rate may be about 100 s^{-1} (45). The tangential force experienced by a spread cell of about 100 μm^2 area is therefore expected to range between $2x10^{-11}$ and $2x10^{-10}$ dyne (using a value of $2x10^{-3}$ poiseuille for the plasma viscosity (46)).

Resting conditions.

In absence of any hydrodynamic flow, a cell such as a granulocyte will move with a velocity of 10-20 µm/s (47). This is the order of magnitude of the rate of displacement of the protrusions sent by these cells (36). The motive force corresponding to a pressure of a few cm H_2O exerted by a cylinder of 0.1 µm radius (such as a microvillus) would be on the order of 10^{-9} newton. The contact time may be several seconds.

It must be pointed out that these conditions may be quite different from those used to perform agglutination experiments, using prolonged incubation after centrifugation.

4 - Quantitative importance of different parameters involved in adhesion.

Since our knowledge of the molecular structure of the cell surface remains quite imprecise, it would be meaningless to make use of sophisticated theoretical methods to achieve an accurate description of adhesion-involved phenomena. It seems more reasonable to estimate the order of magnitude of different parameters likely to play a role in adhesion with procedures simple enough to make apparent the consequences of the assumptions we introduced. The sequential steps of adhesion will be considered separately, although they probably interfere in many circumstances.

4.1 - Cell-to-cell approach.

Before short range intermolecular bonds are formed, several long range interactions are expected to occur when the intercellular distance decreases.

Hydrodynamic forces.

The decrease of the intercellular gap requires the expulsion of water molecules, which generates hydrodynamic repulsive forces. Much theoretical work was devoted to the determination of flow parameters (see 43 for review). Two limiting formulae may be useful. First, the repulsion between two smooth spheres of radius a and intermembrane distance D may be approximated as (43) :

$$F = 3\pi\mu a^2/D \quad dD/dt \qquad (1)$$

where μ is the medium viscosity and dD/dt is the relative cell velocity. Alternatively, if contact surfaces are planar with radius a, the repulsive force F is (16) :

$$F = 3\pi\mu a^4/(2D^3) \ dD/dt \qquad (2)$$

Clearly,the force is critically dependent on the shape of interacting surfaces. Indeed, the interaction between microvilli of 0.1μm radius would be one hundredfold higher if their tips were flattened instead of being spherical ! Considering spherical cells of 5μm diameter colliding in a flow of shear rate $100s^{-1}$, dD/dt would be on the order of 1,000 μm/s. Equation (2) then yields, taking a as 0.1 μm, D as 10 nm and μ as 0.001 poiseuille :

$$F = 4.7\ 10^{-10} \text{ newton} = 4.7\ 10^{-10} \text{ dyne}$$

Hence, hydrodynamic repulsion may play a role under high shear. As pointed out above, the precise shape of cell surface asperities may critically influence hydrodynamic repulsion. Also, it must be emphasized that the above estimate relied on a complete neglect of cell surface deformations, although these might increase repulsion (16). Finally, the local viscosity increase due to the glycocalyx (43) was ignored.

Molecular interactions between cell surface glycocalices.

When the distance between two cell membrane bilayers is on the order of the molecular size of glycocalyx molecules, several basic interactions may occur :

i) electrostatic repulsion between negative groups increases the system free energy.

ii) Electrodynamic attraction is expected, particularly between lipid bilayers.

iii) Polar or hydrophobic interactions are generated by close contact between surface bound molecules.

iv) Conformational changes of flexible surface molecules are expected to occur, with a balance between entropic and enthalpic changes.

v) When the interbilayer distance is lower than molecular size, surface molecules are compressed, with a concomitant reduction of the number of available configurations.

vi) some desorption of macromolecules weakly bound to the cell surface may occur if sufficient time is available.

vii) Finally, solute macromolecules may be expelled from the intercellular space, resulting in osmotic imbalance and intercellular attraction or repulsion depending on geometrical parameters.

Unfortunately, no satisfactory theoretical treatment is available to provide an integrated account of all aforementioned phenomena. The earliest biophysical approaches viewed cell interaction as a balance between electrostatic repulsion and electrodynamic attraction between charged surfaces (48). However, although this concept met with significant success in colloid science, it is too crude to account for actual membrane behaviour. First, the three-dimensional structure of cell boundaries must be incorporated in working models (49). Also, it was pointed out that repulsion between cell surface macromolecules as a consequence of (iii), (iv) and (v) might be the prominent interaction. This is called steric stabilization (50).

The reality of long range-repulsion between surfaces coated with polyelectrolytes (51) or neutral polymers (52) embedded in good solvents was demonstrated by direct experimental measurement with a sophisticated apparatus. The range of repulsion was about eight times the gyration radius of neutral polyethyleneglycol molecules (52). Further, the range of the repulsive forces between mica plates coated with 90,000 molecular weight polylysine was higher than 100 nm (51). Another important point was that when several approach-separation cycles of several hours were performed, marked hysteresis was observed (51). Also, repulsion was increased when the duration of the approach process was varied between 10 and 60 minutes (52). Hence, it is likely that initial intercellular repulsion cannot be viewed as an equilibrium phenomenon. It is therefore not warranted to use available thermodynamic methods (53) to determine the intercellular concentration profile of surface macromolecules in order to derive intercellular forces. Another point (54) is that current theories on polymer structure are not reliable when branched charged molecules are considered. Hence, what seems feasible at the present time is to obtain order-of-magnitude estimates for the importance of different mechanisms, in order to facilitate further choice of workable models of cell interaction.

i) Electrostatic repulsion.

In aqueous electrolytic solutions, the electrostatic interaction between charged particles is screened by both solvent molecules and solute ions. A reasonable formula for the interaction potential between charges q_1 and q_2 at distance r is :

$$W=q_1q_2 \exp(-\kappa r)/(4\pi\varepsilon r) \qquad (3)$$

where ε is the medium dielectric constant (ε=7.07 10^{-10} MKS units it water) and κ is the Debye Huckel reciprocal length ($1/\kappa$ = 0.8 nm in 0.15 M NaCl aqueous solution ; see e. g. 49 for more details). By integrating equation 3, it is found that in a medium of net electric charge density ρ, the energy of an electronic charge q (1.6 10^{-19} coulomb) is $q\rho/\varepsilon\kappa^2$. This quantity is lower than the thermal energy unit kT when ρ is lower than 2.9 10^7 C/m^3. since this value is much higher than the estimated volume charge in the cell glycocalyx, the effect of electrostatic forces on charge distribution may be neglected. The electrostatic energy per unit area of a plane parallel plate with constant charge density ρ and thickness d is :

$$W = (\rho^2/2\varepsilon\kappa^2).(d - (1-\exp(-\kappa d))/\kappa) \qquad (4)$$

If we deal with slices much thicker than $1/\kappa$,the above formula may be approximated as :

$$W = \rho^2 d/(2\varepsilon\kappa^2) \qquad (5)$$

Also, the interaction between adjacent slices may be neglected. Now, formula (5) may be used to estimate intercellular interaction by assuming constant charge density in layers A and B (figure 1). Further, when the interbilayer distance is lower than the unperturbed thickness l of a charged layer of density ρ, uniform compression is assumed, which amounts to replace ρ with $\rho l/d$. Finally, in absence of any reliable information on the shape of cell surface protrusions, interacting surfaces will be considered as planar. As shown on figure 4, five separate domains must be considered. Numerical values are shown on table 1. Clearly, the range of repulsion is much higher than that obtained with conventional charge distribution models.

Electrodynamic attraction.

Neglecting the contribution of low-density A and B region, we estimate electrodynamic attraction by modelling cell surfaces as homogeneous plates of thickness X. The potential at distance d is :

$$W = -(A/12\pi)(1/d^2-2/(d+X)^2+1/(d+2X)^2) \qquad (6)$$

Taking X as 7 nm and Hamaker constant A as 5 10^{-21} J (54), we conclude that electrodynamic attraction may be neglected in a wide range of intercellular distances (table 1).

Steric stabilization.

When the distance between two cells is progressively decreased, glycocalices may overlap. Further approach will result in polymer chain compression. Current polymer theories (55, 56) are of little use to calculate concentration profiles in the intercellular space for two main reasons : first, cell-to-cell approach may be too rapid to allow complete equilibrium. Secondly, surface molecules are probably subjected to strong geometrical constraints (e.g. as a consequence of their branched structure). However, local equilibrium may be obtainable. Hence, it seems reasonable to derive order-of-magnitude estimates for the system energy increase by neglecting long-range conformational changes (overlap regions of figure 4) or assuming uniform compression and calculating local free energy increases with classical expression for the free energy of mixing (57). Using s and w as labels for solute and solvent molecules respectively

$$F = kT\chi n_w v_s - k(n_w \ln v_w + n_s \ln v_s) \qquad (7)$$

where n is the number of molecules per unit volume, v is the volume fraction, and χ is the Flory interaction parameter (57). k is Boltzmann's constant (1.38 10^{-23} J/°K) and T is the absolute temperature. The last term clearly represents the entropy associated to the spatial dispersion of solute molecules independently of their conformation. This should be omitted since membrane-bound molecules have restricted translational freedom. Defining a_w and a_s as the molar volume of a water molecule and a solute monomers (a_s is close to 6.3 10^{-6} cm^3/mole for an hexasaccharide group (58)), we obtain :

$v_s = d_s\ a_s$; $n_w = (1-v_s)/a_w$

$$F = kT\chi d_s a_s(1-d_s a_s)/a_w + kT\ (1-d_s a_s)/a_w\ \ln(1-d_s a_s) \qquad (8)$$

This formula was used to estimate steric repulsion by considering the domains shown on figure 4. Taking a_w as 18 cm^3/mole, and 332 and 3.3 mole/m^3 for the monomer densities in regions B and A shown on figure 1, it is found that steric stabilization may efficiently impair cell-to-cell contact (table 1).

Depletion flocculation.

Biological media may contain substantial amounts of dissolved macromolecules. Due to steric interactions, cell plasma membranes are surrounded by a thin layer where the access of the centre of gravity of these macromolecules is somewhat restricted. When the intercellular distance is decreased enough that the "forbidden layers" surrounding two cells overlap, further decrease of this distance will increase the volume available for soluble macromolecules. This induces an attractive interaction that is fairly proportional to the solute concentration. The relevance of this phenomenon to colloid stability is suggested by theoretical and experimental data (59, 60). A very rough estimate may be obtained by considering that cells are surrounded by a depletion layer of thickness δ on the order of 5 nm (which corresponds to an average protein diameter). The interaction energy per unit area is:

$$W = \Pi\ (2\delta - d) \qquad (9)$$

where Π is the osmotic pressure contributed by excluded proteins and d is the thickness of the intercellular gap. Approximating Π as 5000 N/m^2 (whith is a reasonable order of magnitude of the contribution of serum albumin and immunoglobulins) and δ as 5 nm, numerical values of W may be obtained. As shown on table 1, depletion attraction should be of minor importance in biological media unless it is artificially increased in concentrated polymer solutions (e.g. dextran or polyethyleneglycol).

Short range interactions.

The simplified formulae we described may help understand long- and medium-range cell interactions. However, their validity becomes more and more questionable as the interbilayer distance is decreased below 5-10 nm due to the growing importance of the details of the cell surface. Perhaps other approaches, such as interfacial energy considerations (see the chapter by Pr. van Oss, this volume) may help estimate the energy of interaction between cell surfaces at close contact.

4.2 - Adhesion-related lateral movements of membrane molecules.

Many general arguments strongly suggest that some molecular sorting may occur within intercellular contact zones.

i) First, since the elaboration of the fluid mosaic model of plasma membranes by Singer and Nicolson (61), it has been extensively demonstrated that many membrane molecules are endowed with translational mobility (with a diffusion coefficient on the order of 10^{-10} - 10^{-12} cm^2/s (62)), although 10-40 % of these molecules seem fixed at the micrometer level when studied with photobleaching recovery (62) or post-electrophoresis relaxation (17). This heterogeneity was still observed when the movements of a well defined molecular species, such as membrane immunoglobulin E receptor, was studied (17).

ii) simple thermodynamic reasoning suggests that a significant modification of the local concentration of a given molecular species will occur in presence of an attractive or repulsive potential on the order of kT. Using 12.5 nm^2 as a reasonable value for the molecular area of membrane proteins, it is concluded that a repulsive potential higher than ca. 3.10^{-3} J/m^2 (i.e. $kT/12.5\ 10^{-18}$) may cause substantial molecule depletion in binding areas. As shown on table 1, comparable potentials can be generated by steric interactions. Since these interactions may be substantially influenced by nonspecific molecular properties, significant molecular sorting may be generated resulting in a difference between the membrane composition inside and outside contact areas. This phenomenon might be at least partly responsible for the difference between the structure of bulk plasma membranes and phagosome wall in endocytic cells (63, 64). Reciprocally, even weak nonspecific intermolecular forces (7) may induce substantial concentration of some molecular species in adhesion areas.

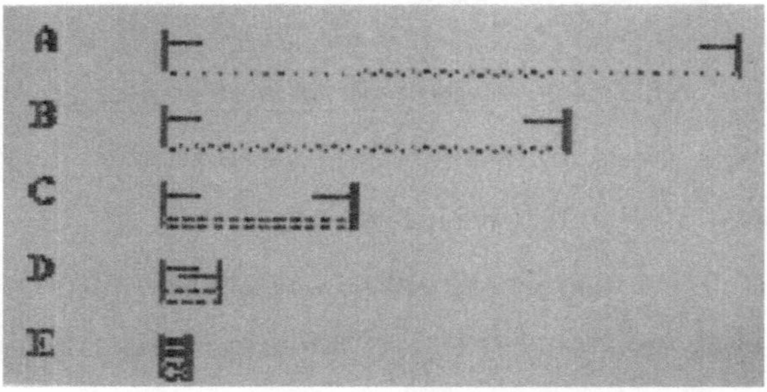

FIGURE 4 : STAGES OF INTERCELLULAR APPROACH
A : overlap of A regions (dotted lines). B: overlap of A regions with A and B regions on the other cell. C : compression of A regions. D : compression of A regions, overlap of B regions. E : compression of A and B regions.

TABLE 1
ESTIMATES FOR VARIOUS INTERCELLULAR POTENTIALS

Type of interaction	Energy (J/m^2) at distance 50 nm	20 nm	10 nm	5 nm
Repulsion				
Electrostatic	6.0 10^{-6}	1.8 10^{-5}	1.3 10^{-4}	3.6 10^{-4}
Steric (χ=0.4)	2.7 10-5	8.1 10-5	6.6 10-4	2.0 10-3
Steric (χ=0)	1.2 10^{-4}	3.8 10^{-4}	2.8 10^{-3}	8.0 10^{-3}
Steric (compression)	1.8 10^{-7}	9.3 10^{-7}	3.5 10^{-6}	1.4 10^{-5}
Attraction				
Electrodynamic	3.8 10^{-9}	8.2 10^{-8}	6.4 10^{-7}	3.8 10^{-6}

iii) Although external forces may trigger a spatial redistribution of membrane molecules, these movements are probably influenced by cytoskeletal elements. This concept is exemplified by the capping process : cross-linking of a population of cell surface molecules may induce passive formation of molecular aggregates, or "patches", followed by active gathering of these clusters into a "cap" (18, 65, 66 ; see the chapter by Dr. Burridge, this book, for more details on membrane-cytoskeletal interactions).

iv) Two additional factors may limit the concentration of ligand molecules into adhesive membrane areas : first, the total number of molecules (of ca. 12.5 nm^2 section area) on a microvillus tip of 0.01 μm^2 area cannot exceed about 800. Hence, steric exclusion sets a drastic limit on the total number of adhesion molecules in contact zones before any deformation occurred. Secondly, when extensive intermembrane contact is achieved (which was shown to occur by numerous electron microscopic studies, see e. g. 37) substantial depletion of some molecular species in free areas may impair further increase of their concentration in contact zones. The thermodynamic basis of these phenomena was described in previous reports (67, 68).

Experimentals.

Recently, the rapid development of quantitative microscopy allowed direct determination of the spatial distribution of surface molecules on cells engaged in adhesive interactions. Rat basophil leukemia cells were coated with fluorescent anti-dinitrophenol (DNP) IgE antibodies and manoeuvered into contact with DNP-coated lipid vesicles. Within a few minutes, the fluorescence intensity increased in the contact zone, with a plateau at 150-200 % of the brightness of cell regions located at 90° to the adhesion area (69). In another study, the interaction between a cytolytic T lymphocyte clone and target tumor cells was studied (70, 71).

Different surface antigens with or without a probable role in adhesion (LFA-1, CD8/Lyt2, class I histocompatibility antigens) were labelled with monoclonal antibodies and cell conjugates were studied with quantitative fluorescence microscopy and electron microscopy. Local fluorescence intensities were determined in order to derive the density of labelled molecules, using electron microscopy to correct for the effect of local membrane deformation on fluorescence distribution. The density of adhesion molecules might be as high as 66 % in adhesive regions, but it could not be determined whether this increase was localized in restricted areas such as the tips of microvilli or the margin of lamellipodia. Interestingly, some redistribution of molecules without any probable role in adhesion was observed. Finally, double labelling experiments revealed a significant correlation between the distribution of adhesion molecules and submembrane polymerized actin. The aforementioned data support previous qualitative evidence (72) and strongly favor the incorporation of lateral molecule diffusion in any model of cell-cell adhesion.

4.3 - Adhesion-related deformations of the plasma membrane.

Definitions.

Clearly, the establishment of intercellular bonds requires some kind of membrane deformation to allow the formation of a sufficient contact area. Some remarks may be useful to clarify this point : imagine a collision between two cells maintained in suspension (figure 5), in a first approximation, corresponding to optical microscopic resolution, these may be viewed as smooth spheres (figure 5A), and some flattening is required to allow the formation of an apparent contact zone. This involves an apparent area increase since the cellular volume remains constant (see above).

Now, as shown on figure 5B, is is likely that i) the actual cell area remains constant and the apparent area increase results from a smoothening of the cell surface (32, 33), ii) in the apparent contact zone, only a fraction of the total cell membrane area is involved in bonding, which defines the <u>actual contact zone</u>. Although there is presently no rigorous way of measuring the <u>actual contact area</u>, electron microscopy usually allows clear identification of a well-defined zone where the electron-dense contours of the membranes of interacting cells are parallel and separated by a gap of less than 50 nm apparent width. It is <u>suggested</u> that these are the actual contact regions. Note that the width of the visible gap is different from the actual intermembranar distance for two main reasons : a) the fixation procedure used to process samples for electron microscopy may modify the glycocalyx thickness. However, this should not drastically alter molecular sizes (73). b) When the plane of section of the sample is not perpendicular to the contact surfaces, the apparent intercellular gap is wider than the actual one. Indeed, the average value of the apparent intercellular gap (as determined on electron micrographs) is expected to be twice higher than the actual gap width (74).

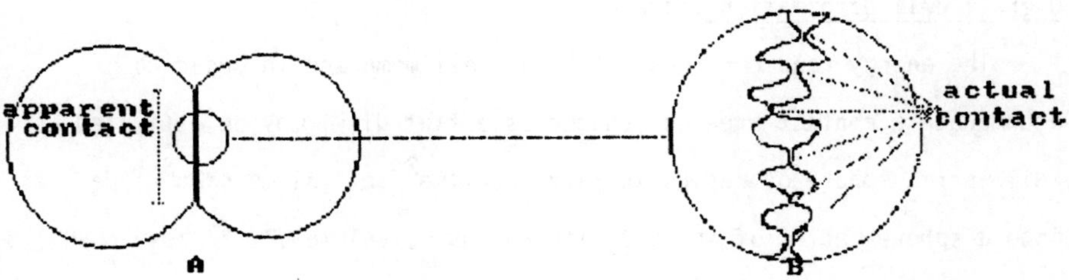

<u>FIGURE 5</u> : INTERCELLULAR CONTACT REGIONS
A : optical microscopy - apparent contact (thick line).
B : electron microscopy (enlargment of a part of the apparent contact zone) : visualization of actual contacts.

Quantitative description of the extent and energetics of cell deformation.

A simple theory of adhesion-induced cell deformation would be to view the membrane shape as a consequence of a balance between adhesive forces and passive mechanical resistance of the cell surface. The intercellular affinity might thus be derived from quantitatve data on cell deformation and deformability. However, the validity of this model cannot be tested. Indeed, this is expected to fail in some cases since external stress (75) or receptor stimulation (76) may induce cell deformation.

Another finding in support of the importance of active cell deformation during adhesion is that in some cases, when cytotoxic T lymphocytes were made to bind target tumor cells, the former were found more resistant to mechanical deformation than the latter (by micropipette aspiration) but they displayed more important smoothing than the latter in adhesive zones (74), thus suggesting that active deformation is a step of adhesion. Hence, biophysical estimates can only tell us what resistance adhesive forces should oppose in absence of active cellular deformation, in contrast with the amount of information obtained by the authors who studied adhesions between artifical vesicles (56).

Overall cell deformation (figure 5A).

The energy required to stretch the cell membrane in order to accomodate a contact area Ac (shaped as a flat disk) may be estimated as follows : a sphere of radius Ro, area $Ao=4\pi Ro^2$ and volume $4\pi Ro^3/3$ deforms into a sphere sector of angle θ with radius R, volume $\pi R^3(2+3\cos\theta-\cos\theta^3)/3$ and area $\pi R^2(2+2\cos\theta+\sin\theta^2)$. Assuming constant volume, we obtain :

$$R/Ro= (4/(2-3\cos\theta+\cos\theta^3))^{1/3} \qquad (10)$$

the relative area increase is therefore :

$$I=(R/Ro^2)(1+\cos\theta+\sin\theta^2/2)-1 \qquad (11)$$

Neglecting bending resistance, the energy required for membrane stretching is :

$$W = \int_0^I Ao.i.T(i)\ di \qquad (12)$$

where T(i) is the membrane tension when the relative area increase is i.T(i) was determined in several experimental models (37, 74). For rat peritoneal macrophages, we obtained :

$$T(i) = 0.0664\ i^2 + 0.00355\ i\ J/m^2 \qquad (13)$$

Using (12) and (13), it is found that the energy required to accomodate a contact area of $5\mu m^2$ (which is a convenient order of magnitude : 37, 74) is $1.7x10^{-16}$ joule, i.e. $3.4x10^{-5}$ J/m^2 of contact area. Hence, in absence of any active cell deformation, about 5,000 weak interactions (5 kcal/mole each) would be required to overcome cell mechanical stiffness and accomodate the apparent contact zone.

Local membrane smoothening.

Obviously, the formation of a binding area between a cell and a flat surface requires some smoothing of cell membrane asperities. This requirement is less apparent when the interaction between the surfaces of two rough cells is considered. This point was adressed experimentally by studying the interaction between target tumor cells and cytotoxic T lymphocytes (13, 73, 77) or LAK cells (77). Conjugates were studied with electron microscopy and the cells contours were digitized and analyzed with a computer program allowing local determination of the cell surface roughness : this was defined as a relative length excess, i.e. the ratio between the length of a membrane segment and the length of a circular line enclosing the same cytoplasm area. Ten series of comparisons were performed between the free and bound areas of cytotoxic lymphoid cells, and targets that were conjugated under different experimental conditions (74, 77). A total of more than 350 cell contours were examined. In three comparisons out of ten, the binding areas were significantly ($P < 0.05$) smoother than free regions. In seven cases, no significant difference was found between free and bound areas. Also, in a series of 103 conjugates, the width of the intercellular gap was measured. It was found that only 30-50 % of the length of cell contours in apparent contact areas (figure 5B) were adjacent to an intercellular gap that was 50 nm wide or less (this limit of 50 nm was used to define close contact zones). Further, when 8 series of comparisons were performed, in a single case a significant negative correlation was found between the roughness parameter of the killer cell surface in binding areas and the percent of this area involved in close contact.

In conclusion, extensive electron microscopic studies of intercellular contacts support the view that local membrane smoothing is associated to the formation of intercellular bonds. However, the importance and generality of this finding cannot be assessed with the methodology we used. More refined morphometric analyses are required to clear this point.

5 - Conclusion and perspectives.

Cell adhesion may be the first important cell function that will be amenable to realistic and testable quantitative modeling. However, much work remains to be done before this goal is reached. Some conclusions and unanswered problems may be emphasized.

First, it seems generally accepted that nonspecific forces are generated during cell-to-cell approach. Several theoretical tools were worked out to deal with electrodynamic attraction, electrostatic repulsion, steric stabilization as well as short range interfacial energies. Some of these are described in other chapters of this book. Also, new experimental methods (particularly Israelachvili's balance apparatus (51, 52) and Evans' micromechanical approach (56)) allowed direct test of numerical calculations. To our view, the main gap to be filled before these advances are relevant to cell-cell interaction is the building of a realistic model of the cell surface and surrounding atmosphere. Also, it would be of obvious interest to extend the aforementioned experimental approaches to artificial surfaces more closely related to cell membranes than polymer-coated mica surfaces or lipid vesicles. However, the estimates summarised on table 1 strongly suggest that significant repulsion may indeed arise from the interaction between cell surface-bound macromolecules.

Secondly, it is well demonstrated that cell adhesion results in extensive changes of the local composition of plasma membranes, as a consequence of lateral movement of surface and cytoskeletal molecules. However, the present picture remains quite imprecise, ant it would be useful to know whether these molecules are actually involved in adhesion or they are only passive bystanders. More experimental information is needed to refine fluorescence data (e. g. confocal microscopy or immuno electron microscopy). Some problems concerning label accessibility and interference between molecule labelling and function may have to be overcome for this purpose.

Thirdly, although much indirect evidence supports the view that substantial changes of local membrane roughness are associated to adhesion, some progress is needed before the importance of these phenomena can be assessed : first, more refined methods are required to <u>describe</u> cell surface microasperities, and the mechanical properties of the cell membrane must be studied at the submicrometer level.

Fourthly, little information is available on the intermembrane distance in adhesion areas (with an accuracy of a few nanometers). At the present moment, only electron microscopy yielded some information on the distance separating the membranes of bound cells. It would be important to develop different methodologies to address this problem in order to evaluate the possibility of artifact generation during sample fixation. Perhaps fluorescence methods might be useful in this respect.

Finally, detailed kinetic studies on the correlation between attachment and structural changes could be useful to assess the functional importance of these changes and check theoretical results and methods.

REFERENCES

1 - Gershey, E. L. & d'Alisa, R. M. (1980). Characterization of a CV1 cell cycle. II - The role of cell substrate attachement. J. Cell Sci. 42 : 357-365.

2 - Chin, Y. H., Carey, G. D. & Woodruff, J. J. (1983). Lymphocyte recognition of lymph node endothelium. J. Immunol. 131 : 1368-1374.

3 - Struncke, R. C., Kunke, K. S., Hollister, J. R. & Musson, R. A. (1982). J. Reticuloendothelial Soc. 32 : 297-309.

4 - Hynes, R. O. & Yamada, K. M. (1982). Fibronectins, multifunctional modular glycoproteins. J. Cell Biol. 95 : 369-377.

5 - Springer, T. A., Dustin, M. L., Kishimoto, T. K. & Marlin, S. D. (1987). The lymphocyte function-associated LFA-1, CD2 and LFA-3 molecules : cell adhesion receptors of the immune system. Ann. Rev. Immunol. 5 : 223-252.

6 - Grumet, M., Rutishauser, U. & Edelman, G. M. (1982). Neural cell adhesion molecule is on embryonic muscle cells and mediates adhesion to nerve cells in vitro. Nature 295 : 693-685

7 - Lindahl, U. & Hook, M. (1978). Glycosaminoglycans and their binding to biological macromolecules. Annu. Rev. Biochem. 47 : 385-417.

8 - Capo, C., Garrouste, F., Benoliel, A. M., Bongrand, P., Ryter, A. & Bell, G. I. (1982). Concanavalin A-mediated thymocyte agglutination - A

model for a quantitative study of cell adhesion. J. Cell Sci. 56 : 21-48.

9 - Sommers, S. D., Whisnant, C. C. & Adams, D. O. (1986). Quantification of the strength of cell-cell adhesion : the capture of tumor cells by activated muring macrophages proceeds through distinct stages. J. Immunol. 136 : 1490-1496.

10 - Grinnell, F. (1974). Studies on the mechanism of cell attachment to a substratum : evidence for three biochemically distinct processes. Arch. Biochem. Biophys. 160 : 304-310.

11 - Van Oss, C. J., Gillman, C. F. & Neumann, A. W. (1975). Phagocytic engulfment and cell adhesiveness. Marcel Dekker, New York.

12 - Capo, C., Bongrand, P., Benoliel, A. M. & Depieds, R. (1979). Nonspecific recognition in phagocytosis : ingestion of aldehyde-treated erythrocytes by rat peritoneal macrophages. Immunology 36 : 501-508.

13 - Mège, J. L., Capo, C., Benoliel, A. M., Foa, C., Galindo, R. & Bongrand, P. (1986) Quantification of cell surface roughness : a method for studying cell mechanical and adhesive properties. J. Theor. Biol. 119 : 147-160.

14 - Chow, I. & Poo, M. M. (1982). Redistribution of cell surface receptors induced by cell-cell contact. J. Cell Biol. 95 : 510-518.

15 - Van Oss, C. J., Chaudury, M. K. & Good, R. J. (1987). Monopolar surfaces. Adv. Coll. Surface Sci. 28 : 35.

16 - Dimitrov, D. S., Stoicheva, N. & Stefanova, D. (1984). A simple interpolation formula for the rate of approach of particles or cells with tension controlled shapes at arbitrary separations. J. Coll. Interface Sci. 98 : 269-271.

17 - Ryan, T. A., Myers, J., Holowka, D., Baird, B. & Webb, W. W. (1988). Molecular crowding on the cell surface. Science 239 : 61-64.

18 - Flanagan, J. & Kack, G. L. E. (1978). Cross-linked surface Ig attaches to actin. Nature 273 : 278-281.

19 - Bongrand, P. & Golstein, P. (1983). Reproducible dissociation of

cellular aggregates with a wide range of calibrated shear forces : application to cytolytic-lymphocyte target cell conjugates. J. Immunol. Methods 58 : 209-224.

20 - Harrison, R. & Lunt, G. G. (1975). Biological Membranes. Blackie, Glasgow.

21 - Levine, Y. K. (1972). Physical studies of membrane structure. Prog. Biophys. Mol. Biol. 24 : 1-74.

22 - Gomperts, B. (1977). The plasma membrane. Models for structure and function. Academic Press, London.

23 - Cook, G. M. W. (1976). Techniques for the analysis of membrane carbohydrates. in : Biochemical analysis of membranes (A. H. Maddy ed.) Chapman & al., London, 283-335.

24 - Vogel, K. G. & Kelley, R. O. (1977). Cell surface glycosaminoglycans : identification and organization in cultures human embryo fibroblasts. J. Cell Physiol. 92 : 469-480.

25 - Kabat, E. A. (1968). Structural concepts in immunology and immunochemistry. Holt, Rinehart & Winston,New York.

26 - Winzler, R. J. (1970). Carbohydrates in cell surfaces. Int. rev. cytol. 29 : 77-125.

27 - Montesano, R., Mossaz, A., Ryser, J. E., Orci, L., & Vassalli, P. (1984). Leukocyte interleukins induce cultured endothelial cells to produce a highly organized glycosaminoglycan-rich pericellular matrix. J. Cell Biol. 99 : 1706-1715.

28 - Kraemer, P. M. (1979). Mucopolysaccharides : Cell Biology and Malignancy. in : Surfaces of normal and malignant cells (R. O. Hynes ed.), Wiley, New York

29 - Mehrishi, J. N. (1972). Molcular aspects of the mammalian cell surface. Prog. Biophys. Mol. Biol. 25 : 3-70.

30 - Bongrand, P., Capo, C. & Depieds, R. (1982). Physics of Cell

Adhesion. Prog. Surface Sci. 12 : 217-285.

31 - Eylar, E. H., Madoff, M. A., Brody, O. V. & Oncley, J. L. (1962). The contribution of sialic acid to the surface charge of the erythrocyte. J. Biol. Chem. 237 : 1992-2000.

32 - Knutton, S., Sumner, M. C. B. & Pasternak, C. A. (1970). Role of microvilli in surface changes of synchronized P815 mastocytoma cells. J. Cell Biol. 66 : 568-575.

33 - Erickson, C. A. & Trinkaus, J. P. (1976). Microvilli and blebs as source of reserve surface membrane during cell spreading. Exp. Cell Res. 99 : 375-384.

34 - Evans, E. & Kukan, B. (1984). Passive material behavior of granulocytes based on large deformation and recovery after deformation tests. Blood 64 : 1028-1035.

35 - Bagge, U. & Branemark, P. I. (1977). White blood cell rheology. Adv. Microcirc. 7 : 1-17.

36 - Schmid-Schonbein, G. W., Sung, K. L. P., Tozeren, H., Skalak, R. & Chien, S. (1981). Passive properties of human leukocytes. Biophys. J. 36 :243-256.

37 - Mège, J. L., Capo, C., Benoliel, A. M. & Bongrand, P. (1987). Use of cell contour analysis to evaluate the affinity between macrophages and glutaraldehyde-treated erythrocytes. Biophys. J. 52 : 177-186..

38 - Tickle, C. & Trinkaus, J. P. (1977). Some clues as to the formation of protrusions by fundulus deep cells. J. Cell Sci. 26 : 139-150.

39 - Harris, A. K., Wild, P. & Stopak, D. (1980). Silicone rubber substrata : a new wrinkle in the study of cell locomotion. Science 208 : 177-179.

40 - Inouye, K. & Takeuchi, I. (1980). Motive force of the migrating pseudoplasmodium of the cellular slime mould dictyostelium discoideum. J. Cell Sci. 41 : 53-64.

41 - Allen, R. D. & Roslansky, J. D. (1959). The consistency of amoeba cytoplasm and its bearing on the mechanism of amoeboid movement. I - An analysis of endoplasmic velocity of chaos chaos. J. Biophys. Biochem. Cytol. 6 : 437-446.

42 - Kamiya, N. & Kuroda, K. (1973). Dynamics of cytoplasm streaming in a plant cell. biorheology 10 : 179-187.

43 - Bongrand, P., Capo, C., Mège, J. L., & Benoliel, A. M. (1988). Use of hydrodynamic flows to study cell adhesion. in : Physical Basis of Cell-Cell Adhesion (P. Bongrand ed.) CRC Press, Boca Raton 125-156.

44 - Franke, R. P., Grafe, M., Schmittler, H., Seiffe, D. & Mittermayer, C. (1984). Induction of human vascular endothelial stress fibres by fluid shear stress. Nature 307 : 648-649.

45 - Asherton, A. & Born, G. V. R. (1973). Relationship between the velocity of rolling granulocytes and that of the blood flow in venules. J. Physiol. 233 : 157-165.

46 - Weiss, L. & Dimitrov, D. S. (1984). A fluid mechanical analysis of the velocity, adhesion and destruction of cancer cells in capillaries during metastases. Cell Biophys. 6 : 9-22.

47 - Lackie, J. M. & Wilkinson, P. C. (1984). Adhesion and locomotion of neutrophil leukocytes on 2D substrate and 3D matrices in : White cell mechanics - Basic Science and Clinical Aspects (H. J. Meiselman, M. A. Lichtman & P. L. LaCelle eds) A. R. Liss, New York.

48 - Curtis, A. S. G. (1967). The cell surface : its molecular role in morphogenesis. Logos Academic Press, New York, 1967.

49 - Bongrand, P. (1988). Intermolecular forces in : Physical Basis of Cell-Cell Adhesion (P. Bongrand, ed.) CRC Press, Boca RAton, 1-37.

50 - Napper, D. H. (1977). Steric stabilization. J. Colloid Interface Sci. 58 : 390-407.

51 - Luckham, P. F. & Klein, J. K. (1984). Forces between mica surfaces bearing adsorbed polyelectrolytes. J. Chem. Soc. Faraday Trans.

1:80 865-878.

52 - Luckham, P. F. & KLein, J. (1985). Interactions between smooth solid surfaces in solutions of adsorbing and non adsorbing polymers in good solvent conditions. Macromolecules 18 : 721-728.

53 - de Gennes, P. G. (1988). Polymer solutions near an interface. 1 - Adsorption and depletion layer. Macromolecules 14 : 1637-1644.

54 - Bongrand, P. & Bell, G. I. (1984). Cell-Cell Adhesion : parameters and possible mechanisms in : Cell Surface Dynamics (A. S. Perelson, C. DeLisi & F. W. Wiegel, eds) Marcel Dekker, New York. 459-493.

55 - de Gennes, P. G. (1988). Model polymers at interfaces in : Physical Basis of Cell-Cell Adhesion (P. Bongrand ed.) CRC Press, Boca Raton, 39-60.

56 - Evans, E. & Needham, D. (1987). Long range interactions between lipid bilayers in salt solutions and solutions of non-adsorbant polymers : comparison of mean field theory with direct measurements. J. Chem. Soc. Faraday Trans.

57 - Flory, P. J. (1953). Principles of Polymer Chemistry. Cornell University Press, Ithaca. 508-512.

58 - Weast, R. C. & Astle, M. J. (1983). Handbook of Chemistry and Physics. CRC Press, Boca Raton. D235.

59 - Feigin, R. I. & Napper, D. H. (1980). Depletion stabilization and depletion flocculation. J. Coll Interface Sci. 75 : 525-541.

60 - Sperry, P. R. (1982). A simple quantitative model for the volume restriction flocculation of latex by water soluble polymers. J. Coll. Inteface Sci. 87 : 375-384.

61 - Singer, S. J. & Nicolson, G. L. (1972). The fluid mosaic model of the structure of cell membranes. Science 175 : 720-731.

62 - Schlessinger, J., Elson, E. L., Webb, W. W., Yahara, I., Rutishauser, U. & Edelman, G. M. (1977). Receptor diffusion on cell

surfaces moedulated by locally bound concanavalin A. Proc. Natl. Acad. Sci. (U.S.A.) 74 : 1110-1114.

63 - Tsan, M. F. & Berlin, R. D. (1972). Effect of phagocytosis on membrane transport of non electrolytes. J. Exp. Med. 134 : 1016-1033.

64 - Berlin, R. D. & Fera, J. P. (1977). Changes in membrane microviscosity associated with phagocytosis : effect of colchicin. Proc. Natl. Acad. Sci. (U.S.A.) 74 : 1072-1076.

65 - Ault, K. A. & Unanue, E. R. (1978). Lymphocyte Surface immunoglobulin and disposition on the membrane in : Theoretical Immunology (G. I. Bell, A. S. Perelson & G. H. Pimbley, eds) Marcel Dekker, New York. 145-169.

66 - Burn, P., Kupfer, A. & Singer, S. J. (1988). Dynamic membrane-cytoskeletal interactions : specific association of integrin and talin arises in vivo after phorbol ester treatment of peripheral blood lymphocytes. Proc. Natl. Acad. Sci. (U.S.A.) 85 : 497-501.

67 - Bell, G. I., Dembo, M. & Bongrand, P. (1984). Cell adhesion - Competition between nonspecific repulsion and specific bonding. Biophys. J. 45 : 1051-1064.

68 - Bell, G. I. (1988). Models of cell adhesion involving specific binding in : Physical Basis of Cell-Cell Adhesion (P. Bongrand, ed.) CRC Press, Boca Raton, 227-259.

69 - McCloskey, M.S A. & Poo, M. M. (1986). Contact-induced redistribution of specific membrane components : local accumulation and development of adhesion. J. Cell Biol. 102 : 2185-2196.

70 - André, P., Capo, C., Benoliel, A. M., Buferne, M. & Bongrand, P. (1989). Analysis of the topological changes induced on cells exposed to adhesive or mechanical stimuli. Cell Biophys. in press.

71 - André, P., Benoliel, A. M., Capo, C., Foa, C., Buferne, M., Schmitt-Verhulst, A. M. & Bongrand, P. (1990). Use of digital image processing to analyze the spatial organization of conjugates made between target tumor cells and a cytotoxic T lymphocyte clone. submitted for

publication.

72 - Kupfer, A. Singer, S. J., Janeway, C. A. & Swain, S. L. (1987). Coclustering of CD4 (L3T4) molecule with the T-cell receptor is induced by specific direct interaction of helper T cells and antigen presenting cells. Proc. Natl. Acad. Sci. 84 : 5888-5892.

73 - Easty, G. C. & Mercer, E. H. (1962). An electron microscope study of model tissues formed by the agglutination of erythrocytes. Exp. Cell Res. 28 : 215-227.

74 - Foa, C., Mège, J. L., Capo, C., Benoliel, A. M., Galindo, J. R. & Bongrand, P.S (1988). T cell-mediated cytolysis : analysis of killer and target cell deformability and deformation during conjugate formation. J. Cell Sci. 89 : 561-573.

75 - Kolega, J. (1986). Effect of mechanical tension on protrusive activity and microfilament and intermdiate filament organization in a epidermal epithelium moving in culture. J. Cell biol. 102 : 1400-1411.

76 - Haston, W. S. & Wilkinson, P. C. (1987). Gradient perception by neutrophil leucocytes. J. Cell Sci. 87 : 373-374.

77 - Foa, C., Galindo, R., Benoliel, A. M., Capo, C., Mege, J. L., Bongrand, P., Leroux, J. Y. & Oth, D. (1987). Study of CTL and LAK contacts to target cells after treatment with mitomycin C and adriamycin. Immunol. Invest 16 : 241-262.

Membrane-cytoskeletal interactions in cell adhesion and locomotion

Keith Burridge and Winthrop T. Jackman
Department of Cell Biology and Anatomy
University of North Carolina at Chapel Hill
Chapel Hill, North Carolina 27599-7090

How cells move in tissue culture continues to command the attention of cell biologists. Although many advances have been made in understanding the biochemistry and organization of various cytoskeletal components, the role these have in generating movement has not been established. Very different theories have been proposed to explain the migration of cultured cells (for example, compare Bretscher 1987, 1988, with Bray and White, 1988). In spite of major disagreements about how the force for cell movement is generated, most theories recognize that cell migration involves three distinct phases: extension of the cell membrane, adhesion of this newly extended region to the underlying substrate and then the bulk translocation of the cytoplasm in the direction of the cell extension. Much of the work in this laboratory has been aimed at understanding cell adhesion. Cells in culture can develop at least two types of adhesion to the underlying substrate. There are broad areas of weak adhesion known as close contacts, in which the plasma membrane is separated from the substrate by about 30-100 nm (Izzard and Lochner, 1976). Some cells also develop smaller, discrete regions that are more tightly adherent and which have a separation of 10-15 nm between the plasma membrane and the substrate (Izzard and Lochner, 1976). These latter have been known as focal adhesions, focal contacts or adhesion plaques and serve to anchor bundles of actin filaments (stress fibers) at their cytoplasmic face (Abercrombie et al., 1971; Heath and Dunn, 1978). By comparison, much more is known about the structure and composition of focal adhesions than close contacts. In this review we will discuss briefly the organization of focal adhesions and their relationship to cell movement as well as to theories of cell migration. The organization and properties of focal adhesions have been discussed in greater detail elsewhere (Burridge, 1986; Burridge et al., 1988; Woods and Couchman, 1988).

Organization of focal adhesions

Cells such as fibroblasts develop focal adhesions when plated on a surface that has adsorbed a component of the extracellular matrix (ECM), such as fibronectin (FN). Normally when cells are grown in the presence of serum, their adhesion to the coverslip or culture dish is promoted by adsorption of either FN or vitronectin (VN), two ECM proteins present in serum. The requirement for an adsorbed layer of ECM for formation of focal adhesions suggests a relationship between these structures and the adhesive junctions cells make to the ECM *in vivo*. This relationship is supported further by their similar morphology and composition. Members of the integrin family of ECM receptors are clustered within focal adhesions, spanning the plasma membrane and providing a link between the ECM and the intracellular cytoskeleton. The particular integrin concentrated in a focal adhesion is specified by the ECM component adsorbed to the substrate (Singer et al., 1988; Dejana et al., 1988; Fath et al., 1989). Thus cells cultured on FN express the FN receptor in their focal adhesions, whereas cells plated on VN or grown in serum concentrate the VN receptor preferentially in their focal adhesions.

Much remains to be learned about how integrins link to the bundles of actin microfilaments. Progress has been made in identifying cytoplasmic components of focal adhesions and a number of interactions have been identified through *in vitro* binding assays. Based on these binding experiments, it is possible to construct a potential chain of proteins linking actin filaments to integrins in the plasma membrane at focal adhesions (figure 1). Undoubtedly this diagram is far from complete since several other proteins have been identified in focal adhesions, but these are not included in the diagram due to the limited information about them. We also suspect that other components remain to be identified. Roles in attaching actin filaments to the membrane at these sites have been suggested for vinculin, talin and α-actinin, although this function has not been proved for any of these proteins. A low affinity interaction between talin and integrin has been identified (Horwitz et al., 1986). Talin also interacts with vinculin (Otto, 1983; Burridge and Mangeat, 1984) and, in turn, vinculin has been shown to bind to α-actinin (Wachsstock et al., 1987),itself an actin-binding protein. The low affinity of some of these interactions suggests the importance of other interactions yet to be identified. Moreover, there are cases where these proteins may associate with sites of actin-membrane interaction independently of each other (Geiger et al., 1985; Kupfer et al., 1986). It seems likely that there may be some redundancy occurring at focal adhesions with multiple modes of attachment linking actin filaments to the membrane. Much effort is currently directed towards this problem.

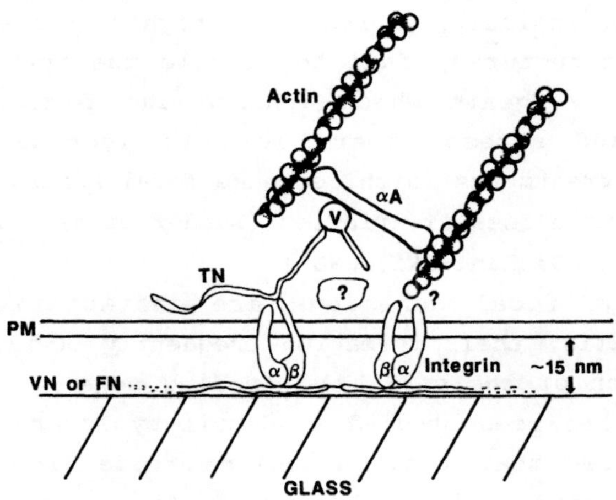

Figure 1. A diagram showing some of the interactions that have been determined *in vitro* for proteins in focal adhesions. Fibronectin (FN) or vitronectin (VN) are shown adsorbed to the glass. These ECM proteins are bound by integrin receptors composed of α and ß chains, both of which span the plasma membrane (PM). At the cytoplasmic face, integrin binds talin (TN), which in turn binds vinculin (V). Vinculin is shown interacting with α-actinin (α-A), a protein which crosslinks actin filaments. Several focal adhesion proteins have not been included because their interactions have not been determined. The figure is reproduced with permission from Burridge et al., 1988.

Focal adhesions and cell migration

Cells that are highly migratory in tissue culture typically reveal close contacts but lack both focal adhesions and stress fibers (Couchman and Rees, 1979; Kolega et al., 1982). Focal adhesions are most prominent in the least motile fibroblasts and it seems that the high adhesion of these discrete regions retards cell migration. This idea is supported by the increased motility of cells in which focal adhesions have been disrupted, for example by tumor promoters (Dugina et al., 1987; Danowski and Harris, 1988). Focal adhesions appear to be the sites where tension generated in stress fibers is transmitted to the substrate. This has been demonstrated by culturing cells on flexible rubber substrates which wrinkle under sufficient tension (Harris et al., 1980). Cells that develop focal adhesions and stress fibers wrinkle these flexible rubber substrates extensively, whereas more highly motile cells, which lack these structures, fail to wrinkle the rubber (Harris et al., 1981). Agents which enhance the formation of focal adhesions and stress fibers lead to increased wrinkling. Conversely, treatments which decrease focal adhesions and stress fibers lead to a loss of wrinkles (Leader et al., 1983; Danowski and Harris, 1988; Danowski, 1989).

Although focal adhesions are characteristic of non-migratory cells, their formation frequently occurs in the most motile region of the cell, the leading edge. Focal adhesion development has been studied in detail by Izzard and coworkers who have shown that focal adhesions arise from a structural precursor, a bundle of actin filaments (a microspike) within the leading edge (DePasquale and Izzard, 1987; Izzard, 1988). This actin bundle contains talin at its tip. Presumably a link between this microfilament bundle and the membrane is triggered in response to integrin receptors being aggregated by the ECM.

Focal adhesions and models of cell migration

From the preceding discussion it should be clear that focal adhesions and their associated stress fibers have little if any role in propelling cells across a substrate. However, focal adhesions and tension generation by microfilaments are relevant to various theories of cell movement. In the rest of this paper we wish to discuss the relationship of focal adhesions to the rearward movement of aggregated ("patched") surface components. The movement of surface aggregates or particles away from the leading edge of a cell is a general phenomenon, often referred to as "capping". Patching and capping were first described in lymphocytes (Taylor et al., 1971), but equivalent movements occur in other motile cells. Patching is a passive process involving the aggregation of surface components in the plane of the membrane, brought about by an antibody, lectin or other multivalent ligand. Capping is an active process, which in lymphocytes involves translocation of the patches to one pole of the cell. It is generally agreed that the rearward movement of patches on the surface of cells is related closely to the forward movement of cells across a substrate. Consequently, capping provides a model for studying cell movement. It is in this context that focal adhesions are relevant to cell migration. Their formation involves the aggregation of integrins by the ECM and tension is exerted at these sites.

Although numerous theories have been advanced to explain capping, they can be divided into two classes according to the source of the motive force generating the movements. Most of the theories envisage a link between surface patches and the actin cytoskeleton, with the movement of patches being driven by a sliding filament, actomyosin-type mechanism, similar to that found in muscle (Bourguignon and Singer, 1977; Bourguignon and Bourguignon, 1984). At the opposite extreme, Bretscher has proposed a membrane flow model where the motive force comes not from the interaction of actin and myosin but rather from the exocytosis/endocytosis pathway (Bretscher 1976; 1984; 1987). In the latter model, a flow of membrane components is generated as a result of polarized insertion of new and recycled membrane at the leading edge of the cell and membrane uptake all over the cell surface via coated pits. Most membrane proteins will maintain a random distribution because their rapid diffusion in the plane of the membrane will tend to overcome the rearward flow. Following crosslinking, however, the rate of diffusion of components will decrease significantly and, as a result, they will be caught up in the rearward flow and carried backwards.

In support of the membrane flow model proposed by Bretscher, insertion of new membrane components has been observed at the leading edge of migrating cells (Bretscher, 1983; Bergmann et al., 1983; Singer and Kupfer, 1986). This model predicts that anything should cap in a motile cell if it has a sufficiently slow diffusion rate. Consistent with this prediction, foreign antigens that have been introduced into cell membranes cap in response to their aggregation (Stern and Bretscher, 1979; Wolf et al., 1980). Similarly, capping of aggregated lipids in the outer bilayer of the plasma membrane has been noted (Stern and Bretscher, 1979; Kellie et al., 1983; Spiegel et al., 1984), as has the capping of various truncated membrane proteins with short or absent cytoplasmic domains (Laub et al., 1981; Murre et al., 1984; Edidin and Zuniga, 1984). The capping of these various components is easily accommodated by Bretscher's membrane flow model, but it is difficult, although not impossible, to envisage how such components may link to the cytoskeleton.

A further prediction of this membrane flow model is that surface components that are being endocytosed rapidly and recycled back to the surface will tend to be more concentrated on large cells close to their site of insertion, i.e. at the leading edge. Such a distribution has been observed for both the transferrin and LDL receptors, two components that are recycled rapidly (Bretscher, 1983). Conversely, proteins that do not recycle but remain on the surface for prolonged periods would be expected to show a reverse gradient being most concentrated towards the rear of the cell, away from the site of new membrane insertion. Again, such a distribution has been observed (Ishihara et al., 1988), but it should be noted that these authors suggested a very different explanation for this gradient (see also the article in this volume by Lee et al.)

The absence of a role for actin and myosin in a model of cell movement seems counterintuitive, but gains credence from motile cells which lack these proteins or contain only mutant forms. The amoeboid sperm of C. elegans, for example, contain negligible amounts of actin and myosin (Nelson et al., 1982) and yet are highly motile. Evidence has been presented for a rearward flow of membrane in the pseudopod of this sperm, together with insertion of new membrane components at the tips of the pseudopodial projections (Roberts and Ward, 1982a; Pavalko and Roberts, 1987). These results and the lack of actin and myosin have led those working on this unusual cell to suggest that it may use membrane flow for its propulsion (Roberts and Ward, 1982b).

A second example of cell migration continuing despite a defective or absent myosin heavy chain has provoked much interest. This situation was generated experimentally in the amoebae of Dictyostelium discoideum, either by antisense RNA (Knecht and Loomis, 1987) or by transfection and homologous recombination with a truncated gene encoding just part of the myosin heavy chain (DeLozanne and Spudich, 1987). In both cases, the resulting amoebae were unable to undergo normal cleavage of daughter cells during cell division, but, perhaps surprisingly, were still capable of some migration and of phagocytosis. These observations have been interpreted as further evidence for motility being driven by a mechanism that does not involve actin and myosin (Bretscher, 1988). However, a subsequent paper investigating the migration of these mutant amoebae in detail found that their motility was abnormal in several ways (Wessels et al., 1988). This would suggest a modified conclusion that myosin is required for these movements to occur normally. It should be noted that a second form of myosin has been detected in D. discoideum (Coté et al. 1985) and it will be interesting to determine what role this protein has in their movement.

Various lines of evidence support cytoskeletal models of capping. There have been many reports describing the co-localization of actin, myosin and various actin-associated proteins with patches and caps (for review, see Bourguignon and Bourguignon, 1984). Some differences, however, have been reported in the literature with, for example, myosin associating with caps of some lymphocyte surface proteins but not with others (Braun et al., 1978a). For some components it has been shown that crosslinking by antibodies converts these surface proteins into a detergent-insoluble state, presumed to reflect an induced association with the cytoskeleton (Braun et al., 1982; Woda and McFadden, 1983; Turner and Shotton, 1987). Such an association has been demonstrated in a few cases biochemically (Flanagan and Koch, 1978; Condeelis, 1979; Bourguignon et al., 1985). Additional evidence supporting cytoskeletal models for capping has derived from the inhibition of capping of many surface molecules with drugs such as the cytochalasins which inhibit actin polymerization (Taylor et al., 1971; Schreiner and Unanue, 1976; Jackman and Burridge, 1989). However, some of the drug studies have been confusing and inhibition of capping has not been a universal observation (Sundqvist and Ehrnst, 1976; Williams et al., 1977).

For cytoskeletal models of capping, focal adhesions are very relevant. Not only can these be considered as patches of integrin crosslinked by their binding to the ECM, but they are obvious regions of interaction with actin filaments (for review, see Burridge et al., 1988). Tension is transmitted to focal adhesions and their rearward movement would seem to be prevented only because they adhere too strongly to the substrate. Indeed, there are circumstances when these patches are actively moved rearward together with the associated ECM. This is observed when cells are plated on FN-coated coverslips in the presence of serum. Under these conditions, the FN is less adherent and many cells will move it in association with the FN receptor and other components of focal adhesions towards the cell body away from the motile edge (Avnur and Geiger, 1981; Grinnell, 1986; Fath and Burridge, unpublished observations). Here is a case where the translocation of a particular type of patch appears to be driven by a cytoskeletal mechanism.

Focal adhesions contain integrins that have been aggregated presumably as a consequence of their binding to the ECM, but patching and capping of these receptors has also been induced experimentally with antibodies and cells in suspension (Burn et al., 1987; Pavalko et al., 1989). Surprisingly, talin was not associated with these aggregated integrins unless the cells were treated with phorbol esters, agents which stimulate protein kinase C (Burn et al., 1987; Pavalko et al., 1989). This suggests that the association of some of the cytoskeletal proteins with integrins in focal adhesions may involve signals besides the aggregation of integrins in the plane of the membrane. In general, proteins such as talin and vinculin do not appear to be associated with patches or caps of most surface molecules. However, certain other microfilament-associated proteins, such as fodrin (nonerythrocyte spectrin), have been identified with many types of patched and capped surface molecules (Levine and Willard, 1983; Nelson et al., 1983; Bourguignon et al., 1985; Turner et al., 1988). Patches of different surface components may utilize different intermediates for linking to the actin cytoskeleton. There is much evidence indicating that actin interacts with the plasma membrane in multiple ways.

Strong support for a cytoskeletal basis of capping comes from the rearward flow of actin that has been observed in many cell types (Dunn, 1980; Heath, 1983a; Fisher et al., 1988; Forscher and Smith, 1988). Some models for capping and cell movement envisage a membrane flow, but one that is driven by the rearward movement of the cortical microfilament network rather than by the pathway of endocytosis and exocytosis (Bray and White, 1988). In migrating fibroblasts, this rearward movement of actin is seen as a movement of "arcs", bundles of microfilaments that form just behind the leading edge and which sweep back in waves to a region in front of the nucleus (Heath, 1983a,b). Patches have been observed to move rearwards in association with these arcs. With cells in suspension, including lymphocytes, constriction rings rich in actin filaments have also been observed to move backwards (Haston and Shields, 1984). Polymerization of actin at the leading edge (Wang, 1985; Small, 1981) may contribute to the rearward flow of cortical microfilaments. However, although this might extend individual actin filaments a few microns, it fails to account for the distance covered, for example, by arcs in fibroblasts, nor does it account for the movement of actin filaments in arcs that are oriented at right angles to their direction of movement. Furthermore, the inhibition of actin polymerization by cytochalasin failed to stop the rearward movement of actin filaments in growth cones (Forscher and Smith, 1988). It seems probable that, after their polymerization, actin filaments are driven rearwards in the cortex by their interaction with myosin. It has been suggested that myosin generates this flow by creating tension throughout the cortex that is strongest at the rear and weakest at the leading edge, where myosin is usually absent (Bray and White, 1988). Counterbalancing the rearward flow of cortical actin filaments, there must be a forward flow of actin monomers or short oligomers in the more central regions of the cytoplasm. The corollary of actin polymerization at the leading edge must be its disassembly at more distal sites.

In many of the cytoskeletal models, polymerization of actin at the leading edge provides the motive force for cell extension. This may be accompanied by other forces which extend the membrane such as osmotic or hydrostatic pressure within the cell (Oster and Perelson, 1987). It should be noted that myosin plays little, if any, role in cell extension in many of these models, and this may provide one explanation for why the myosin-defective mutants of D. discoideum are still capable of some movement, particularly extension. The lack of a need for myosin in the migratory movements of fibroblasts has been demonstrated by microinjection of antibodies which perturb myosin filament formation and function (Honer et al., 1988). Many of the cells injected with anti-myosin showed increased migration and developed bizarre morphologies, with the motile front of the cell often becoming very extensive and migrating away from the rest of the cell body. Similar morphologies have also been seen in response to tumor promoters (Schliwa et al., 1984; Dugina et al., 1987; Danowski and Harris, 1988). These agents have been shown to decrease cell contractility (Danowski and Harris, 1988) consistent with the observation in vitro that phosphorylation of myosin by protein kinase C decreases its activity (Nishikawa et al. 1985). This suggests to us that a major function of myosin in these cells is to maintain the tension in the cell cortex that keeps the migratory part of the cell, the motile leading lamella, together with the cell body. Inhibiting this function in fibroblasts leads to the migratory front of the cell moving away from the rest of the cell in an uncoordinated fashion. Normally the tension between the cell body and the motile front of the cell will contribute to translocating the bulk of the cell, but it will also tend to restrain the migration of the leading edge. Consistent with this interpretation is the observation that the strength of contractility of cultured cells, as judged by their ability to wrinkle flexible substrates, is inversely proportional to their migratory behavior (Harris et al., 1981).

From the above experiments, it can be concluded that myosin activity is unnecessary for the extension phase of cell migration. It will be important to explore the consequences of myosin deletion or inactivation on both the cortical flow of microfilaments and the rearward movement of patches. Such experiments should reveal whether myosin has a role in generating these movements.

Summary and Concluding remarks

Focal adhesions are of interest not only to those studying the interaction of cells with the ECM, but also to those concerned with the mechanisms of cell movement, in spite of being present in the least motile of cells. They are sites where cells transmit force to the substrate and their lack of movement reflects the strength of their adhesion to the coverslip or culture dish. With respect to the models of capping of aggregated surface components, they provide a compelling example of attachment of microfilaments to clustered receptors (integrins) in the plasma membrane. However, the details of this attachment have not been worked out fully and are probably complex. In addition, some of the proteins involved in this attachment appear to be specialized for this family of receptors and are not associated with other aggregated surface molecules.

It is perhaps surprising that the debate between the different theories for cell movement and capping has not been resolved. Some of the protagonists in this debate hold the view that a single mechanism must underly all amoeboid movement, from amoebae to fibroblasts (for example, see Bretscher, 1988). We find, however, that there is evidence to support both theories and come to the conclusion that both mechanisms may operate in some cells. Given the variability of cells we would not be surprised if some had harnessed one mechanism for their movement, whereas others had adopted a different strategy. The amoeboid sperm of C. elegans would seem a likely candidate for movement being driven by membrane exocytosis and recycling, whereas the movement of A. proteus would seem to depend predominantly on the activity of its cytoskeletal and contractile proteins. In many cells both mechanisms may operate simultaneously but to varying degrees. The existence of two mechanisms for lymphocyte capping was suggested previously (Braun et al., 1978a,b), based on the patching and capping of different surface antigens. These could be classified into two groups which differed in their association with the cytoskeleton and in their response to various inhibitory drugs. Two mechanisms operating simultaneously within some cells would also account for the apparently conflicting evidence that has made this field so confusing. With the increasing ability to inhibit the activity of specific proteins in cells by molecular genetics or microinjection of appropriate probes, it should be possible in the near future to settle this question and to determine whether two mechanisms for generating movement co-exist within the same cell.

Acknowledgements

We thank our colleagues for their helpful advice and comments. We are particularly grateful to Dr. Barbara Danowski for many valuable discussions. The patience of Ms. Betty Stewart who prepared this manuscript is gratefully acknowledged. The authors were supported by NIH grants GM 29860 and CA 42771.

References

Abercrombie, M., Heaysman, J., Pegrum, S.M. (1971) The locomotion of fibroblasts in culture. Exp Cell Res 67:359-367

Avnur, Z., Geiger, B. (1981) The removal of extracellular fibronectin from areas of cell-substrate contact. Cell 25:121-132

Bergmann, J.E., Kupfer, A., and Singer, S.J. (1983) Membrane insertion at the leading edge of motile fibroblasts. Proc Natl Acad Sci 80:1367-1371

Bourguignon, L.Y.W. and Bourguignon, G.J. (1984) Capping and the cytoskeleton. Int Rev Cytol 87:195-224

Bourguignon, L.Y.W. and Singer, S.J. (1977) Transmembrane interactions and the mechanism of capping of surface receptors by their specific ligands. Proc Nat Acad Sci 74:5031-5035

Bourguignon, L.Y.W., Suchard, S.J., Nagpal, M.L. and Glenney, J.R. (1985) A T-lymphoma transmembrane glycoprotein (gp180) is linked to the cytoskeletal protein, fodrin. J Cell Biol 101:477-487

Braun, J., Fujiwara, K., Pollard, T.D. and Unanue, E.R. (1978a) Two distinct mechanisms for redistribution of lymphocyte surface macromolecules. 1. Relationship to cytoplasmic myosin. J. Cell Biol 79:409-418

Braun, J., Fujiwara, K., Pollard, T.D. and Unanue, E.R. (1978b) Two distinct mechanisms for redistribution of lymphocyte surface macromolecules. 2. Contrasting effects of local anesthetics and a Calcium ionophore. J Cell Biol 79:419-426

Braun, J., Hochman, P.S. and Unanue, E.R. (1982) Ligand-induced association of surface immunoglobulin with the detergent-insoluble cytoskeletal matrix of the B-lymphocyte. J. Immunol 128:1198-1204

Bray, D. and White, J.G. (1988) Cortical flow in animal cells. Science 259:883-888

Bretscher, M.S. (1976) Directed lipid flow in cell membranes. Nature 260:21-23

Bretscher, M.S. (1983) Distribution of receptors for transferrin and low density lipoprotein on the surface of giant HeLa cells. Proc Natl Acad Sci 80:454-458

Bretscher, M.S. (1984) Endocytosis: Relation to capping and cell locomotion. Science 224:681-686

Bretscher, M.S. (1987) How animal cells move. Sci Am 257:72-90

Bretscher, M.S. (1988) Fibroblasts on the move. J Cell Biol 106:235-237

Burn, P., Kupfer, A., Singer, S.J. (1988) Dynamic membrane-cytoskeletal interactions: specific association of integrin and talin arises <u>in vivo</u> after phorbol ester treatment of peripheral flood lymphocytes. Proc Natl Acad Sci USA 85:497-501

Burridge, K. (1986) Substrate adhesions in normal and transformed fibroblasts: organization and regulation of cytoskeletal, membrane and extracellular matrix components at focal contacts. Cancer Reviews 4:18-78

Burridge, K., Fath, K., Kelly, T., Nuckolls, G., Turner, C. (1988) Focal adhesions: transmembrane junctions between the extracellular matrix and the cytoskeleton. Ann Rev Cell Biol 4:487-525

Burridge, K., Mangeat P. (1984) An interaction between vinculin and talin. Nature 308:744-746

Condeelis, J. (1979) Isolation of Concanavalin A caps during various stages of formation and their association with actin and myosin. J Cell Biol 80:751-758

Couchman, J.R., Rees, D.A. (1979) The behavior of fibroblasts migrating from chick heart explants: changes in adhesion, locomotion and growth, and in the distribution of actomyosin and fibronectin. J Cell Sci 39:149-165

Coté, G.P., Albanesi, J.P., Ueno, T., Hammer, J.A. and Korn, E.D. (1985) Purification from Dictyostelium discoideum of a low-molecular-weight myosin that resembles myosin I from Acanthamoeba castellanii. J Biol Chem 260:4543-4546

Danowski, B.A. (1989) Fibroblast contractility and actin organization are stimulated by microtubule inhibitors. J Cell Sci, in press

Danowski, B.A. and Harris, A. (1988) Changes in fibroblast contractility, morphology and adhesion in response to a phorbol ester tumor promoter. Exp Cell Res 177:47-59

Dejana, E., Colella, S., Conforti, G., Abbadin, M., Gaboli, M. and Marchisio, P.C. (1988) Fibronectin and vitronectin regulate the organization of their respective arg-gly-asp adhesion receptors in cultured human endothelial cells. J Cell Biol 107:1215-1223

DeLozanne, A. and Spudich, J.A. (1987) Disruption of the Dictyostelium myosin heavy chain gene by homologous recombination. Science 236:1086-1091

DePasquale, J.A., Izzard, C.S. (1987) Evidence for an actin-containing cytoplasmic precursor of the focal contact and the timing of incorporation of vinculin at the focal contact. J Cell Biol 105:2803-2809

Dugina, V.B., Svitkina, T.M., Vasiliev, J.M. and Gelfand, I. (1987) Special type of morphological reorganization induced by phorbol ester: reversible partition of cells into motile and stable domains. Proc Natn Acad Sci USA 84:4122-4125

Dunn, G.A. (1980) Mechanisms of fibroblast locomotion. In Cell Adhesion and Motility, pp 409-423. Curtis, A.S. and Pitts, J.D., Eds. (Cambridge Univ Press New York)

Edidin, M. and Zuniga, M. (1984) Lateral diffusion of wild-type and mutant L^d antigens in L cells. J Cell Biol 99:2333-2335

Fath, K.R., Edgell, C-J.S. and Burridge, K. (1989) The distribution of distinct integrins in focal contacts is determined by the substratum composition. J Cell Sci 92:67-75

Fisher, G.W., Conrad, P.A., DeBiasio, R.L. and Taylor, D.L. (1989) Centripetal transport of cytoplasm, actin, and the cell surface in lamellipodia of fibroblasts. Cell Motil Cytoskel 11:235-247

Flanagan, J. and Koch, G.L.E. (1978) Cross-linked surface Ig attaches to actin. Nature 273:278-281

Forsher, P. and Smith, S.J. (1988) Actions of cytochalasins on the organization of actin filaments and microtubules in a neuronal growth cone. J Cell Biol 107:1505-1516

Geiger, B., Volk, T., Volberg, T. (1985) Molecular heterogeneity of adherens junctions. J Cell Biol 101:1523-1531

Grinnell, F. (1986) Focal adhesion sites and the removal of substratum-bound fibronectin. J Cell Biol 103:2697-2706

Harris, A.K., Stopak, D., Wild, P. (1981) Fibroblast traction as a mechanism for collagen morphogenesis. Nature 290:249-251

Harris, A.K., Wild, P., Stopak, D. (1980) Silicone rubber substrata: a new wrinkle in the study of cell locomotion. Science 208:177-179

Haston, W.S. and Shields, J.M. (1984) Contraction waves in lymphocyte locomotion. J Cell Sci 68:227-241

Heath, J.P. (1983a) Direct evidence for microfilament-mediated capping of surface receptors on crawling fibroblasts. Nature 302:532-534

Heath, J.P. (1983b) Behavior and structure of the leading lamella in moving fibroblasts. J Cell Sci 60:331-354

Heath, J.P., Dunn, G.A. (1978) Cell to substratum contacts of chick fibroblasts and their relation to the microfilament system. A correlated interference-reflexion and high voltage electron microscope study. J Cell Sci 29:197-212

Honer, B., Citi, S., Kendrick-Jones, J. and Jockusch, B.M. (1988) Modulation of cellular morphology and locomotory activity by antibodies against myosin. J Cell Biol 107:2181-2189

Horwitz, A., Duggan, K., Buck, C., Beckerle, M.C., Burridge, K. (1986) Interaction of plasma membrane fibronectin receptor with talin - a transmembrane linkage. Nature 320:531-533

Ishihara, A., Holifield, B. and Jacobson, K. (1988) An analysis of the lateral redistribution of a major plasma membrane glycoprotein which occurs during locomotion. J Cell Biol 106:329-343

Izzard, C.S. (1988) A precursor of the focal contact in cultured fibroblasts. Cell Motil Cytoskel, 10:137-142

Izzard, C.S., Lochner, L.R. (1976) Cell-to-substrate contacts in living fibroblasts: an interference reflexion study with an evaluation of the technique. J Cell Sci 21:129-159

Jackman, W.T. and Burridge, K. (1989) Polymerization of additional actin is not required for capping of surface antigens in B-lymphocytes. Cell Motil Cytoskel 12:23-32

Kellie, S., Patel, B., Pierce, E.J. and Critchley, D.R. (1983) Capping of cholera toxin-ganglioside GMI complexes on mouse lymphocytes is accompanied by co-capping of -actinin. J Cell Biol 97:447-454

Knecht, D.A. and Loomis, W.F. (1987) Antisense RNA inactivation of myosin heavy chain gene expression in Dictyostelium discoideum. Science 236:1081-1086

Kolega, J., Shure, M.S., Chen, W., Young, N.D. (1982) Rapid cellular translocation is related to close contacts formed between various cultured cells and their substrata. J Cell Sci 54:23-34

Kupfer, A., Singer, S.J., Dennert, G. (1986) On the mechanism of uni-directional killing in mixtures of two cytotoxic T-lymphocytes. J Exp Med 163:489-498

Laub, F., Kaplan, M. and Gitler, C. (1981) Actin polymerization accompanies Thy-1 capping on mouse thymocytes. FEBS Lett. 124:35-38

Leader, W.M., Stopak, D., Harris, A.K. (1983) Increased contractile strength and tightened adhesions to the substratum result from reverse transformation of CHO cells by dibutyryl cyclic adenosine monophosphate. J Cell Sci 64:1-11

Levine, J. and Willard, M. (1983) Redistribution of fodrin (a component of the cortical cytoplasm) accompanying capping of cell surface molecules. Proc Natn Acad Sci USA 80:191-195

Murre, C., Reiss, C.S., Bernabeu, C., Chen, L.B., Burakoff, S.J., and Seidman, J.G. (1984) Construction, expression and

recognition of an H-2 molecule lacking its carboxyl terminus. Nature 307:432-436
Nelson, G.A., Roberts, T.M. and Ward, S. (1982) Caenorhabditis elegans spermatozoan locomotion: ameboid movement with almost no actin. J Cell Biol 92:121-131
Nelson, W.J., Colaco, C.A.L.S. and Lazarides, E. (1983) Involvement of spectrin in cell-surface receptor capping in lymphocytes. Proc Natl Acad Sci 80:1626-1630
Nishikawa, M., Sellers, J.R., Adelstein, R.S. and Hidaka, H. (1984) Protein kinase C modulates in vitro phosphorylation of the smooth muscle heavy meromyosin by myosin light chain kinase. J Biol Chem 259:8808-8814
Oster, G.F. and Perelson, A.S. (1987) The physics of cell motiliity. J Cell Sci Suppl 8:35-54
Otto, J.J. (1983) Detection of vinculin-binding proteins with an ^{125}I-vinculin gel overlay technique. J Cell Biol 97:1283-1287
Pavalko, F.M. and Roberts, T.M. (1987) Caenorhabditis elegans spermatozoa assemble membrane proteins onto the surface at the tips of pseudopodial projections. Cell Motil Cytoskel 7:169-177
Pavalko, F.M., Turner, C.E. and Burridge, K. (1989) Regulation of actin-membrane interactions at focal adhesions of cultured cells. Adv Prot Phosphatases 5:279-298
Roberts, T.M. and Ward, S. (1982a) Membrane flow during nematode spermatogenesis. J Cell Biol 92:113-120
Roberts, T.M. and Ward, S. (1982b) Centripetal flow of pseudopod surface components could propel the ameboid movement of Caenorhabditis elegans spermatozoa. J Cell Biol 92:132-138
Schliwa, M., Nakamura, T., Porter, K.R., Euteneuer, V. (1984) A tumor promoter induces rapid and coordinated reorganization of actin and vinculin in cultured cells. J Cell Biol 99:1045-1059
Schreiner, G.F. and Unanue, E.R. (1976) Membrane and cytoplasmic changes in B lymphocytes induced by ligand-surface immunoglobulin interaction. Adv Immunol 24:37-165
Singer, I.I., Scott, S., Kawka, D.W., Kazazis, D.M., Gailit, J. and Ruoslahti, E. (1988) Cell surface distribution of fibronectin and vitronectin receptors depends on substrate composition and extracellular matrix accumulation. J Cell Biol 106:2171-2182
Small, J.V. (1981) Organization of actin in the leading edge of cultured fibroblasts: influence of osmium tetroxide and dehydration on the ultrastructure of actin meshworks. J Cell Biol 91:695-705
Speigel, S., Kassis, S., Wilchek, M. and Fishman, P.H. (1984) Direct visualization of redistribution and capping of fluorescent gangliosides on lymphocytes. J Cell Biol 99:1575-1581
Stern, P.L. and Bretscher, M.S. (1979) Capping of exogenous Forssman glycolipid on cells. J Cell Biol 82:829-833
Sundqvist, K. and Ehrnst, A. (1976) Cytoskeletal control of surface membrane mobility. Nature 264:226-231
Taylor, R.B., Duffus, W.P.H., Raff, M.C. and dePetris, S. (1971) Redistribution and pinocytosis of lymphocyte surface immunoglobulin molecules induced by anti-immunoglobulin antibody. Nature New Biol 233:225-229

Turner, C.E., Newton, M.R. and Shotton, D.M. (1988) Cytoskeletal involvement in the sequential capping of rat thymocyte surface glycoproteins. J Cell Sci 89:309-319

Turner, C.E. and Shotton, D.M. (1987) Isolation and initial biochemical characterisation of caps of two major rat thymocyte glycoproteins: Evidence for the involvement of a 205 K Con A binding protein and cytoskeletal components in capping. Cell Motil Cytoskel 8:37-43

Wachsstock, D.H., Wilkins, J.A., Lin, S. (1987) Specific interaction of vinculin with alpha-actinin. Biochem Biophys Res Comm 146:554-560

Wang, Y-L. (1985) Exchange of actin subunits at the leading edge of living fibroblasts: possible role of treadmilling. J Cell Biol 101:597-602

Wessels, D., Soll, D.R., Knecht, D., Loomis, W.F., DeLozanne, A. and Spudich, J. (1988) Cell motility and chemotaxis in Dictyostelium amebae lacking myosin heavy chain. Dev Biol 128:164-177

Williams, D.A., Boxer, L.A., Oliver, J.M. and Baehner, R.L. (1977) Cytoskeletal regulation of Concanavalin A capping in pulmonary alveolar macrophages. Nature 267:255-257

Woda, B.A. and McFadden, M.L. (1983) Ligand-induced association of rat lymphocyte membrane proteins with the detergent-insoluble lymphocyte cytoskeletal matrix. J Immunol 131:1917-1919

Wolf, D.E., Henkart, P. and Webb, W.W. (1980) Diffusion, patching and capping of stearoylated dextrans on 3T3 cell plasma membranes. Biochem 19:3893-3904

Woods, A. and Couchman, J.R. (1988) Focal adhesions and cell-matrix interactions. Collagen Rel Res 8:155-183

The behavior of cell surface components during cell locomotion: A quantitative approach

J. Lee, A. Ishihara[+], B. Holifield and K. Jacobson
Department of Cell Biology and Anatomy
University of North Carolina at Chapel Hill,
Chapel Hill, NC 27599. USA

INTRODUCTION

The precise mechanism of cell locomotion is poorly understood despite extensive research into this area. The main reason for this is that it is a very complex phenomenon involving many interrelated cellular events. In order to move, a cell must first adhere to the substratum and then generate enough force to drive the cell forward. In addition, sustained movement in a particular direction always requires the polarisation of cells into a 'front' and 'rear' end. Despite experimental advances and progress in understanding these events, the molecular mechanism of cell locomotion and its control remain largely unknown. The purpose of this article is to describe how the experimental approach to cell locomotion has changed and to illustrate this with examples of recent work in this area from our laboratory.

PREVIOUS APPROACH TO THE STUDY OF CELL LOCOMOTION

Earlier investigations of cell locomotion were based extensively on the movement of fibroblasts on a glass coverslip. Numerous morphological observations were made of these cells using the light microscopy and time-lapse cinemicroscopy. These studies were important in describing the changes in gross morphology and especially the complex

---

[+]MRC Cell Biophysics Unit
26-29 Drury Lane
London WC2B 5 RL

movements of the cell membrane that occur during locomotion (Abercrombie 1970,; Harris, 1973; Trinkaus, 1984).

A more detailed view of moving cells has been obtained by the combined use of interference reflection microscopy (IRM) and differential interference contrast microscopy (DIC). This has revealed the nature and distribution of cell-substratum contacts. For the first time, it was possible to see that the formation and dissolution of cell-substratum contacts is a continuous process during locomotion (Lochner & Izzard, 1973; Abercrombie & Dunn, 1977; Izzard & Lochner, 1980).

Although EM studies have also revealed cell-substratum contacts and other fine structural details of motile cells, they provide no direct information about the dynamics of these structures during locomotion. Biochemical studies have determined the molecular composition of focal contacts (Burridge et al., 1988,) but our understanding of how they are assembled and disassembled during locomotion is just developing (DePasquale and Izzard, 1988).

The increasing use of fluorescence and biochemical techniques have led studies of cell motility to the molecular level. They have begun to reveal the molecular composition of cytoskeletal elements, focal adhesions and the cell membrane. However, one of the most important applications of these techniques is to allow the observation of these molecules in living cells. Fluorescently-labelled antibodies and fluorescent analogs of cytoskeletal proteins can be used to visualise single classes of molecules, allowing their distribution and movements to be observed in living cells over extended periods of time (Wang, 1984; DeBiasio et al., 1988; Duband et al., 1988; Forscher and Smith, 1988). Antibodies can be used to disrupt the normal interactions or the function of specific proteins in order to deduce their role in locomotion. These techniques represent the finer tools with which to probe the molecular mechanism of cell locomotion and form the basis of the next level of investigation.

CURRENT APPROACH TO THE STUDY OF CELL LOCOMOTION

A key to understanding the molecular basis of cell mobility lies in investigating the dynamics of specific molecules during locomotion. Current work performed in this laboratory is directed towards documenting the movements of protein and lipid molecules in the cell membrane during locomotion. This is of particular interest because the cell membrane is the site at which events crucial to locomotion take place. The interaction of the cell membrane with the substratum or with other cells is essential in determining whether a cell will move or not. Furthermore, it is the site at which cell-substratum contacts and other membrane specialisations such as filopodia form.

This laboratory employs quantitative fluorescence microscopy to study the movements of membrane components during locomotion. This has the advantage of allowing detailed analyses to be made of molecular movements within the plasma membrane of living cells. The two major techniques used in this laboratory are fluorescence recovery after photobleaching (FRAP) and digitized fluorescence microscopy (DFM).

The FRAP procedure involves photobleaching a small region of fluorescently-labelled molecules on the cell surface with a pulse of laser light. The fluorescence in this area is briefly destroyed but recovers due to the diffusion of unbleached molecules into this area. The fluorescence intensity is measured before photobleaching and at short intervals during recovery. The rate of fluorescence recovery depends on the diffusion coefficient and the flow velocity of the bleached molecules (Jacobson, et al., 1976; Axelrod et al., 1976). The proportion of mobile fluorescent molecules is indicated by the degree to which fluorescence reaches its prebleach value.

Digitized florescence microscopy (DFM) employs ultra low light level video detectors, video image digitizers and digital microcomputers (DiGiuseppi, et al., 1985). This instrumentation allows the movements of specific classes of molecules to be traced over extended periods of time without significant fading of fluorescence or loss of cell viability. The phase contrast image, a fluorescence image whose focal plane is close to the ventral cell surface and another fluorescence image whose focal plane is close to the dorsal cell surface (Fig. 1).

The movement of several individual patches was tracked in different regions of the cell surface over a period of 1-2 hours. This was facilitated by the development of a computer program which collects a stores the x,y, coordinates of specific patches. This information can then be used to plot trajectories of these patches (Fig. 1). Time-lapse movement could also be simulated in this way allowing patch movement to be compared simultaneously over different regions of the cell surface.

We found that movement of patches in the cell membrane depends on their position on the cell surface (Holifield et al., 1989, submitted). Patches located on both dorsal and ventral surfaces of the leading lamella move rearwards away from the leading edge. However, movement of patches on the ventral surface is often obstructed a short distance back from the leading edge. In contrast, more posterior patches located on the dorsal surface over the nucleus move forward relative to the substratum. This can be correlated with the forward movement of the nucleus resulting from retraction of the rear end of the cell. Thus, it seems that movement of patches on the leading lamella is independent of that on the dorsal, supranuclear surface. These two regions appear to be separated by a boundary or 'null border'. When rearward moving patches reach this boundary their movement stops and they begin to move forward. However, unlike patch movement at the front of the cell, rearward patch movement only occurs during retraction of the trailing edge and so is dependent on deformations of the cell body. This striking position-dependent movement of patched GP80 on the cell surface reflects the distinct regionalisation of mechanisms driving active transport.

The capping of membrane proteins is believed to be mediated via links with the cytoskeleton. One possibility is that patch movement is directed by the cortical actin cytoskeleton (Heath, 1983a). To investigate this, double-label

B. Active Transport

This refers to the active (ATP dependent) movement of molecules in the plane of the membrane in addition to simple diffusion. The best known example of this is 'capping.' This involves the movement of groups of 'patched' or cross-linked proteins away from extending cell margins towards the cell center. In contrast to lateral diffusion, this is an active, ATP-requiring process. It is thought that this type of movement is mediated by the cytoskeleton via direct or indirect linkages with certain membrane proteins. Capping only occurs in motile cells and so is regarded to be a reflection of the same mechanism(s) that leads to locomotion. Since the rearward movement of membrane proteins on the dorsal surface during capping may also occur on the ventral surface, anchorage of such molecules to the substrate near the front of the cell may cause it to be "pulled" forwards. The possible relationship between lateral diffusion and active transport to a locomotory mechanism has prompted a deeper investigation of these processes. The first study is related to the movement of patched or aggregated protein in the membrane of motile cells. The second is concerned with the movements of single or non-aggregated antibody complexes of the same protein during locomotion. Both studies have independently provided new information about the molecular nature of cell locomotion.

The redistribution of 'patches' of cross linked protein in locomoting cells.

The purpose of this experiment was to observe the movement of single patches of membrane protein over different regions of the cell surface during locomotion. Patching of a predominant surface glycoprotein (GP80) was induced by cross-linking GP80 with a monoclonal GP80 antibody. The addition of a rhodamine-labelled, secondary antibody leads to capping of the patched GP80. Throughout this process, movement of individual patches could be followed by collecting sets of high resolution digitized images at 3 minute intervals. Each set consists of a phase contrast image, a fluorescence image whose focal plane is close to the ventral cell surface and another fluorescence image whose focal plane is close to the dorsal cell surface (Fig. 1).

The movement of several individual patches was tracked in different regions of the cell surface over a period of 1-2 hours. This was facilitated by the development of a computer program which collects a stores the x,y, coordinates of specific patches. This information can then be used to plot trajectories of these patches (Fig. 1). Time-lapse movement could also be simulated in this way allowing patch movement to be compared simultaneously over different regions of the cell surface.

We found that movement of patches in the cell membrane depends on their position on the cell surface (Holifield et al., 1989, submitted). Patches located on both dorsal and ventral surfaces of the leading lamella move rearwards away from the leading edge. However, movement of patches on the ventral surface is often obstructed a short distance back from the leading edge. In contrast, more posterior patches located on the dorsal surface over the nucleus move forward relative to the substratum. This can be correlated with the forward movement of the nucleus resulting from retraction of the rear end of the cell. Thus, it seems that movement of patches on the leading lamella is independent of that on the dorsal, supranuclear surface. These two regions appear to be separated by a boundary or 'null border'. When rearward moving patches reach this boundary their movement stops and they begin to move forward. However, unlike patch movement at the front of the cell, rearward patch movement only occurs during retraction of the trailing edge and so is dependent on deformations of the cell body. This striking position-dependent movement of patched GP80 on the cell surface reflects the distinct regionalisation of mechanisms driving active transport.

The capping of membrane proteins is believed to be mediated via links with the cytoskeleton. One possibility is that patch movement is directed by the cortical actin cytoskeleton (Heath, 1983a). To investigate this, double-label

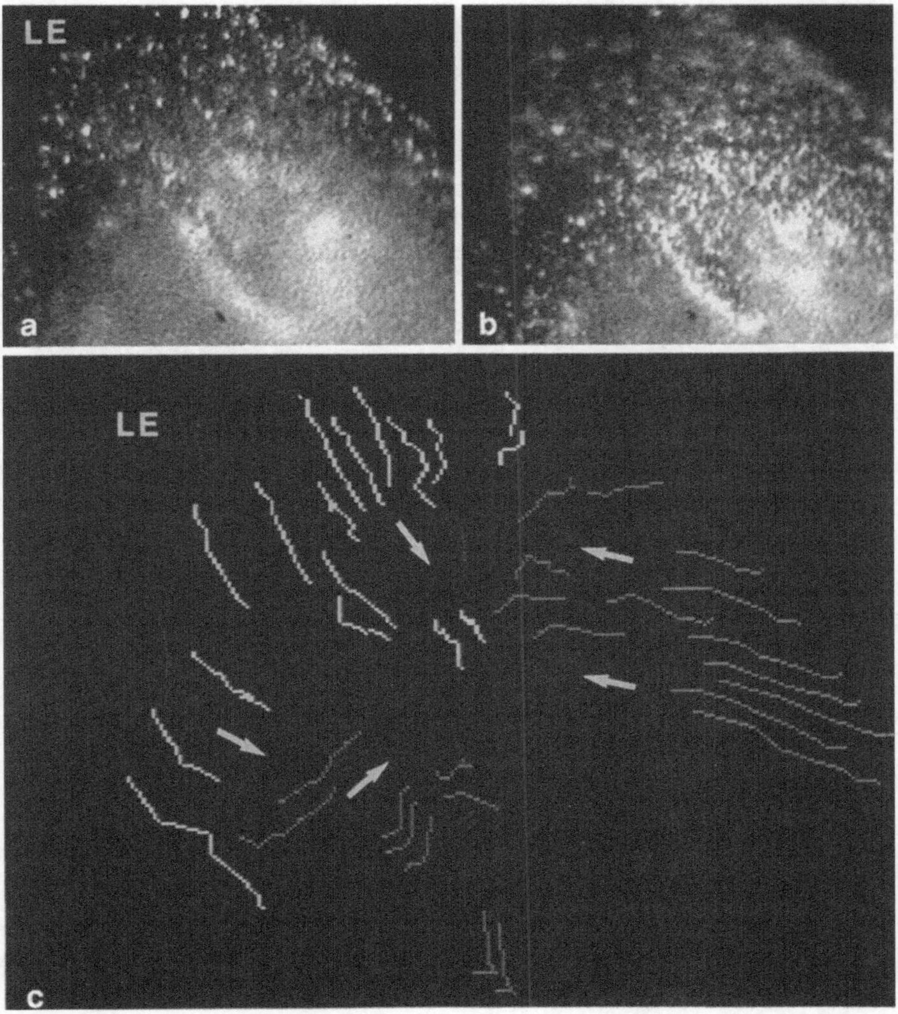

Figure 1. Digitized fluorescence images of patched GP80. The image in panel (a) is taken at a higher focal plane and (b) a lower one. Panel (c) shows the same field of view with patch movements reconstructed from the x and y coordinates of successive images. White lines represent patch movements rearward, away from the leading edge (LE) of the cell, and gray lines represent movements in a predominantly forward direction. Arrows indicate directions in which nearby patches moved.

fluorescence studies were performed to localise the patched GP80 in relation to the F-actin distribution. Observations made at different focal planes showed an F-actin meshwork lying beneath the dorsal cell membrane. The actin filaments appear to be organised in bundles parallel to the leading edge. These correspond to the phase-dense, arc-shaped structures that are seen to move rearward in concert with patched protein on the dorsal surface (Heath, 1983b).

The results lead us to suggest that patches on the dorsal surface of the leading lamella are actively transported rearwards as a result of a localized rearward driving force generated by the cortical actin meshwork. However, forward movement of patches on the posterior cell surface appears to occur together with the general forward movement of cytoplasm and organelles which occurs during retraction. It has been shown that retraction involves both ATP-dependent and independent processes (Chen, 1981). Since capping is thought to reflect the activity of an underlying locomotory mechanism, these results support the idea that the cytoskeleton is a crucial component in this process.

Analysis of GP80 redistribution in the cell membrane during locomotion

Previous studies into the dynamics of unpatched GP80-monoclonal antibody (GP80-mAB) complexes have been directed towards understanding the factors which limit their lateral mobility in the plasma membrane and the biological significance of this. The lateral mobility of GP80-mAB complexes in motile fibroblasts has been studied extensively using the FRAP technique. One interesting observation in relation to cell locomotion is that fluorescent staining of GP80 was greatest at the rear of the cell and decreased towards the leading edge (Jacobson et al., 1984a). Furthermore, this gradated distribution of GP80-mAB complexes was found in locomoting but not stationary cells (Jacobson et al., 1984b). However, this phenomenon appears to be specific to GP80 since other proteins such as Thy-1, H-2 and HA (the flu virus hemagglutinin) show a uniform distribution in locomoting fibroblasts (Holifield, et al., 1989, submitted). An explanation for why the GP80-antibody complex forms a gradient in motile cells while other protein-antibody complexes do not, remains to be found. Since a gradated distribution of GP80 is only found in moving cells, a detailed quantitative study would be expected to provide more clues to the molecular mechanism underlying locomotion.

Several models could in theory account for a gradated distribution of GP80 in motile cells. To investigate these possibilities the observed mAB-complex distribution in locomoting cells was compared with predicted distributions generated by computer simulation of these models. The experimental distribution of GP80 was obtained by observing changes in distribution of this rhodamine-labelled molecule during locomotion. This information was quantified using ultra low-light level digitized fluorescence microscopy. The staining of GP80 persisted for about 100 h without being appreciably internalised and it does not appear to affect cell locomotion (Jacobson, et al., 1984b).

The predicted GP80 distribution was obtained by computer simulation. This is a new feature of our experimental approach and has proven to be a powerful tool in analyzing quantitative results and interpreting them in relation to current theories of cell locomotion. It allows one to independently vary parameters such as lipid flow rate or diffusion coefficient to provide the best fit to experimental data. The degree to which such parameters are biologically reasonable provides a test of a model's plausibility.

Experimental Observations

The major observation from this study is that a 'steady-state' gradient of GP80-mAB complex is maintained on the surface of locomoting fibroblasts (Ishihara et al., 1988). Although the fluorescence staining was generally diffuse, a distinct gradient could be seen. Very little fluorescence exists on the leading edges, whereas a marked increase in fluorescence could be seen towards the rear of the cell. This information was quantitated by obtaining fluorescence intensity

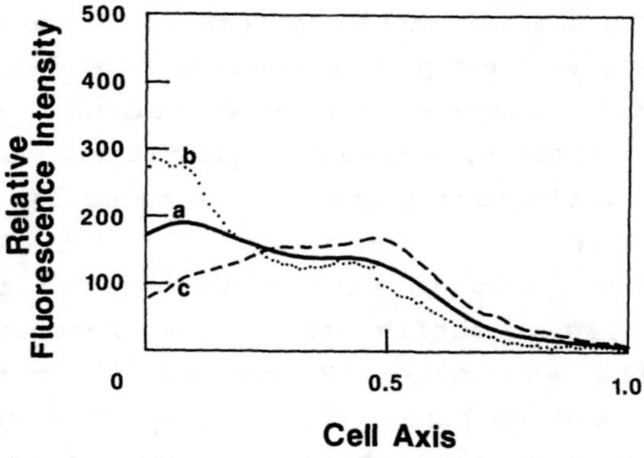

Figure 2. The mean GP80 gradient. A mean profile of the gradated fluorescence distributions was obtained from profiles of seven cells whose shapes were approximately rectangular. Abscissa indicates the relative location on the cell axis from the rear end to the front end of cell. Ordinate indicates the relative fluorescence intensity in arbitrary units. The intensity profile was corrected first by the subtraction of the background and then by normalization to give each of the profiles an equal weight before the averaging. Line a shows the average of the seven profiles. Lines b and c show the average of the three steepest profiles and of the three shallowest profiles, respectively.
Figs. 2-5 reproduced with permission from the Journal of Cell Biology.

profiles over a period of 1½ hours. The profiles were obtained by averaging the fluorescence intensity across a rectangle defined parallel to the long axis of several cells (Fig. 2). The ability of several models to account for the experimental GP80-mAB complex distribution was tested by computer simulation. However, only one of these is described here.

The second observation concerns the behavior of the GP80 gradient on warming prechilled cells to 37° C. Cooled, nonmotile cells exhibit a uniform distribution of GP80. When warmed, cell locomotion and the extension of new leading lamellae resumes. However, these lamellae are almost devoid of GP80. Thus a marked fluorescence boundary is created between the leading edge and the perinuclear region. Over time, this boundary erodes as the fluorescence intensity increases in the newly extended lamellae and decreases in the preexisting area. Fluorescence intensity profiles taken during this process show that labelled GP80 is gradually transported into the newly extended area. Computer simulations were used to test whether this transport could be accounted for by lateral diffusion.

Computer simulation of the current models of cell locomotion

A detailed description of these simulations has already been given (Ishihara et al., 1988). Therefore, only a brief overview will be given here. It should be noted that the theoretical models have been reduced to one dimension to facilitate simulation. Therefore, they are only a rough approximation of the complex morphological changes and of the discontinuous nature of cell locomotion. Nevertheless, computer simulations allow preliminary quantitative comparisons to be made between theoretical models and experimental observations of such a complex phenomenon.

In each simulation, the cell is represented by a 50 μm-long rectangle consisting of an array of 100 'bins.' Membrane proteins are represented by a large constant number of 'particles' (10,000). Prior to any simulation, these particles are equally distributed into the bins since the initial GP80 distribution is assumed to be uniform. During the simulation, these particles are redistributed into the bins according to a set of rules to simulate random diffusion and the effects of lipid flow. Lateral diffusion is simulated by transferring particles into two adjacent bins. Lipid flow is simulated by moving a particle into the next bin in the direction of flow. Flow operations are inserted intermittently into the diffusion operations to simulate realistic flow rates. Cell movement is assumed to be constant in speed and direction. Extension of the leading lamella is represented by adding one bin to the 'front' of the array. The average of 10 - 20 simulations was taken as the final result.

Comparison of Simulated and Experimental Fluorescence Boundary Erosion in Extending Lamellae

Redistribution of GP80 molecules was simulated using three different diffusion coefficients (4 x 10^{-9}, 4 x 10^{10}, 4 x 10^{-11} cm^2/S). Extension of the leading lamella was assumed to be constant in speed and direction. It can be seen that a diffusion coefficient of 4 x 10^{-10} cm^2/S comes closest to matching the experimental fluorescence intensity profile after 35 mins (Fig. 3). It is of interest that this value agrees closely with that obtained for labeled GP80 by fluorescence recovery after photobleaching (D = 3.2 x 10^{-10} cm^2/S). This result suggests that the redistribution of GP80 into newly extended lamellae is governed by the lateral diffusion of these molecules within the plane of the membrane.

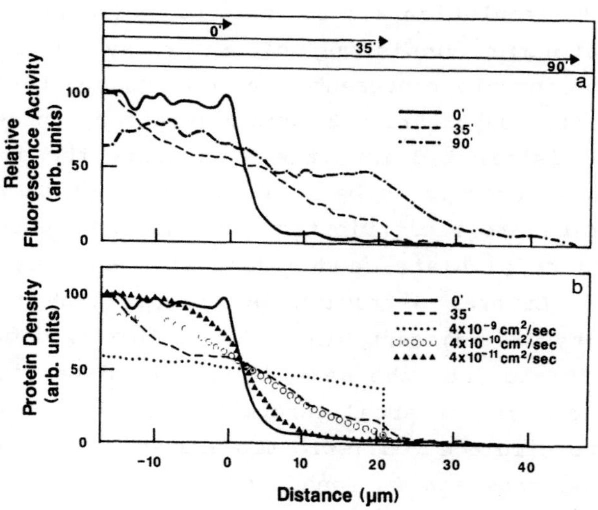

Figure 3. Redistribution of GP80 in spreading cells. Prechilled cells were labeled with R-anti-GP80 (95 μg/ml and warmed to 37°C at time zero. Panel a gives the mean intensity profiles of relative fluorescence intensity obtained by averaging individual profiles. The arrows at the top indicate the positions of the cell's leading edge. Panel b shows the theoretical predictions of GP80 distribution at 35' when the distribution at time zero relaxed due to pure lateral diffusion. The calculation was carried out by computer simulation for various diffusion coefficients. The leading

edge was assumed to advance at a constant rate of 0.6 μ/min. Reasonable agreement with the experimental result was obtained by taking $D = 4 \times 10^{-10}$ cm^2/s.

<u>Simulation of the Retrograde Lipid Flow (RLF) Model: A comparison with experimental observations.</u>

The basic premise of the RLF model is that the membrane lipid flows from the front to the rear of the cell on both the ventral and dorsal cell surfaces (Bretscher, 1984). Other membrane components such as monomeric or aggregated proteins are thought to be swept rearwards by this flow. Extension of the leading lamella is thought to be achieved by insertion of membrane lipid at the tip of the leading edge which is balanced by endocytosis of lipid over the remaining cell surface. This is suggested to generate a rearward flow of lipid. A schematic representation of this model is shown (Fig. 4)

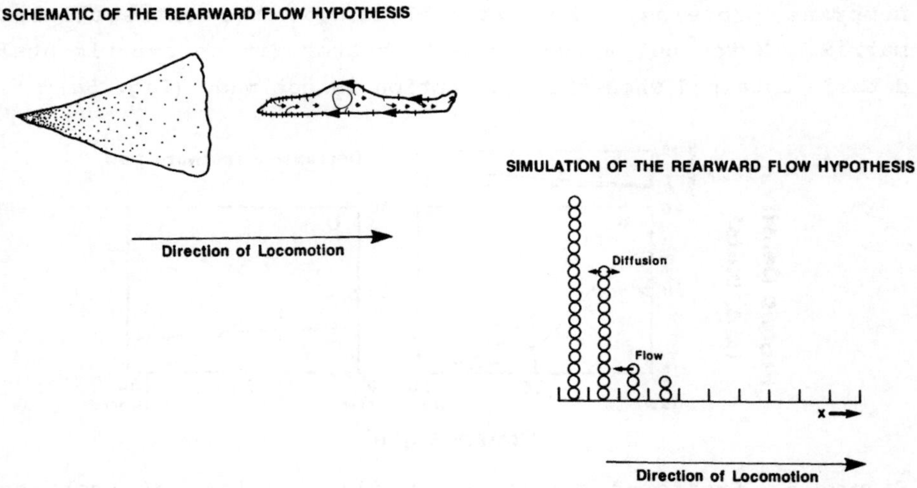

<u>Figure 4.</u> Schematic diagram of the rearward flow hypothesis and its simulation. (a) The model postulates the recycling of membrane lipids. Membrane lipids with some recycling proteins are internalized as endocytosis occurs over the cell surface; recycled lipids are translocated to the front of the cell and inserted to form a new leading edge. On the cell surface lipids flow rearward (←) and membrane proteins tend to be

carried with the flow, resulting in the formation of surface concentration gradient. The gradient is steeper when the flow velocity is faster or when the diffusion coefficient is smaller. (b) Scheme of one-dimensional simulation. A cell was represented by a linear array of bins, and a number of particles representing membrane proteins were distributed in each of the bins. Actual simulation was carried out with 100 bins and 10,000 particles. The particles were transferred from bin to bin under the following rules. Diffusion operation: each particle is transferred randomly to one of the two adjacent bins. Flow operation: each particle is transferred to the adjacent bin to the left to represent retrograde flow. The diffusional operation was repeated successively with intermittent insertion of the flow operation to simulate realistic flow rates.

The RLF model was simulated using several different assumptions. The results of some of these are shown (Fig. 5). Both simulations assume a rearward lipid flow velocity that decreases from front to rear. However, the results shown in Fig. 5a are based on the assumption that GP80-mAB complexes are moved rearwards with the lipid flow whereas Fig. 5b shows the result based on assuming that the GP80-mAB complexes are 'protected' from lipid flow by transient links with other membrane proteins, the cytoskeleton or the extracellular matrix. Note that a considerably better fit to experimental data is obtained when this assumption is not made (Fig. 5a).

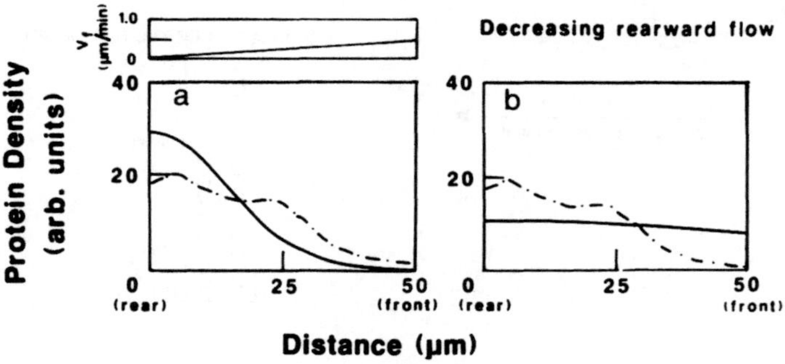

Figure 5. Predicted steady state distributions of membrane proteins based on the retrograde lipid flow model. The figures show the membrane protein density from the front (the right end) to the rear (the left end) along the 50-μm-long cell axis. Broken lines indicate the mean steady state distribution obtained experimentally. Solid lines give the simulation results. The area under the distributions represents the total amount of the membrane protein and is arranged to be identical

for each plot. The flow velocity (V_f) was defined relative to the substratum. In all cases $D = 3.2 \times 10^{-10}$ cm^2/s was assumed. In a and and b, V_f was assumed to decrease linearly from the front end (0.5 μm/min) to the rear end (0 μm/min) as indicated in the panel at the top of 5a. In 5a no transient anchoring was assumed. In 5b transient anchoring of 95% of the protein population was assumed.

However, previous experiments indicate that the lateral diffusion of GP80-mAB complexes is hindered. This could conceivably be due to transient linkages of GP80 with immobile components of the cell membrane (Jacobson et al., 1984a). Support for this idea comes from measurements of the lateral diffusion coefficient of GP80 in motile fibroblasts. This value is much lower than would be expected if this protein were reconstituted into liposomal membranes (Jacobson et al., 1983; Vaz et al., 1984). Thus it appears that if the assumption of transient protein linkages is not made, then any theoretical model cannot represent reality. Even though the RLF model agrees well with experiment when membrane proteins are assumed mobile with a low diffusion coefficient (Fig. 5a) experimental evidence contradicts this idea. Furthermore, when the experimental diffusion coefficient of GP80 is assumed, the RLF model fails to simulate the observed GP80 distribution (Fig. 5b).

More recent work in our laboratory indicates another fundamental discrepancy between model and experiment. This has been strikingly shown in experiments where motile fibroblasts were double labeled with anti-GP80 and antibodies against HA. (Holifield et al., 1989, submitted). In this case the GP80-antibody complex formed a gradated distribution while the antibody labeled HA molecules remained evenly distributed over the same leading lamella. This result is particularly surprising since both HA and GP80 antibody complexes have similar diffusion coefficients. Therefore, according to the RLF model both these proteins should have a markedly gradated distribution in motile cells.

In the light of these results the RLF model no longer seems tenable. Yet the idea of a retrograde flow driving GP80 redistribution remains attractive especially if it is ascribed to the cytoskeletal elements beneath the leading lamella (Abercrombie, 1980; Dunn, 1980). Thus, it is possible that movement of the unpatched GP80-mAB complex may be directed by a moving cortical actin meshwork in a similar way as is proposed for patched GP80.

It has been suggested that a rearward flow of cortical cytoskeleton beneath the dorsal surface of the leading lamella is responsible for the rearward movement of patched membrane proteins (Abercrombie, 1980; Dembo and Harris, 1981; Dunn 1980; Heath, 1983b). In addition this is thought to provide the driving force for locomotion. One may imagine that transient attachment of GP80 molecules to the cortical actin cytoskeleton would tend to move these molecules rearwards during locomotion. However, if the GP80 molecules become detached from the cytoskeleton, their random diffusion would work against an 'active' redistribution mechanism. Thus the 'steady-state' GP80 distribution in moving cells would be the product of an equilibrium between rearward transport and lateral diffusion.

The direction of lipid flow in locomoting cells

So far this article has been confined to describing the movements of proteins within the plasma membrane during cell locomotion. We have compared experimental observations with computer simulations to test the ability of one major model of locomotion to account for the behaviour of GP80 in motile cells. The predicted behavior of membrane proteins was based on an assumption about the direction of lipid flow. Therefore, it would seem that the next logical step in the investigation of cell motility is to determine the direction of lipid flow in locomoting cells.

Current studies in our laboratory use human polymorphonuclear leukocytes (PMN's) to investigate the direction of lipid flow during locomotion. These cells have the advantage of being highly motile on a glass substratum and are resilient to experimental manipulation. The experiment involves marking a discrete region of the cell membrane and observing the subsequent movement of this mark during locomotion. This was achieved by staining the membrane lipid of the PMN's with a lipid analogue dihexadecanoyl indocarbocyanine (DiI) and then photobleaching a line on the cell surface, perpendicular to the direction of cell motion. The line remains visible for about 3 seconds before it is completely erased due to diffusional recovery. However, if the cell is moving rapidly then there will be enough time to detect line movement before it is completely erased. We have shown that cell locomotion is not impaired by staining with DiI together with repeated photobleaching. Furthermore, no significant staining could be detected in the cytoplasm by confocal microscopy during the time taken to perform the experiment. Low light level fluorescence microscopy and digital image processing techniques were used to analyze a series of images acquired at short intervals after photobleaching. Examination of the fluorescence intensity profiles obtained from these images allowed a comparison to be made between the velocity of the bleached line and the extending lamella (Fig. 6).

Preliminary results suggest that the velocity of the bleached line is the same as the leading edge during locomotion. Thus a line bleached at the front of the cell moves forward by the same amount as the leading lamella. However, according to the RLF Model this line would be expected to move rearwards relative to the leading edge. Thus our results are consistent with the idea that the cell membrane moves as a single unit during locomotion. The significance of this result is that it demonstrates for the first time that lipid flow can be detected in moving cells using this method. Furthermore, since the results are quantitative they are amenable to computer simulation. Using this approach we are presently simulating bleaching experiments according to each of three major models of cell locomotion and comparing the simulated results with the experimental result. Together these data should provide another vital step towards understanding the mechanism of cell locomotion.

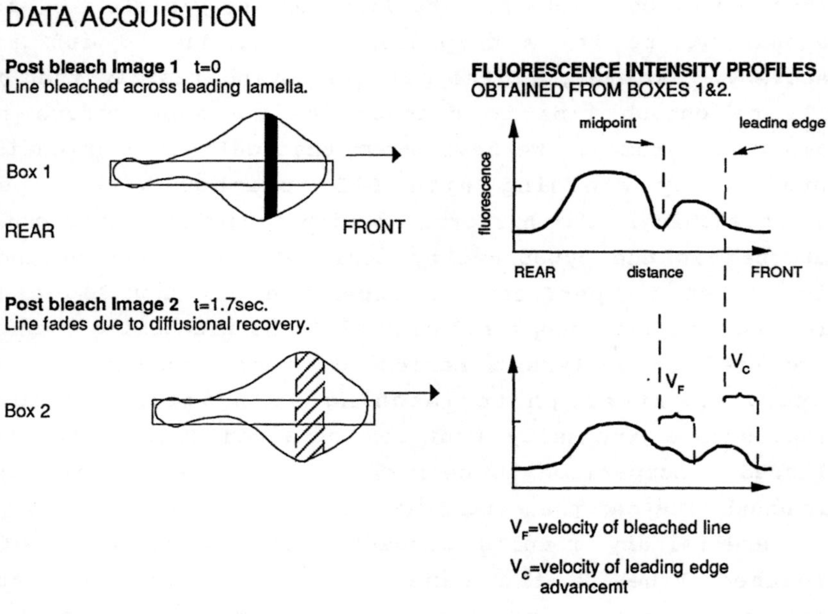

Fig. 6. Digitized fluorescence images of a bleached PMN are acquired during extension of the leading lamella before diffusional recovery is complete. Post bleach images 1 and 2 are used to generate fluorescence intensity profiles 1 and 2 by averaging the fluorescence intensity across the defined rectangle for each pixel along its long axis. The velocity of the bleached line and the leading lamella is calculated from the distance moved by these regions of a profile, with respect to a fixed point on the substratum.

SUMMARY AND CONCLUSIONS

The increasing use of video microscopy and biophysical techniques have brought studies of cell locomotion from the level of gross morphology to the molecular level. The movements of individual proteins can now be traced in single locomoting cells. Image processing techniques and computer simulation have proven to be valuable tools in analyzing experimental observations. Ultimately, this has allowed our current concepts of cell locomotion to be refined.

Investigation into the movements of both aggregated and non-aggregated GP80-mAB complexes indicate a cytoskeletal involvement in the redistribution of these molecules during locomotion. This contradicts the RLF theory which supposes that the movement of membrane proteins during locomotion is entirely due to the effects of a rearward lipid flow. Further opposition to this model is provided by PMN photobleaching experiments which suggest that membrane lipid flows forwards during locomotion, not rearwards.

Our current concept of the behaviour of cell surface components during locomotion is summarized as follows. During locomotion aggregated or crosslinked membrane proteins on the dorsal surface of the leading lamella are actively transported rearwards to the "null border". In special cases, non-aggregated proteins such as the GP80-mAB complex, are also transported rearwards. In contrast, other membrane proteins are not transported rearwards unless they are crosslinked into patches. The rearward movement of the cortical actin cytoskeleton is presently regarded as the most likely driving force for the rearward transport of proteins on the leading lamella. Photobleaching experiments with PMN's indicate that the membrane lipid of the leading lamella moves forwards at the same velocity as the advancing cell. Thus it appears that some intracellular force is acting to drive rearward transport of certain membrane proteins while at the same time the lipid bilayer undergoes bulk forward flow keeping pace with the extending lamella. However, the degree to which these proteins are moved rearwards or their ability to form a gradated distribution during locomotion will depend on their affinity to the actin cytoskeleton. In addition rearward transport will be counteracted by the effects of random diffusion and forward lipid flow.

Although some progress has been made in unraveling the finer details of cell locomotion many questions still remain. For example we do not fully understand how membrane proteins interact with cytoskeletal elements and the substratum to bring about cell movement. Presumably the forces which lead to the rearward movement of some membrane proteins are the same as those that drive cells forwards during locomotion. However, the magnitude of such forces and the mechanism(s) by which they are generated remain to be discovered.

REFERENCES

Abercrombie M (1980) The crawling Movement of Metazoan Cells. Proc Roy Soc B 207 P 129

Abercrombie M, Dunn GA, JP Heath (1977) The Shape and Movement of Fibroblasts in Culture. In Cell and Tissue Interactions, JW Lash and MM Burger eds (Rabier Press NY) p 57-70

Abercrombie M, JEM Heaysman and SM Pegrum (1970) The locomotion of fibroblasts in culture. I. Movements of the leading edge. Exp Cell Res 59:393-398

Axelrod D, Koppel DE, Schlessinger E Ebson and W Webb (1976) Mobility measurement by analysis of fluorescence photobleaching recovery kinetics. Biophysical Journal 16:1055-1069

Bretscher MS (1984) Endocytosis: Relation to capping and cell locomotion. Science 224:681-686

Burridge K, Fath K, Kelly T, Nuckolls G and C Turner (1988) Focal Adhesions: Transmembrane junctions between the extracellular matrix and the cytoskeleton. Ann Rev Cell Biol 4:487-525

Chen W-T (1981) Mechanism of retraction of the trailing edge during fibroblast movement. J Cell Biol 90:187-200

DeBiasio RL, Wang L-L, Fisher GW and Taylor DL (1988) The dynamic distribution of fluorescent analogues of actin and myosin in protrusions at the leading edge of migrating Swiss 3T3 fibroblasts. J Cell Biol 107:2631-2645

Dembo M and Harris AK (1981) Motion of particles adhering to the leading lamella of crawling cells. J Cell Biol 91:528-536

DePasquale J.A. and CS Izzard (1987) Evidence for an actin-containing cytoplasmic precursor of the focal contact and the timing of incorporation of vinculin at the focal contact. J Cell Biol 105:2803-2809

DiGuiseppi J, R Inman, A Ishihara, K Jacobson and B Herman (1985) Applications of digitized fluorescence microscopy to problems in cell biology. Biotechniques 3:394-403

Duband JL, Nuckolls GH, Ishihara A, Hasegawa T, Yamada KM, Thiery JP and K Jacobson (198) Fibronectin receptor exhibits high lateral mobility in embryonic locomoting cells but is immobile in focal contacts and fibrillar streaks in stationary cells. J Cell Biol 107:1385-1396

Dunn GA (1980) Mechanisms of Fibroblast Locomotion In Cell Adhesion and Motility (Third Symposium British Society for Cell Biology, Cambridge University Press) Eds ASG Curtis and JD Pitts, p 409-423

Fisher GW, Conrad PA, DeBiasio RL and Taylor DL (1988) Centripetal transport of cytoplasm, actin and the cell surface in lamellipodia of fibroblasts. Cell Motil Cytoskel 11:235-247

Forscher P and Smith SJ (1988) Actions of cytochalasins on the organization of actin filaments and microtubules in a neuronal growth cone. J Cell Biol 107:1505-1516

Harris AK (1973) Cell Surface Movements Related to Cell Locomotion. In Locomotion of Tissue Cells, Ciba Foundation Symposium No 14, Eds R Porter and DW Fitzsimmons, p 3-20

Heath JP (1983) Behaviour and structure of the leading lamella in moving fibroblasts. J Cell Sci 60:331-354

Heath JP (1983b) Direct evidence for microfilament-mediated capping of surface receptors on crawling fibroblasts. Nature (Lond) 302:532-534

Holifield B, Ishihara, A and Ken Jacobson (1989) Behavior of membrane protein-antibody complexes on motile fibroblasts: Implications for a mechanism of capping (submitted)

Ishihara A, B Holifield and K Jacobson (1988) Analysis of lateral redistribution of a plasma membrane glycoprotein-monoclonal antibody complex which occurs during cell locomotion. J Cell Biol 106:329-343

Izzard CS and Lochner LR (1980) Formation of cell-10-substrate contacts during fibroblast motility: an interference-reflexion study. J Cell Sci 42:81

Jacobson K (1983) Lateral diffusion in membranes. Cell Motil 3:367-373

Jacobson K, D O'Dell and JT August (1984a) Lateral diffusion of an 80,000-dalton glycoprotein in the plasma membrane of murine fibroblasts: relationships to cell structure and function. J Cell Biol 99:1624-1633

Jacobson K, O'Dell D, Holifield B, TL Murphy and JT August (1984b) Redistribution of a major cell surface glycoprotein during cell movement. J Cell Biol 99:1613-1623

Jacobson K, E Wu and G Poste (1976) Measurement of the translational mobility of concanavalin A in glycerol-saline solutions and on the cell surface by fluorescence recovery after photobleaching. Biochim Biophys Acta 433:215

Lochner L and CS Izzard (1973) Dynamic aspects of cell-substrate contact in fibroblast motility. J Cell Biol 58:199a

Trinkaus JP (1984) Cells into organs. The forces that shape the embryo. 2nd ed Prentice Hall Inc New Jersey

Vaz WLC, F Goodsaid-Zaldaondo and K Jacobson (1984) Lateral diffusion of lipids and proteins in bilayer membranes. FEBS (Fed Eur Biochem Soc) Lett 174:199-207

Wang Y-L (1984) Reorganization of actin filament bundles in living fibroblasts. J Cell Biol 99:1478-1485

Index

GPSR Compliance

The European Union's (EU) General Product Safety Regulation (GPSR) is a set of rules that requires consumer products to be safe and our obligations to ensure this.

If you have any concerns about our products, you can contact us on ProductSafety@springernature.com

In case Publisher is established outside the EU, the EU authorized representative is:

Springer Nature Customer Service Center GmbH
Europaplatz 3
69115 Heidelberg, Germany

Batch number: 10620143

Printed by Printforce, the Netherlands